Radicals, Ion Radicals, and Triplets

Radicals, Ion Radicals, and Triplets

The Spin-Bearing Intermediates of Organic Chemistry

Nathan L. Bauld

Department of Chemistry and Biochemistry
The University of Texas at Austin
Austin, Texas

New York · Chichester · Weinheim · Brisbane · Singapore · Toronto

This book is printed on acid-free paper.♾

Published simultaneously in Canada.

Library of Congress Cataloging-in-Publication Data:
Bauld, Nathan L.
Radicals, ion radicals, and triplets : the spin-bearing intermediates of organic chemistry / Nathan L. Bauld.
p. cm.
Includes bibliographical references.
ISBN 0-47119-035-7
1. Intermediates (Chemistry) 2. Organic compounds—Synthesis.
I. Title.
QD476.B355 1997
547′.1224—dc21 96-37072

Printed in the United States of America.

10 9 8 7 6 5 4 3 2 1

For Jane Scoggins Bauld

Contents

Preface

The author's purpose in writing this book is to present fundamental aspects of the chemistry of all the major spin-bearing intermediates of organic chemistry in a single volume, thereby emphasizing their integrity and hopefully highlighting their importance in chemistry and biology.

The presentation assumes a 1-year, introductory course in organic chemistry, but it strives to develop the key concepts to the level of current usage. Structural, mechanistic, synthetic, and theoretical aspects are all considered. Molecular orbital theory is used primarily at the simplest level (HMO). Problem sets are included to test and extend the reader's understanding of the material, and appropriate literature references that provide answers are included with the problems wherever possible. The material is considered appropriate for upper division and graduate chemistry courses and for professionals wishing to renew or advance their familiarity with this chemistry. For the biologically oriented reader, applications in biological systems are exemplified in several contexts and are especially strongly emphasized in the problem sets.

The author has been active in research involving spin-bearing intermediates for approximately 40 years, beginning with his doctoral research project with Professor E. J. Corey at the University of Illinois (1956–1959), and continuing with postdoctoral research in the laboratories of Professor P. D. Bartlett of Harvard University (1959–1960). His research interests were also stimulated and strongly influenced by Professor Cheves Walling, through his classic book *Free Radicals in Solution*. Much of the author's own research in the area of spin-bearing intermediates has involved anion radicals and cation radicals. His broader interests are in the area of physical-organic chemistry.

Acknowledgment

I am deeply indebted to my wife, Jane, without whose encouragement, support, and example this manuscript could not have been written. I also thank Janet Macdonald for her considerable expertise in word processing and generating the numerous graphics contained in the figures.

CHAPTER

1

Basic Concepts of Free Radicals

Free radicals can be defined as chemical species that have a single unpaired electron. In the important case of the methyl radical, the radical center is trivalent and trigonally hybridized (Figure 1.1). The sp^2 hybridized carbon atom and the three hydrogens are coplanar and the unpaired (odd) electron occupies a $2p$ carbon atomic orbital (here referred to as $2p_z$). This singly occupied molecular orbital is of especial importance to free radical chemistry and is often abbreviated as SOMO. The odd electron can have either an α or β spin, but the two spin states are isoenergetic in the absence of a magnetic field. Free radicals are sometimes referred to as *doublets* because of the existence of these two discrete states, which are energetically distinguishable in the presence of a magnetic field. In the same sense, chemical species with no unpaired electrons are referred to as *singlets*, and those with two unpaired electrons are called *triplets*. As will be developed in Chapter 3, the preference for a trigonal radical site is by no means general, even for simple alkyl radicals.

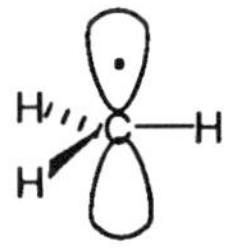

Figure 1.1 Methyl radical.

1.1 Nomenclature

Specific organic free radicals are assigned the IUPAC name of the corresponding substituent (e.g., methyl) and the generic term *radical* is then appended.

1.2 Generation of Free Radicals

The most fundamental process for generating free radicals is homolytic cleavage (homolysis) of a covalent bond (Figure 1.2). The standard enthalpy change for this reaction is defined as the dissociation energy of the A–B bond [$D(A$–$B)$]. This is also referred to as the bond dissociation energy (BDE). Since ions are not formed in this process, a polar solvent is neither required nor especially advantageous over a nonpolar one (or indeed over the vapor phase). The thermal generation of radicals in solution at moderate temperatures and at convenient rates requires bonds of relatively low D, and peroxides are frequently the compounds of choice for this purpose (Figure 1.3).

Although most O–O bond homolyses occur in the simple fashion described in Figure 1.3, a number of instances are recognized in which homolysis and fragmentation are concerted, that is, in which two bonds undergo homolytic cleavage simultaneously. A classic example is the homolytic cleavage of *tert*-butylperoxyphenylacetate (Figure 1.4).[1] The cleavage of a hypothetical phenylacetoxy radical ($PhCH_2CO_2^•$) to carbon dioxide and a resonance-stabilized benzyl radical is evidently activationless and even provides substantial driving force for the O–O bond cleavage. The decomposition occurs at a rate approximately 300 times

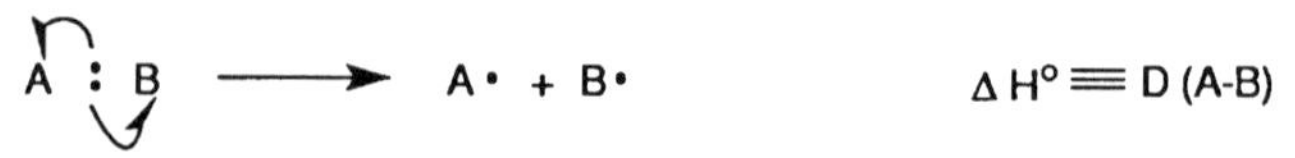

$$A:B \longrightarrow A\bullet + B\bullet \qquad \Delta H^\circ \equiv D\,(A\text{-}B)$$

Figure 1.2 Homolysis.

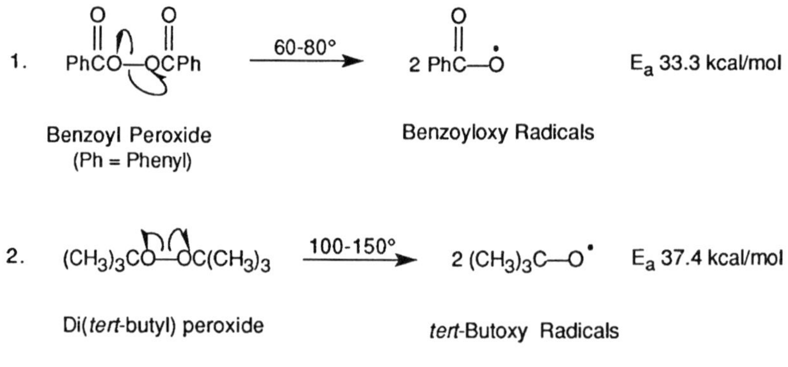

1. $PhC(=O)O–OC(=O)Ph \xrightarrow{60\text{-}80^\circ} 2\ PhC(=O)–O^\bullet$ E_a 33.3 kcal/mol

Benzoyl Peroxide (Ph = Phenyl) → Benzoyloxy Radicals

2. $(CH_3)_3CO–OC(CH_3)_3 \xrightarrow{100\text{-}150^\circ} 2\ (CH_3)_3C–O^\bullet$ E_a 37.4 kcal/mol

Di(*tert*-butyl) peroxide → *tert*-Butoxy Radicals

Figure 1.3 Generation of radicals by the thermal decomposition of organic peroxides.

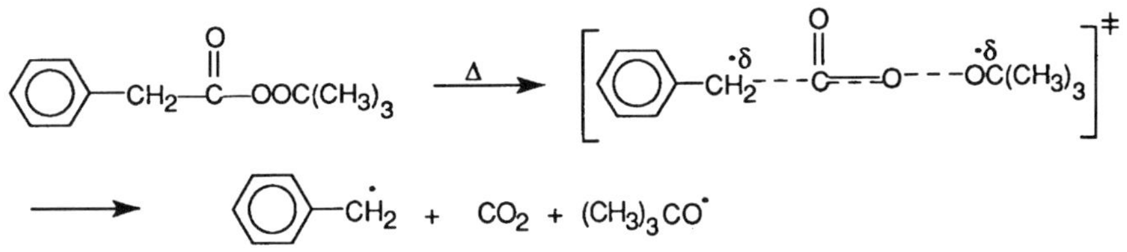

Figure 1.4 Concerted two-bond cleavages in the homolysis of *tert*-butylperoxyphenylacetate.

as fast as the cleavage of the corresponding *tert*-butylperoxyacetate (at 60°C) and has an activation enthalpy *ca.* 9 kcal less. On the other hand, the activation entropy is *ca.* 13 entropy units less positive than that for the peroxyacetate ester, suggesting that the decrease in $\Delta H\ddagger$ occasioned by concerted homolysis/fragmentation is achieved at the expense of "freezing" rotation around the phenyl-to-benzylic carbon bond, which is required in order to maximize the resonance stabilization of the benzyl radical.

Azo compounds, especially azobis(isobutyronitrile) [AIBN], are also effective radical sources. In the case of AIBN, two bonds are dissociated simultaneously to gain the thermodynamic advantage of forming the highly stable nitrogen molecule (Figure 1.5).

Another important factor facilitating this concerted decomposition is the resonance stabilization of the two 2-cyano-2-propyl radicals (Figure 1.6). That the unpaired electron is delocalized onto both carbon and nitrogen has a direct

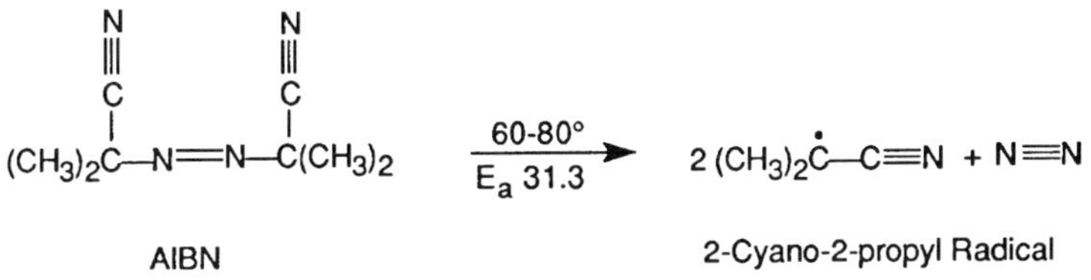

AIBN

2-Cyano-2-propyl Radical

Figure 1.5 Thermal decomposition of AIBN.

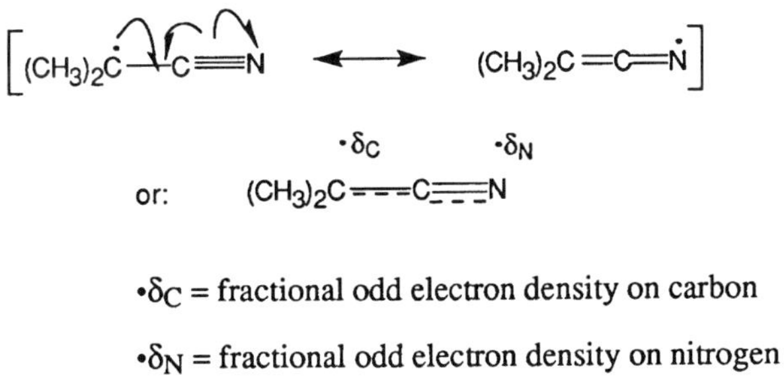

$\bullet\delta_C$ = fractional odd electron density on carbon

$\bullet\delta_N$ = fractional odd electron density on nitrogen

Figure 1.6 Resonance stabilization of the 2-cyano-2-propyl radical.

consequence in that subsequent coupling of two of these radicals occurs both between two carbon centers and between one carbon and one nitrogen center (Figure 1.7).[2] The latter product (a ketene imine) is unstable and rearranges to the former (tetramethylsuccinonitrile) upon further heating. Apparently, coupling between two nitrogens either does not occur or the product is too unstable to be detected. It should be noted that concerted two-bond cleavages of azoalkanes are by no means general, but are usually observed only when *both* radical fragments are substantially resonance stabilized.

1.2.1 Redox Cleavage

Radicals can be smoothly generated, often even at room temperature or below, by reductive or oxidative cleavage of a covalent bond. Addition of a third electron to a bond (via the antibonding MO) or removal of an electron from a bond sharply weakens the bond and greatly facilitates cleavage. Reductive cleavage of peroxidic bonds is an especially useful instance. Fenton's reagent uses ferrous ion as the reductant and either hydrogen peroxide or an organic hydroperoxide as the reducible substrate (Figure 1.8).[3,4] It will be noted that in redox cleavage only one radical is produced, in contrast to the radical pairs generated by thermal cleavage.

1.2.2 Photochemical Cleavage

The absorption of light quanta can also provide the necessary energy for cleaving covalent bonds at virtually any desired temperature. Homolytic cleavage of an excited state of dibenzoyl peroxide, for example, is virtually 100% efficient in producing benzoyloxy radicals (Figure 1.9).[5]

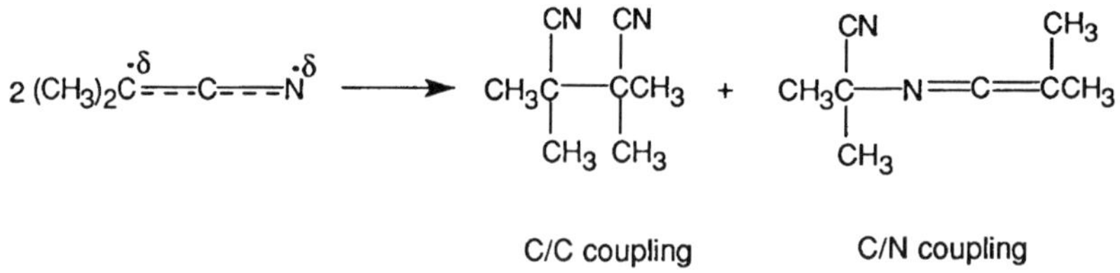

$$2\,(CH_3)_2\overset{\cdot\delta}{C}{=}{=}C{=}{=}\overset{\cdot\delta}{N} \longrightarrow CH_3C(CN)(CH_3){-}C(CN)(CH_3)CH_3 + CH_3C(CN)(CH_3){-}N{=}C{=}C(CH_3)CH_3$$

C/C coupling C/N coupling

Figure 1.7 Products of the thermal decomposition of AIBN.

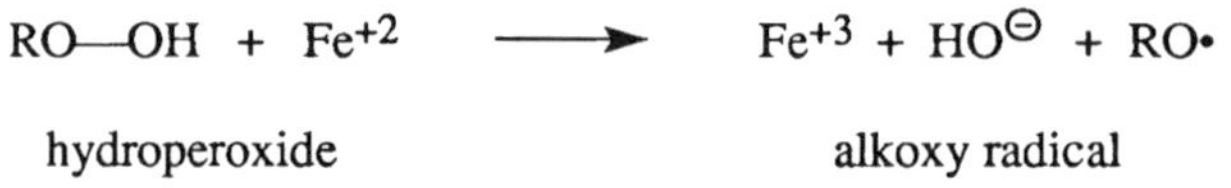

$$RO{-}OH + Fe^{+2} \longrightarrow Fe^{+3} + HO^{\ominus} + RO\cdot$$

hydroperoxide alkoxy radical

Figure 1.8 Fenton's reagent.

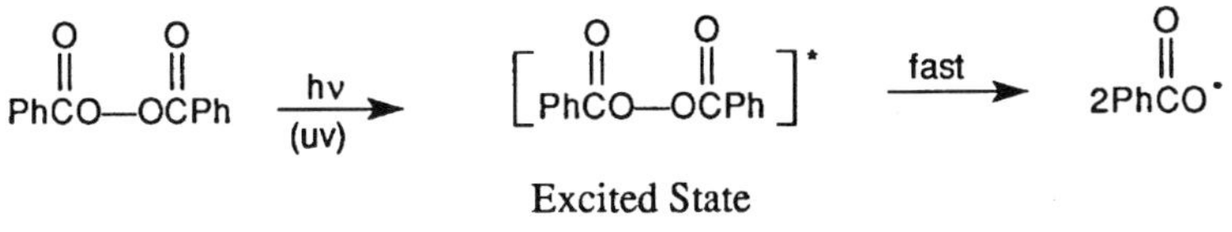

$$PhC(=O)O{-}OC(=O)Ph \xrightarrow[(uv)]{h\nu} [PhC(=O)O{-}OC(=O)Ph]^{*} \xrightarrow{fast} 2PhC(=O)O^{\bullet}$$

Excited State

Figure 1.9 Photochemical cleavage.

1.3 The First Identified Free Radical

The triphenylmethyl ("trityl") radical, discovered by Moses Gomberg at the University of Michigan in 1900, was the first known organic free radical (Figure 1.10).[6,7] Overcoming strong skepticism, essentially all Gomberg's views concerning this radical and its chemistry (e.g., reaction with molecular oxygen to give trityl peroxide) have been supported *except* his formulation of the dimer as hexaphenylethane. Apparently as a result of steric resistance to coupling between two highly hindered benzylic positions, coupling occurs between one benzylic and one *para* position (Figure 1.11)[8,9]

Whereas this dimer is only slightly dissociated into trityl radicals, substituted trityl radicals having all three *para* positions blocked by a substituent (even methyl will do) are much more highly dissociated in solution.[10,11] Incidentally, radicals that are relatively long-lived in solution but not readily isolable (in some cases because of the tendency to couple in the solid state) are termed "persistent" radicals, as distinguished from stable (i.e., isolable) radicals or reactive (transient) radicals.[12]

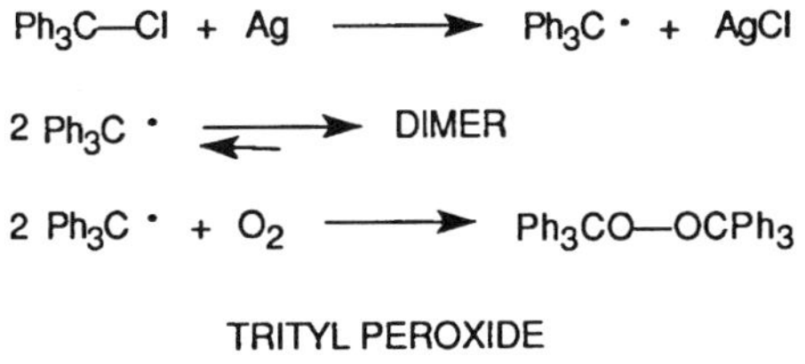

$$Ph_3C{-}Cl + Ag \longrightarrow Ph_3C^{\bullet} + AgCl$$

$$2\,Ph_3C^{\bullet} \rightleftharpoons \text{DIMER}$$

$$2\,Ph_3C^{\bullet} + O_2 \longrightarrow Ph_3CO{-}OCPh_3$$

TRITYL PEROXIDE

Figure 1.10 The trityl radical.

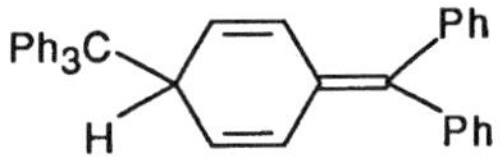

Ph3C, H, Ph, Ph

Figure 1.11 The trityl radical dimer.

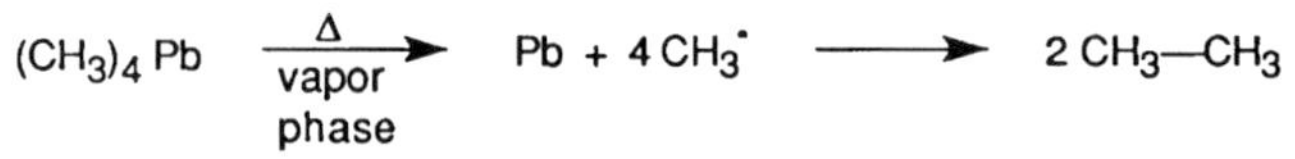

$$(CH_3)_4\,Pb \xrightarrow[\text{vapor phase}]{\Delta} Pb + 4\,CH_3^{\bullet} \longrightarrow 2\,CH_3{-}CH_3$$

Figure 1.12 Radicals as reactive intermediates in the vapor phase.

1.4 Detection of Reactive Radicals

The first conclusive demonstration of free radicals as reactive intermediates was accomplished by Paneth and Hofeditz in 1929 in their now classic studies of the pyrrolysis of tetramethyl lead (Figure 1.12).[13] Methyl radicals were suspected because of the production of ethane and were confirmed as transient intermediates by the observation that the presence of a previously deposited lead mirror further down the reaction chamber diminished the yield of ethane and regenerated tetramethyl lead.

1.5 Radical Intermediates in Solution

The involvement of free radicals as reactive intermediates in solution chemistry was not decisively established until 1937.[14] In that year, Kharasch discovered the now-familiar radical chain mechanism for the addition of hydrogen bromide to alkenes (discussed in Chapter 2). In the same year, Hey and Waters published a review proposing a radical mechanism for this same reaction as well as several others,[15] and Flory interpreted vinyl polymerization in terms of a radical chain mechanism.[16] As Walling has pointed out, "1937 is a convenient date to mark the beginning of our present recognition of free radicals as reactive intermediates in ordinary temperature, liquid phase reactions."[17]

1.6 Hyperconjugative Stabilization of Radicals

The relative stabilities of a series of carbon centered radicals, $R^{\bullet}$, is conventionally evaluated from a comparison of the bond dissociation energies of the appropriate R–H bonds. The implicit simplifying assumption is that differences in $D(R\text{–H})$ arise exclusively from effects that stabilize (or destabilize) the radicals ($R^{\bullet}$) as opposed to their precursor substrates (R–H). The progressive decrease in $D(\text{C–H})$ in entries 1–4 in Table 1.1 is then considered to reflect the stability order $CH_3^{\bullet} < CH_3CH_2^{\bullet} < (CH_3)_2CH^{\bullet} < (CH_3)_3C^{\bullet}$, that is, the progressive stabilization of the radical center by successive alkyl group substitutions for hydrogen at the radical site. This is attributed primarily to hyperconjugative stabilization (delocalization of the odd electron) of the radical by alkyl groups attached to the carbon center (C_{α}) but that are not available for a hydrogen atom attached to C_{α}.

Table 1.1 Bond dissociation energies and radical stabilization

Substrate	*D(C–H) (kcal/mol)*	*Relative stabilization*
1. CH_3—H	104	0
2. CH_3CH_2—H	98	6
3. $(CH_3)_3CH$—H	95	9
4. $(CH_3)_2C$—H	92	12
5. CH_2=$CHCH_2$—H	85	19
6. $PhCH_2$—H	85	19

The resonance structure **A** of the ethyl radical (Figure 1.13), which has all three C–H bonds intact, is substantially lower in energy than structure **B**, which replaces the formally broken C_β–H bond with a much weaker C–C π bond. Consequently the true ethyl radical structure more closely resembles **A** than **B**. ESR studies of the ethyl radical (see Chapter 3) in fact suggest that about 85% of the odd electron density resides on carbon, with about 5% on each of the three β hydrogens. Since the methyl radical is not subject to hyperconjugative stabilization (delocalization), the entire energy difference (6 kcal/mol) between entries 1 and 2 is often attributed to the hyperconjugative stabilization of the ethyl radical by a single alkyl substituent (*viz.* methyl) attached to C_α. The replacement of additional H_α's by alkyl groups (entries 3 and 4) provides further stabilization, but the effects are somewhat smaller than for the first substitution. Other simple primary, secondary, and tertiary C–H bonds are found to have dissociation energies and stabilization energies very close to those observed for ethane, propane, and isobutane, respectively, and it is usually considered that alkyl substituents other than methyl when attached to a radical center exert very nearly the same stabilizing effect as does methyl. Consequently the stabilization energies of Table 1.1 are considered to apply rather generally to typical primary, secondary, and tertiary C–H bonds.

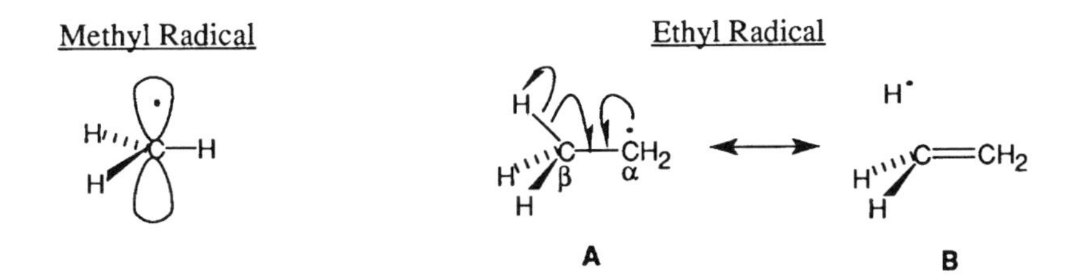

Figure 1.13 Hyperconjugative stabilization of radicals.

1.7 Conjugative Stabilization

Conjugative stabilization of radicals by attached unsaturated functions is illustrated by propene (allyl radical, entry 5) and toluene (the benzyl radical, entry 6). These stabilizations are, of course, much larger than the hyperconjugative stabilizations exerted by alkyl groups. In the case of the allyl radical (Figure 1.14), both canonical structures (**A** and **B**) are of equal energy, and the odd electron density is delocalized equally to C_1 and C_3. In the benzyl case, structure **A**′, which retains the aromaticity of the ring, is more stable than structures **B**′, **C**′, and **D**′ and the odd electron density at the benzylic position ($\bullet\delta_B$) is greater than that of the *ortho* ($\bullet\delta_o$) and *para* ($\bullet\delta_p$) positions.

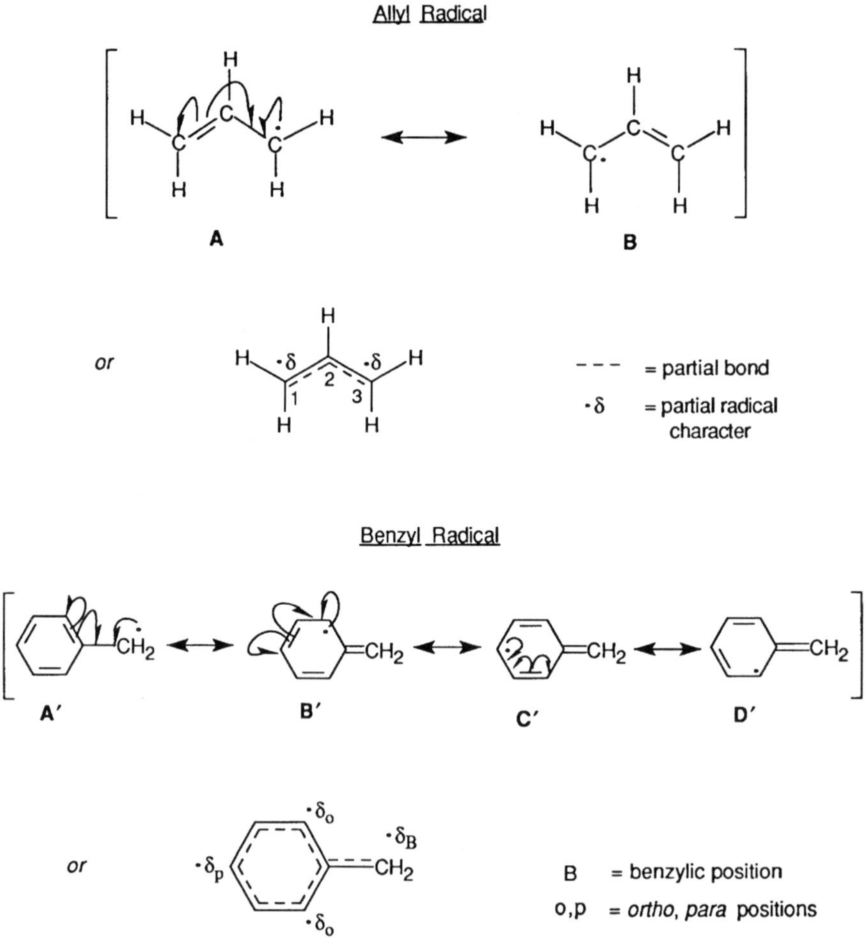

Allyl Radical

A B

or

--- = partial bond
•δ = partial radical character

Benzyl Radical

A′ B′ C′ D′

or

B = benzylic position
o,p = *ortho, para* positions

Figure 1.14 Conjugative stabilization of radicals.

1.8 Three-Electron Bonds

A stabilization effect that is unique to radicals seems to be operative in providing the more modest stabilizations of radical centers by attached substituents having nonbonded electron pairs. This "three-electron bonding" can be described in terms of resonance theory including ionic structures (Figure 1.15). An especially clear description of this stabilization is available in terms of molecular orbital theory (Figure 1.16). The existence of overlap between the carbon $2p_z$ orbital and the p orbital on the substituent S establishes a π bond between C and S. However, three electrons must be accommodated by the two MOs (the bonding MO and the antibonding MO) corresponding to the π bond. This requires that one electron be placed in the high-energy ABMO, but since two electrons occupy the BMO, modest net bonding can result.

1.9 Radical Reaction Modes

Organic free radicals are typically highly reactive both toward other radicals (i.e., radical–radical reactions) and toward most singlet molecules (radical–molecule reactions). The most common radical–radical reaction mode is coupling, which is the reverse of homolytic dissociation (Figure 1.17). As such, these reactions are highly exothermic [ΔH° = –D(A–B)]. For relatively unstabilized radicals such as

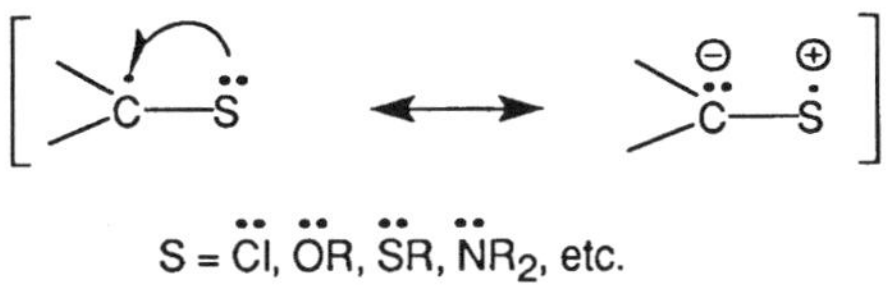

S = Cl, OR, SR, NR_2, etc.

Figure 1.15 Resonance theoretical description of three-electron bonding.

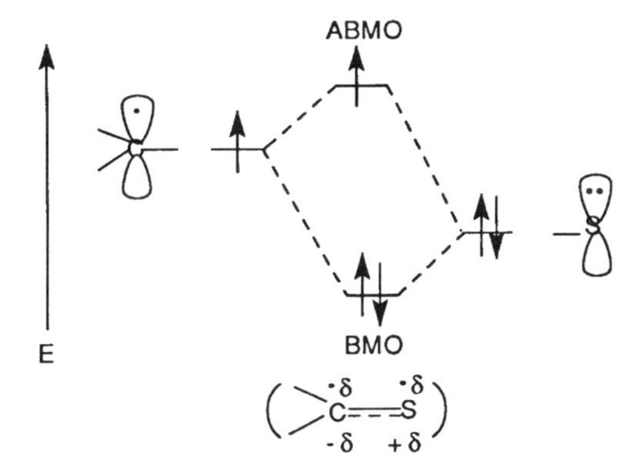

Figure 1.16 MO diagram of the three-electron bond.

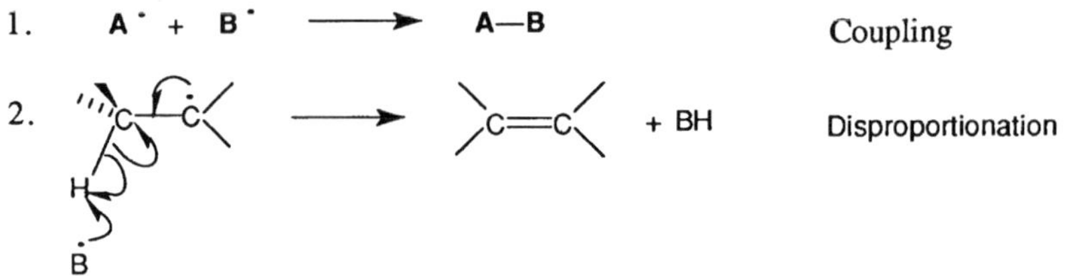

1. A$^{\cdot}$ + B$^{\cdot}$ ⟶ A—B Coupling

2. ⟶ C═C + BH Disproportionation

B

Figure 1.17 Radical–radical reaction modes.

simple alkyl radicals ($R^{\bullet}$) and halogen atoms ($X^{\bullet}$), both the coupling and cross-coupling reactions are activationless and essentially diffusion controlled. This high level of "homophilic" reactivity distinguishes radicals from carbocations or carbanions and represents the essential challenge to the preparation of an isolably stable free radical. A somewhat less common radical–radical reaction mode is disproportionation, which is essentially a homolytic β *elimination* reaction.

The reactions of radicals with singlet molecules are of two basic types, *abstraction* and *addition* (Figure 1.18). Abstraction refers to the homolytic removal

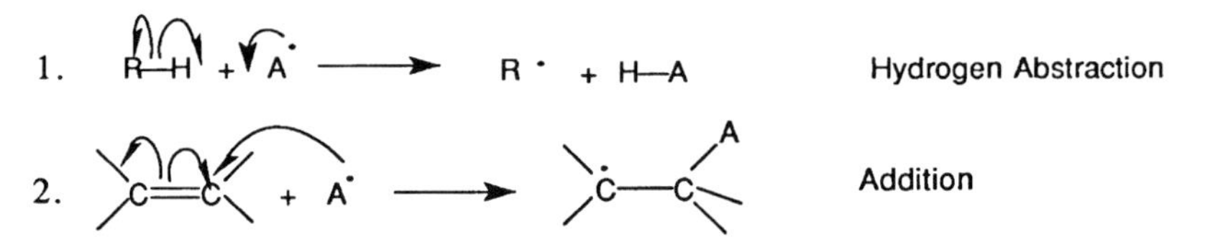

Figure 1.18 Radical–molecule reaction modes.

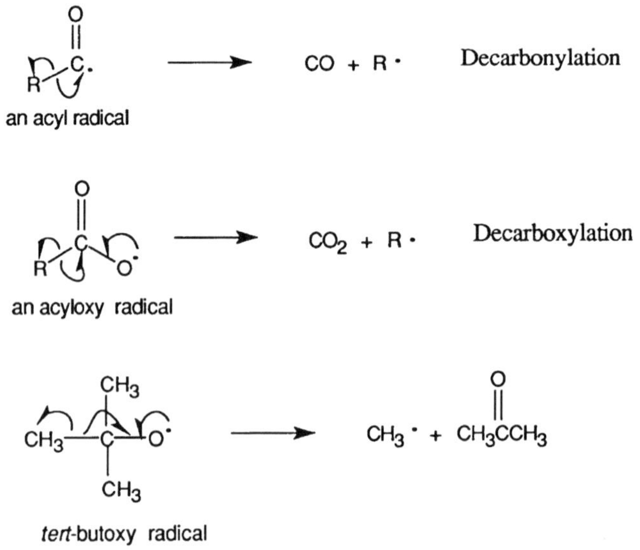

an acyl radical ⟶ CO + R · Decarbonylation

an acyloxy radical ⟶ CO_2 + R · Decarboxylation

tert-butoxy radical ⟶ CH_3 · + CH_3CCH_3 (O)

Figure 1.19 Fragmentation reactions of radicals.

of an atom (or more rarely a group) from a molecule by a radical, thereby generating a new radical and a new singlet molecule. Hydrogen and the heavier halogens (except fluorine) are the atoms most readily removed. When the molecule is unsaturated, radical addition is a feasible reaction mode.

Addition and abstraction reactions may, of course, occur in either an intermolecular or an intramolecular context. This is also true for elimination, which can occur via a radical–radical reaction (disproportionation) or via a unimolecular radical reaction (the reverse of addition; Figure 1.18). When the bond homolyzed in a unimolecular elimination is a C–C bond, the reaction is often described as a fragmentation (Figure 1.19). Finally, rearrangements (or sigmatropic shifts) are far less common in radicals than in carbocations, but are occasionally observed (vide infra).

1.10 Stable Radicals

The ultimate challenge in attaining an isolably stable free radical is preventing coupling of two radicals to afford a dimer (Figure 1.20). Stabilizing a free radical relative to its dimer can be approached through effects that exclusively or primarily stabilize the radical (especially delocalization effects), those that primarily

$$2R^{\bullet} \rightleftharpoons \underset{\text{DIMER}}{R\text{—}R}$$

Figure 1.20 Radical coupling.

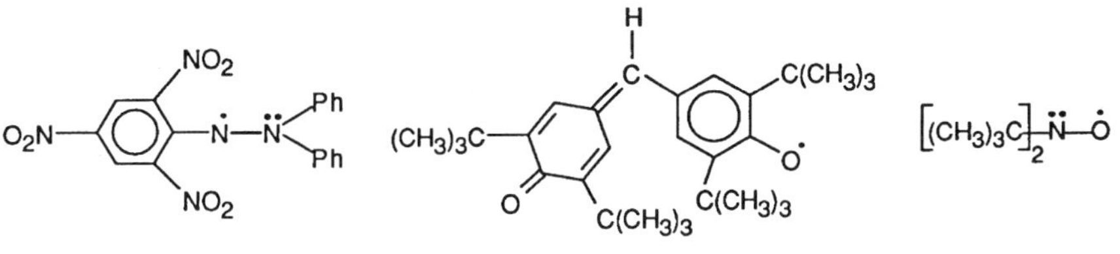

Diphenylpicrylhydrazyl[18] (DPPH)

Galvinoxyl[19]

di-tert-Butylnitroxyl[20] (a distillable liquid)

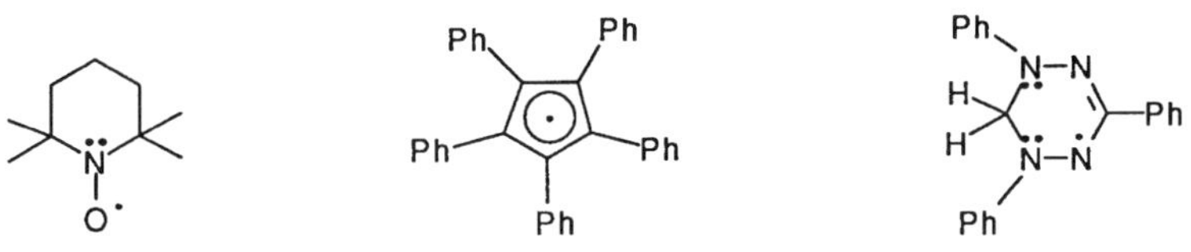

TEMPO[21]

Pentaphenylcyclopentadienyl[22]

Triphenylverdazyl[23]

Figure 1.21 Stable free radicals.

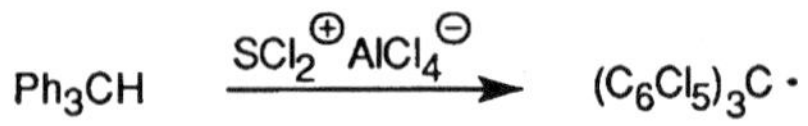

$$Ph_3CH \xrightarrow{SCl_2^{\oplus}AlCl_4^{\ominus}} (C_6Cl_5)_3C\cdot$$

Figure 1.22 The perchlorotrityl radical.

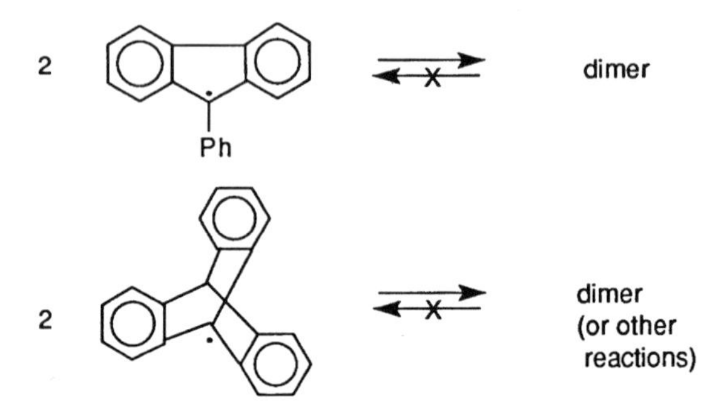

Figure 1.23 Relatively unstable trityl-type radicals.

destabilize the dimer (especially steric effects), or those that are strictly relative (an inherently weak R–R bond, e.g., an O–O bond). The high degree of relative stabilization required to attain a stable free radical usually requires a combination of these effects, as illustrated for the six stable radicals depicted in Figure 1.21. It should be noted that the first, third, fourth, and sixth of these radicals probably have some stabilization from a three-electron bond.

Although the six radicals shown above are impressively stable, the perchlorotrityl radical is quite possibly the most stable organic radical known.[24] This bright red, crystalline solid is 100% dissociated both in solution and in the solid state, is stable in air, and has a half-life estimated to be *ca.* 100 *years*. It is thermally stable up to 300°C (in air) and is unreactive toward NO, NO_2, sulfuric acid, concentrated nitric acid, and chlorine. This novel radical is prepared by perchlorination of triphenylmethane (Figure 1.22).

Radicals that are much *less* stable than the trityl radical result when steric effects are diminished, as in the 9-phenylfluorenyl radical,[25] and especially when the conjugation of the radical site with the aryl rings is sharply diminished or even removed, as in the case of the triptycyl radical (Figure 1.23).[26]

1.11 An "Aromatic" Radical

The phenalenyl radical (Figure 1.24),[27] which is stabilized essentially only by cyclic delocalization, is a "persistent" radical in solution (similar to trityl). It is rare indeed for a radical to have this level of stability without the benefit of steric retardation of dimerization or an inherently weak dimer bond. The impressible stability of the phenalenyl radical is easily rationalized by noting that both the phenalenyl cation

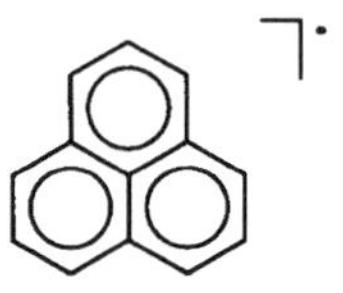

Figure 1.24 Phenalenyl radical.

and anion are aromatic (the LUMO of the cation, which is the HOMO of the anion, is a nonbonding MO).[28] Thus the radical where the SOMO is this same NBMO would be expected to have a similarly high stabilization derived from cyclic delocalization. Consequently, the phenalenyl radical could reasonably be construed as "aromatic," although this terminology is not usually applied to radicals.

1.12 Radical Scavengers

Stable radicals resist dimerization but remain highly reactive toward typical, reactive free radicals. Such radicals as DPPH and Galvinoxyl are therefore efficient *radical scavengers* and can be used to intercept reactive radical intermediates (Figure 1.25). In the thermal decomposition of AIBN, for example, the normally efficient conversion to dimer is reduced to a constant 34% by a variety of radical scavengers.[29] Consequently, only about 66% of the radicals are considered to become free and are trappable by the scavenger. The other 34% are considered to couple before escaping the solvent cage, a phenomenon known as *geminate* (or *cage) recombination.* In contrast, the benzoyloxy radicals produced by the thermal decomposition of dibenzoylperoxide are 100% trappable by I_2 or Galvinoxyl.[30] Evidently, the fragmentation of benzoyloxy radicals to phenyl

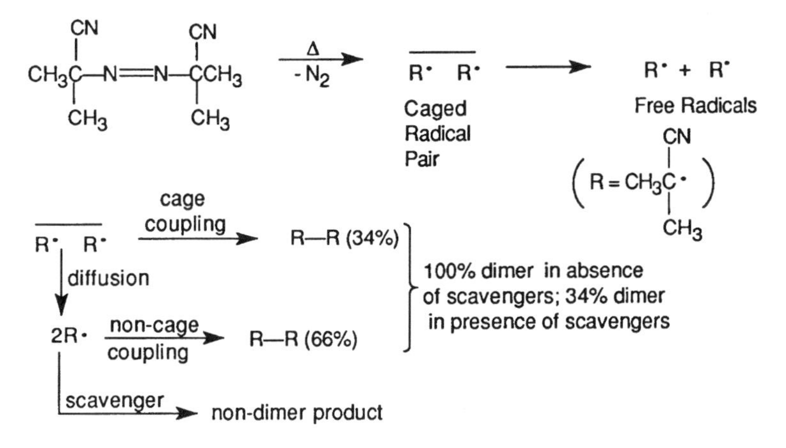

Figure 1.25 Radical scavengers.

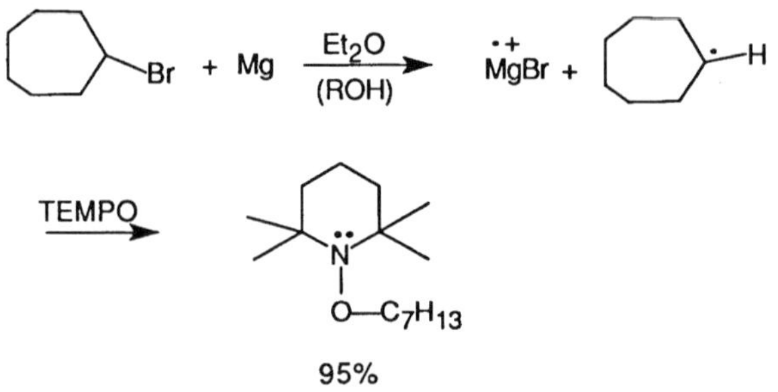

Figure 1.26 Trapping of cycloheptyl radicals by TEMPO in the reaction of cycloheptyl bromide with magnesium.

radicals and carbon dioxide is slower than diffusion of the benzoyloxy radicals out of the cage. As is illustrated by I_2, efficient radical scavengers need not themselves be radicals, but the best scavengers, *viz.* those that couple with transient radicals at a virtually diffusion-controlled rate, are often stable radicals (e.g., TEMPO). TEMPO has been used, for example, to trap cycloheptyl radicals efficiently in the reaction of cycloheptyl bromide with magnesium in diethyl ether (Figure 1.26).[31] The experiment, in its simplest form, is potentially complicated by the subsequent reaction of the Grignard reagent with TEMPO. Consequently, excess (5.0 M) *tert*-pentyl alcohol was included to protonate any Grignard reagent produced as rapidly as it is formed, giving cycloheptane. The product of the radical trapping, N-cycloheptoxy-2,2,6,6-tetramethylpiperidine, was obtained in 95% yield. The experiment illustrates the importance of assuring the compatibility (i.e., the nonreactivity) of the radical scavenger with both reactants and products.

1.13 Spin Traps

Radical scavengers permit evaluation of the extent of *free radical* as opposed to *caged radical* pair formation in a reaction by capturing the free radicals and diverting them from their normal product channels. The characterization of the structure of the specific radicals being trapped is often accomplished by means of *spin traps* (Figure 1.27), that is, molecules that react efficiently with radicals to form a stable or persistent radical, the structure of which can be examined via electron spin resonance (ESR).[32]

The ESR structural analysis is made more secure by comparing the observed spectrum with the spectrum of the authentic nitroxyl radical. This can be synthesized, for example, by addition of the appropriate Grignard reagent (RMgX) to the nitrone and then air oxidizing the resulting anion. The specific features of the ESR spectrum that permit characterization of the R group are considered in Chapter 3.

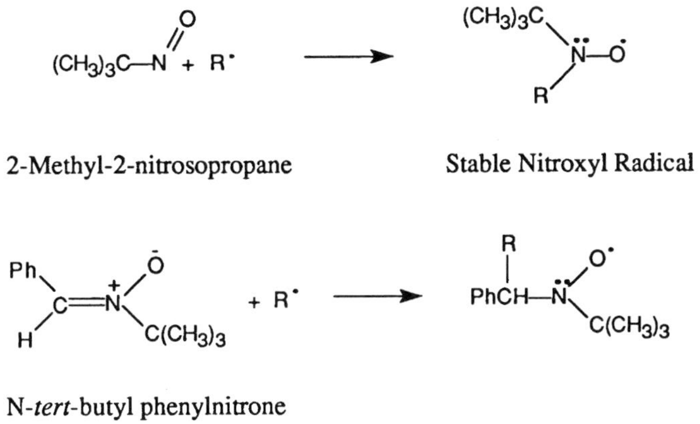

2-Methyl-2-nitrosopropane — Stable Nitroxyl Radical

N-*tert*-butyl phenylnitrone

Figure 1.27 Spin traps.

1.14 Radical Inhibitors

Substances used in small amounts (e.g., 0.5–5 mol %) to inhibit (read *prevent*) radical chain reactions are termed *inhibitors*. They operate, like radical scavengers, by consuming efficiently any radical intermediates that are generated, but their function is to prevent further reaction via radical chains. Stoichiometric amounts of an inhibitor are unnecessary since radical chain reactions are necessarily propagated via a relatively small number of radicals provided through initiation. Stable radicals are neither required nor are they desirable for inhibition since the level of reactivity towards radicals provided by even such simple molecules as hydroquinone and a variety of other phenol derivatives is sufficient, and since even stable radicals may be able to initiate some radical chains.

1.15 Radical Probes and Clocks

Radical probes are molecules that, if they are converted to free radical intermediates in the course of any reaction, undergo a unique and rapid radical reaction (called the probe reaction), which decisively establishes the involvement of a free radical mechanism. Equally important, failure to observe the probe reaction can often be construed as strong evidence against the operation of a free radical mechanism. Radical probes are thus powerful tools for discerning free radical mechanisms. A familiar example of a radical probe reaction is the cyclization of the 5-hexenyl radical to the cyclopentylmethyl radical, which is an intramolecular radical addition (Figure 1.28).[33–36]

The use of this probe reaction as a mechanistic tool can be illustrated by considering the reduction of alkyl bromides (*R*Br) by tributyltin hydride (Bu_3SnH) in the presence of benzoyl peroxide, a reaction that is postulated to occur via the radical chain mechanism of Figure 1.29.[36,37]

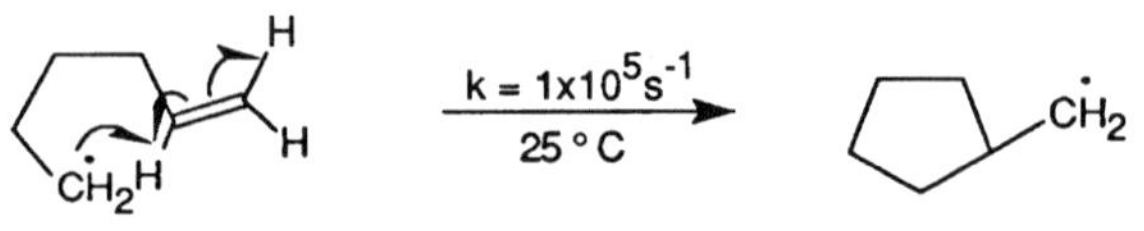

$k = 1\times10^5 s^{-1}$, 25 °C

5-Hexenyl Radical → Cyclopentylmethyl Radical

Figure 1.28 Cyclization of the 5-hexenyl radical.

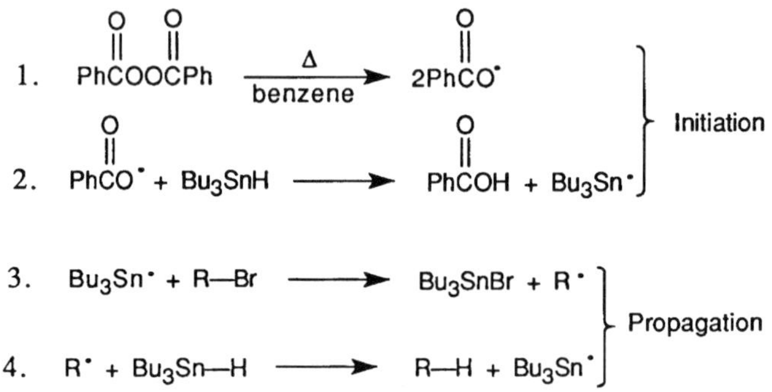

1. $PhC(=O)OOC(=O)Ph \xrightarrow[\text{benzene}]{\Delta} 2PhC(=O)O^{\bullet}$
2. $PhC(=O)O^{\bullet} + Bu_3SnH \longrightarrow PhC(=O)OH + Bu_3Sn^{\bullet}$

Initiation

3. $Bu_3Sn^{\bullet} + R\text{—}Br \longrightarrow Bu_3SnBr + R^{\bullet}$
4. $R^{\bullet} + Bu_3Sn\text{—}H \longrightarrow R\text{—}H + Bu_3Sn^{\bullet}$

Propagation

Figure 1.29 Reduction of alkyl bromides by tributyltin hydride.

If this reaction is carried out using the probe molecule 5-bromo-1-hexene as the alkyl bromide, the radical produced should be the 5-hexenyl radical. In view of the rapidity of the previously described probe reaction (i.e., cyclization), hydrogen atom abstraction from Bu_3SnH should give both 1-hexene and methylcyclopentane in relative amounts that depend directly on the concentration of Bu_3SnH (Figure 1.30). The experimental confirmation of this prediction represents strong support for the proposal of alkyl radical intermediates generally in these tributyltin hydride reductions. Further, since the rate constant of the probe reaction is known (1×10^5 s^{-1} at 25°C), the (second-order) rate constant for the hydrogen abstraction can be calculated from the ratio of the two products and the concentration of Bu_3SnH. Probe reactions of known rate constant can be used to determine rate constants of a variety of potentially competing reactions and are referred to in this context as

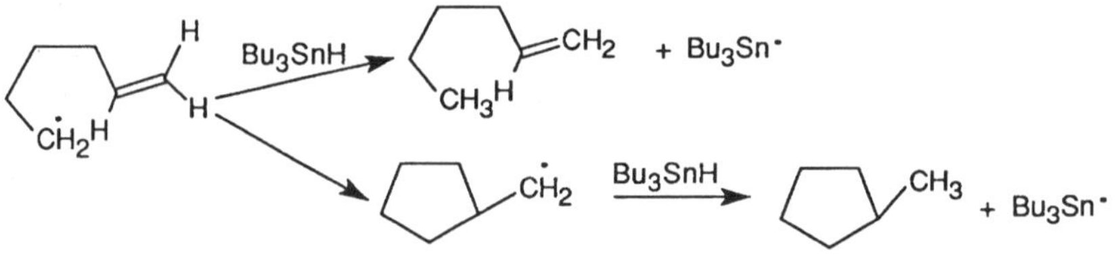

Bu_3SnH → 1-hexene + $Bu_3Sn^{\bullet}$

→ cyclopentylmethyl radical $\xrightarrow{Bu_3SnH}$ methylcyclopentane + $Bu_3Sn^{\bullet}$

Figure 1.30 Competition between the probe reaction and hydrogen atom abstraction.

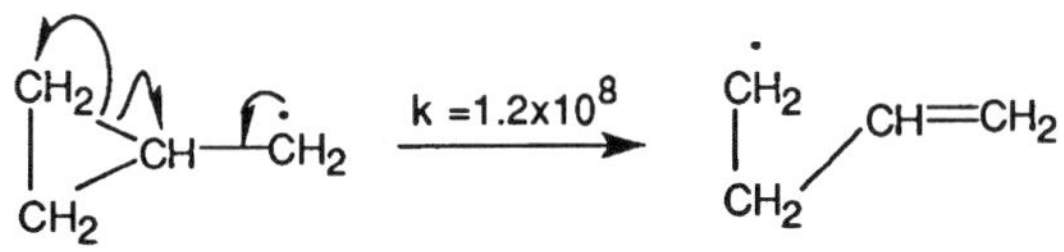

CH_2 CH—$\dot{C}H_2$ CH_2 $\xrightarrow{k = 1.2\times10^8}$ $\dot{C}H_2$ CH_2 $CH{=}CH_2$

Cyclopropylcarbinyl Radical

Allylcarbinyl Radical

Figure 1.31 The cyclopropylcarbinyl radical probe reaction.

"clock reactions." For reactions involving exceptionally short-lived radical intermediates, or to more rigorously rule out radical intermediates in a given reaction, still more sensitive (i.e., faster) probe reactions are required. A considerably faster probe reaction is the cleavage of the cyclopropylmethyl radical, which has a rate constant of $1.2 \times 10^8\ s^{-1}$ (Figure 1.31).

1.16 Hypersensitive Mechanistic Probes

Radical probe reaction products can often be detected rather sensitively. For example, methylcyclopentane, the probe product of the 5-hexenyl radical probe, could easily be detected in 0.1–1% yield, where the main product is 1-hexene. Consequently, radical intermediates would be detected even if the rate of abstraction of a hydrogen atom from Bu_3SnH or other hydrogen donors were 10^2 or 10^3 times the rate of the probe reaction (at 1 M Bu_3SnH, e.g.). That is to say, probe reactions do not have to be as fast as the reactions they are competing against in order to provide evidence for a radical intermediate. Under the circumstances specified above, radical intermediates (as evidenced by the formation of probe products) would be detected by the 5-hexenyl probe even if the abstraction rate constants were as much as $10^8\ M^{-1}\ s^{-1}$ (at 1 M hydrogen donor). Hypothetically, if the probe product were not observed, two possibilities would exist; that is, either radical intermediates are not involved or the competing (abstraction) reaction has a rate constant $k_{comp} > 10^8\ M^{-1}\ s^{-1}$. Since many radical reactions have rate constants larger than this, it is obviously important to develop still faster radical probe reactions. A large number of such hypersensitive mechanistic probes have been developed and their probe reaction rate constants measured.[38,39] The two examples illustrated in Figure 1.32 are fast enough to permit the detection of a radical intermediate in competition with *any* competing reaction, since a species which reacts at a rate of $10^{13}\ s^{-1}$ is presumably a transition state. Incidentally, the "radical clock" concept proves to be extremely useful in measuring the rates of these ultrafast probe reactions. It was previously noted that rates of hydrogen abstraction (e.g., from Bu_3SnH) can be measured using radical clock reactions of known rate constant. These hydrogen abstraction rate constants can then be used to obtain rate constants of still faster probe reactions, which can be used to calibrate faster hydrogen

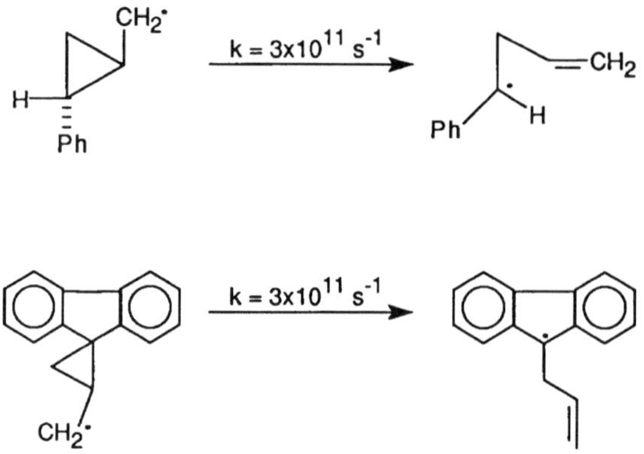

Figure 1.32 Hypersensitive mechanistic radical probes.

abstraction reactions. Faster hydrogen abstraction is provided by thiophenol (k_{abs} = $1.1 \times 10^8 M^{-1} s^{-1}$ at 25°C) and especially by benzeneselenol ($k = 2 \times 10^9 M^{-1} s^{-1}$ at 25°C).[40] The latter is especially useful in clocking the rates of hypersensitive probe reactions. The rates of hydrogen transfer from benzeneselenol to primary alkyl radicals are fast enough to be partially diffusion controlled.

1.17 The Uniqueness of the Radical Probe Reaction

As was indicated previously, a radical probe reaction must not only be rapid but it should ideally represent a unique reaction mode, specific to radical chemistry. For example, a rapid cyclopropylcarbinyl to allylcarbinyl cleavage would not be expected of a carbocation intermediate, because of the great stability of the cyclopropylcarbinyl cation in comparison to the allylcarbinyl cation (if the latter even exists). A more serious problem is the possible conversion of the normal (e.g., a cyclopropylcarbinyl alcohol) reaction product to the isomeric probe product (allyl carbinyl alcohol) subsequent to the reaction step of primary mechanistic interest, possibly via a carbocation mechanism. This is especially problematic in enzyme-catalyzed reactions, where the product may remain bound to the enzyme for a considerable time after the primary reaction event. A case in point is the hydroxylation of bicyclo[2.2.1]pentane, a rather fast radical probe that undergoes cleavage at a rate of $2.1 \times 10^9 s^{-1}$. Enzymatic hydroxylation produces, in addition to the normal product (the bicyclic alcohol), substantial amounts of the probe product (Figure 1.33).[41] This result tends to suggest a radical mechanism (abstraction of hydrogen by the $Fe^V = O$ intermediate) in preference to a concerted oxygen insertion into the C–H bond. However, the much greater thermodynamic stability of the probe alcohol than the normal alcohol, coupled with the existence of a plausible acid or electrophile-catalyzed interconversion mechanism engenders concern that the supposed probe product may actually be

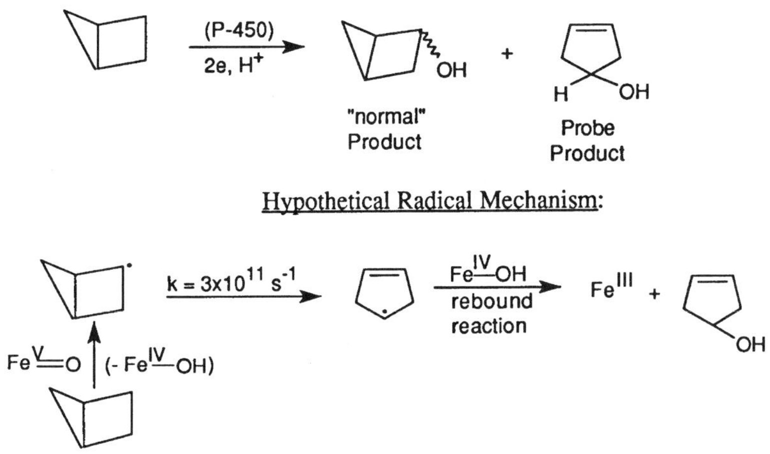

Figure 1.33 The 2-bicyclo[2.1.0]pentyl radical probe in enzymatic hydroxylation.

formed indirectly, that is, by isomerization of the normal product. Consequently, no definitive conclusions should be drawn from these data concerning the mechanism of the hydroxylation step. A more rigorous approach to the application of radical probes, especially in enzyme-catalyzed reactions, is to build into a single-probe reaction system regiochemically different radical and carbocation reaction modes (Figure 1.34). Such probes have now been made available,[42] and a careful study of their application to P-450 hydroxylation has recently been reported.[43] Incubation of the appropriate methylcyclopropane derivative (Figure 1.35) with microsomal P-450 (a mixture of isozymes) or with the CYP2B1 isozyme leads to a mixture of products, but the predominant product corresponds to hydroxylation at the methyl group *without* cleavage of the cyclopropane ring.

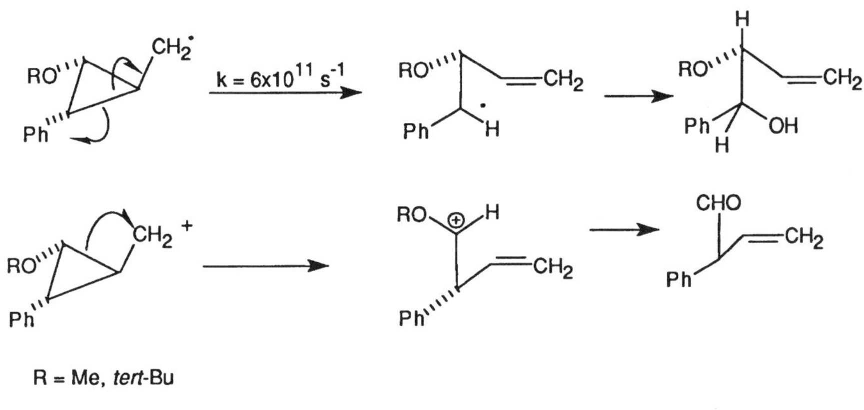

Figure 1.34 Probes with distinct radical and carbocation reaction modes.

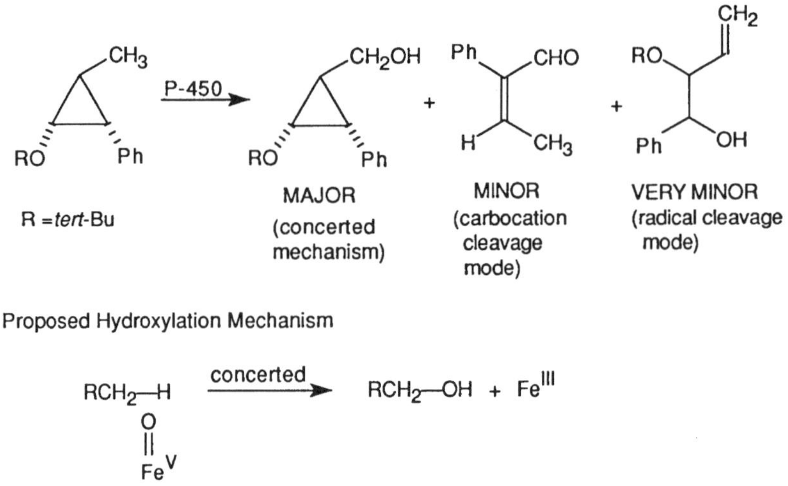

Figure 1.35 P-450 Hydroxylation of an ultrasensitive probe that distinguishes radical and carbocation cleavage modes.

Ring cleaved products are relatively minor in extent and correspond primarily to the *carbocation* regiochemical mode. However, very small amounts of the radical probe product are found.

From the known rate of the radical probe reaction ($6 \times 10^{11} s^{-1}$) and the ratio of the uncleaved cyclopropylcarbinyl alcohol to the allylcarbinyl alcohol that corresponds to radical cleavage, the lifetime of the radical species that gives rise to the radical cleavage product is found to be just 70 fs. Consequently, this radical species is not considered to be a true intermediate, which would have a lifetime greater than that of a vibration ($>10^{-13}$s). The hydroxylation mechanism is therefore most appropriately described as concerted, since the main reaction path apparently does not involve a true intermediate.

1.18 Radical Rearrangements

Although 1,2 shifts of alkyl groups and hydrogen are a familiar aspect of carbocation chemistry, such rearrangements are completely unknown in the radical series, even where very substantial driving force exists (Figure 1.36). Because neither radicals nor carbanions undergo 1,2-alkyl or hydrogen shifts, the occurrence of such shifts can be considered diagnostic for carbocation intermediates. The higher activation energies required for radical than for carbocation rearrangements could potentially have its basis in the generally much smaller thermodynamic driving force for the radical rearrangements. However, carbocation rearrangements are readily observable when no, or even negative, driving force exists, so that the

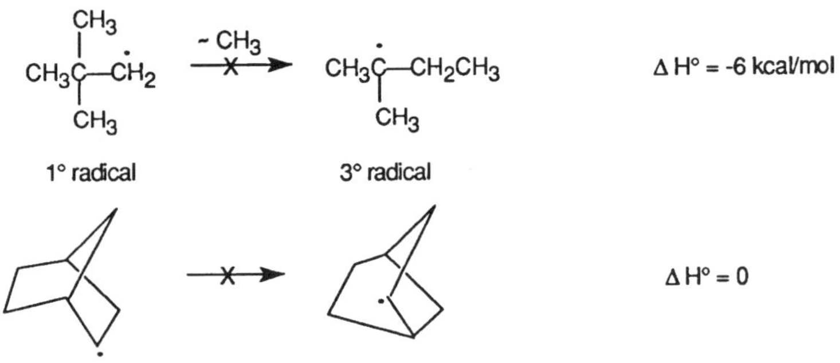

$CH_3\dot{C}(CH_3)_2$—$\dot{C}H_2$ $\xrightarrow[\times]{-CH_3}$ $CH_3\dot{C}(CH_3)$—CH_2CH_3 $\Delta H° = -6$ kcal/mol

1° radical 3° radical

$\Delta H° = 0$

Figure 1.36 Nonrearrangement of the neopentyl and norbornyl radicals.

Figure 1.37 (a) TS for carbocation rearrangement; (b) TS for radical rearrangement.

contrasting radical/carbocation behavior must have a purely kinetic basis. A simple rationalization of this, based upon the concept of transition-state (TS) aromaticity, notes that the transition state for a 1,2 rearrangement has a *pericyclic* orbital system consisting of three AOs (Figure 1.37).[44] In the case of a carbocation system, the electron population in this pericyclic system is two, making it “aromatic,” whereas the TS for a radical rearrangement has three electrons in the same pericyclic system and is nonaromatic.

Rearrangements of phenyl and other unsaturated groups, on the other hand, are feasible, but these probably occur in a stepwise fashion involving addition/elimination. The pioneering study of Winstein and Seubold established that both *tert*-butylbenzene and isobutylbenzene are formed in the radical chain decarbonylation of 3-methyl-3-phenylbutanal (Figure 1.38).[45] Further, the amount of rearranged product (isobutylbenzene) is found to be inversely proportional to the concentration of the aldehyde reactant, suggesting that the reaction giving rearranged product is suppressed by an increased aldehyde concentration, as would be expected for a radical rearrangement that competes with hydrogen atom abstraction from the aldehyde. Consequently the formation of both *tert*-butylbenzene and isobutylbenzene is not the simple result of nonregiospecific hydrogen abstraction at both the primary and tertiary carbons of the cyclic intermediate $A^{\bullet}$, which is the radical analogue of a Cram phenonium ion.

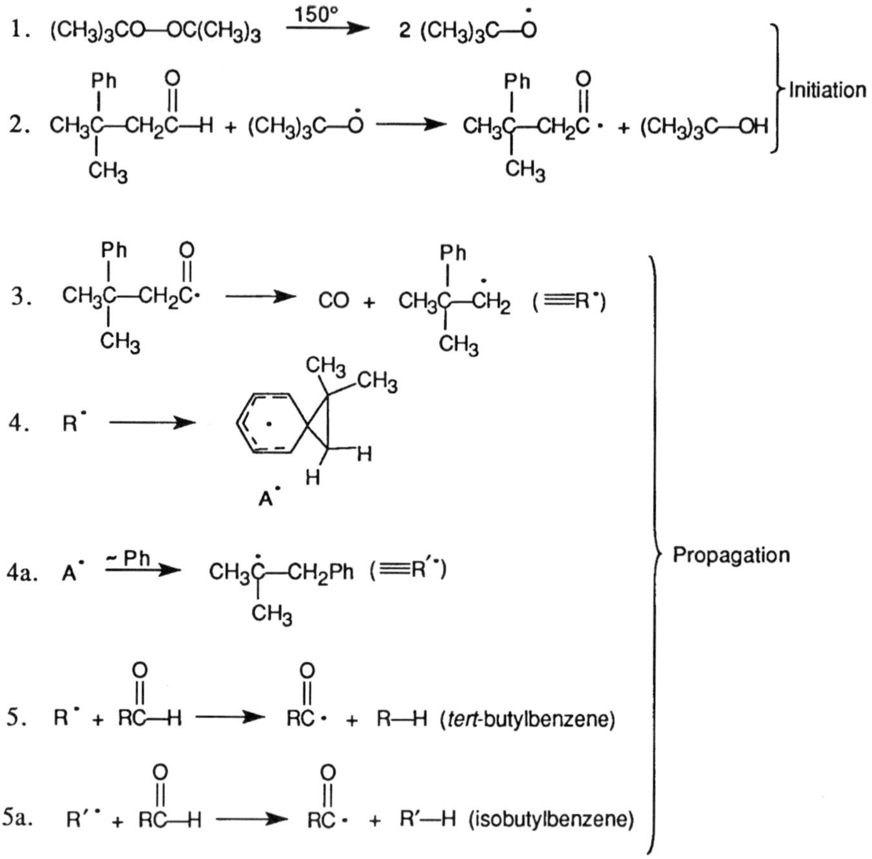

1. $(CH_3)_3CO{-}OC(CH_3)_3 \xrightarrow{150^\circ} 2\ (CH_3)_3C{-}\dot{O}$

2. $CH_3C(Ph)(CH_3){-}CH_2C(=O){-}H + (CH_3)_3C{-}\dot{O} \longrightarrow CH_3C(Ph)(CH_3){-}CH_2\dot{C}(=O) + (CH_3)_3C{-}OH$

Initiation (steps 1–2)

3. $CH_3C(Ph)(CH_3){-}CH_2\dot{C}(=O) \longrightarrow CO + CH_3C(Ph)(CH_3){-}\dot{C}H_2 \ (\equiv R^\bullet)$

4. $R^\bullet \longrightarrow A^\bullet$

4a. $A^\bullet \xrightarrow{\sim Ph} CH_3\dot{C}(CH_3){-}CH_2Ph \ (\equiv R'^\bullet)$

5. $R^\bullet + RC(=O){-}H \longrightarrow R\dot{C}(=O) + R{-}H$ (*tert*-butylbenzene)

5a. $R'^\bullet + RC(=O){-}H \longrightarrow R\dot{C}(=O) + R'{-}H$ (isobutylbenzene)

Propagation (steps 3–5a)

Figure 1.38 Rearrangement of phenyl groups in the neophyl radical.

1.19 Nonclassical Radicals

In sharp contrast to the norbornyl cation, the norbornyl radical not only does not adopt a symmetrical nonclassical structure; it does not rearrange at all. 5-Norbornenyl radicals do undergo rearrangement but not to give a nonclassical radical (Figure 1.39).[46] The rearrangement, which is actually an intramolecular radical addition to the $C_5{-}C_6$ double bond to give a tricyclic radical, is slow enough to permit competition with hydrogen abstraction from various hydrogen donors. Again, the extent of rearrangement is inversely dependent upon the concentration of the hydrogen donor. These results, however, should not necessarily be taken as excluding a small amount of delocalization of the electron in the 5-norbornenyl radical onto the double bond and thus a small stabilization derived from a nonclassical $(C_2{-}C_6)$ interaction.

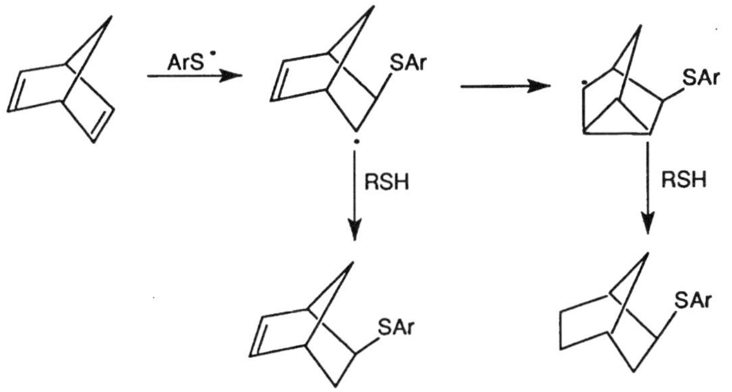

Figure 1.39 Rearrangement of 5-norbornenyl radicals.

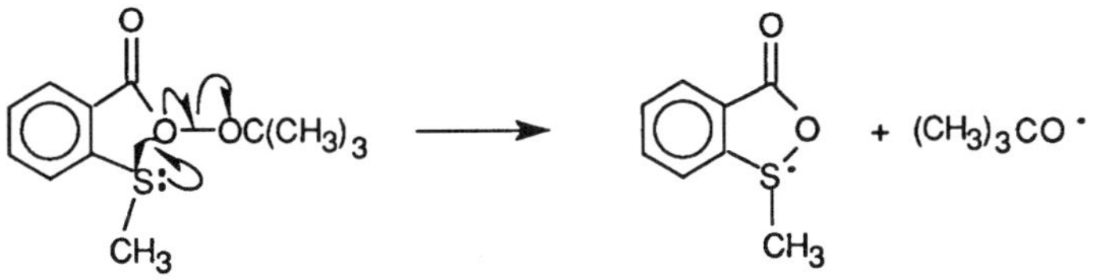

Figure 1.40 Anchimeric assistance to homolysis.

1.20 Anchimeric Assistance for Homolysis

Anchimeric (or neighboring group) assistance is much less common for homolytic than heterolytic cleavages, but the feasibility of homolytic assistance is nicely demonstrated by the homolysis of *tert*-butylperoxybenzoate esters having *ortho* methylthio and, to a lesser extent, *ortho* iodo substituents (Figure 1.40).[47] The *o*-methylthio derivative undergoes O–O cleavage at a rate 140,000 times that of unsubstituted compound, while the *o*-iodo derivative reacts 43 times as rapidly as the parent. Steric effects are evidently not responsible for the rate acceleration, since the *o-tert*-butyl derivative homolyzes at a rate only three times that of the parent.

1.21 Polar Effects in Radical Reactions

To a degree that is initially somewhat surprising, the rates of many radical reactions are found to be subject to the strictly polar effects of substituents. For example, the relative rates of benzylic hydrogen abstraction from *m*- and *p*-substituted toluene derivatives by bromine atoms (Figure 1.41) are well correlated by the Hammett–Brown equation ($\log k/k_0 = \rho\sigma^+$) with $\rho = -1.36$.[48] The existence of a good correlation with σ^+ (instead of the Hammett sigma) indicates the generation of

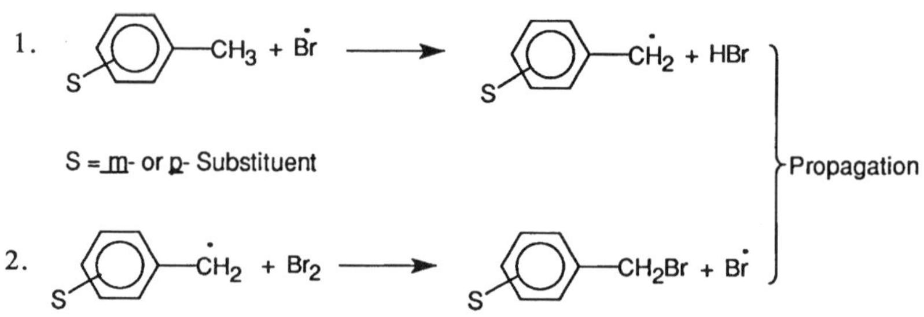

Figure 1.41 Radical chain bromination of toluene derivatives.

partial positive charge at the benzylic position in the TS for abstraction. The value of ρ is an indication of the (modest) magnitude of this positive charge. As a point of calibration, *p*-methoxytoluene reacts about ten times as fast as toluene. The simple dissociation of a hydrogen atom from toluene, generating the benzyl radical, does not engender benzylic positive charge, although the conversion of a benzylic methyl group (which is hyperconjugatively electron donating) to a benzylic methylene (i.e., radical) site might slightly decrease the electron density at the benzylic position and provide a small negative ρ value. This effect, if it exists at all, is far too small to explain the bromination results. Consequently the substituent effects on these abstractions must represent a purely kinetic effect on TS stability and certainly do not arise from benzylic radical character in the TS. Apparently, the TS, but interestingly not the reactants or products, is somewhat polar, specifically with benzylic carbocation character, as indicated in the resonance theoretical treatment of the TS (Figure 1.42).

The development of this polar character, of course, is a consequence of the relatively high electrophilicity of bromine atoms. An especially interesting aspect of the Hammett–Brown correlation observed in the bromination of toluene derivatives is that the σ^+ substituent parameter represents a purely polar substituent effect, that is, the substituent's effect on a benzylic positive charge. The purely radical

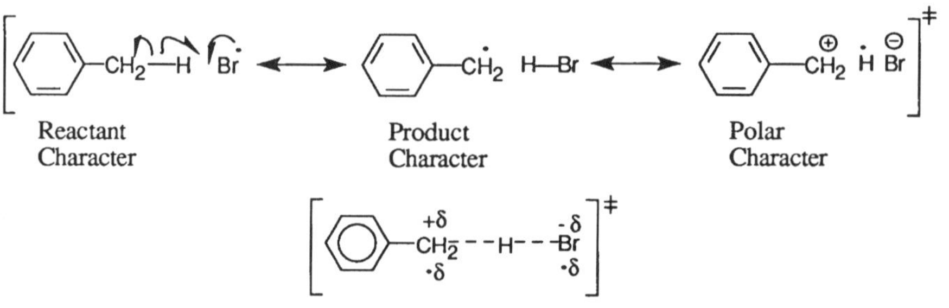

Figure 1.42 Polar character in the TS for hydrogen abstraction by $Br^{\bullet}$.

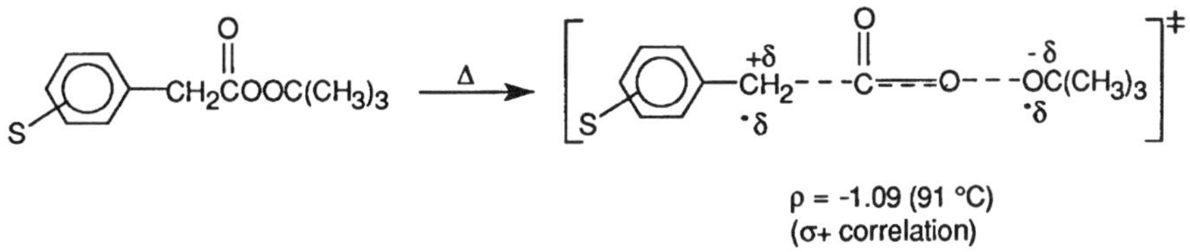

ρ = -1.09 (91 °C)
(σ+ correlation)

Figure 1.43 Homolytic two-bond cleavages in *tert*-butylperoxyphenylacetates.

stabilizing effect of substituents ($\sigma^{\bullet}$) is apparently dominated by the polar effects, at least in this case. The corresponding ρ value for hydrogen abstraction from toluene by chlorine atoms is substantially less than for bromination (–0.66), undoubtedly because the transition state is much more reactantlike in this strongly exergonic abstraction.[49] On the other hand, polar effects are comparable to those in bromination when the abstracting radical is trichloromethyl ($\rho = -1.4$).[50] The methyl and phenyl radicals apparently have very small polar effects ($\rho = -0.1$) and the ρ value for abstraction by triethylsilyl radicals is +0.3. Sizeable polar effects are also observed in the two-bond-concerted cleavages of *tert*-butylperoxy esters of arylacetic acids (Figure 1.43).[51]

Polar effects are also expressed in radical addition reactions. An impressive example is the strong polarization found for the transition states for the addition of *tert*-butyl radicals to ethene and substituted ethenes.[52] High-quality *ab initio* calculations indicate that the transition states for *tert*-butyl radical additions are stabilized by 4.8–6.0 kcal/mol (in the gas phase) more than the corresponding transition states for the methyl radical additions via a polar effect (Figure 1.44). As a consequence of polar effects, the much more highly hindered and more stabilized radical is actually more reactive in additions to not only electron-deficient alkenes but even to ethene itself. For comparison, *tert*-butyl radicals appear to be more nucleophilic than even hydroxymethyl radicals ($\bullet CH_2OH$), and their ionization energy is less.

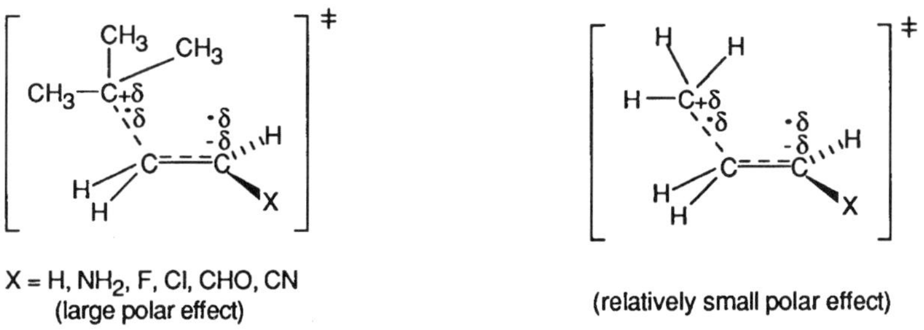

X = H, NH_2, F, Cl, CHO, CN
(large polar effect)

(relatively small polar effect)

Figure 1.44 The nucleophilic character of the *tert*-butyl radical.

1.22 Frontier-Orbital Interpretation of Polar Effects in Radical Additions

The addition of an alkyl radical to an alkene is substantially exothermic, since it involves the conversion of a carbon–carbon π bond to a σ bond. Consequently, the transition state should be relatively early, that is, reactantlike. This would be especially true for additions to alkenes that have radical-stabilizing substituents such as cyano, carbethoxy, or phenyl. Under these conditions a simple perturbational approach such as frontier-orbital (FO) theory should provide a useful model of the transition state.[53] The relevant frontier orbitals are, for the attacking radical, the SOMO and, for the alkene, the HOMO and LUMO. Interaction between the SOMO and the HOMO has the effect of delocalizing the electron pair of the HOMO onto the radical ($R^{\bullet}$) and the odd electron of the SOMO onto the alkene. Since the HOMO is doubly occupied and the SOMO singly occupied, the net effect is transfer of electron density from the alkene to the radical moiety. In an analogous manner, the interaction of the SOMO with the alkene LUMO transfers electron density from the radical moiety to the alkene. If the SOMO/HOMO and SOMO/LUMO interactions are equally strong, as they would be if the SOMO is midway between the HOMO and the LUMO in energy, there is no net electron transfer, and the transition state is essentially nonpolar. This is approximately the case for the addition of the methyl radical to ethene (Figure 1.45). However, if the SOMO is much closer in energy to the LUMO than to the HOMO, the former interaction (SOMO/LUMO) is dominant, and there is electron transfer from the radical moiety to the alkene. Electron-withdrawing groups (EWG) on the alkene moiety lower the energy of both the HOMO and LUMO and thus favor the SOMO/LUMO interaction. The same interaction is favored when the radical moiety is electron rich, a circumstance corresponding to a higher-energy SOMO.

The combination of both an electron-rich radical and an electron-deficient alkene yields an especially strong stabilizing interaction and an extensive transfer of electron density from the radical moiety to the alkene moiety. Conversely, when the radical is electron deficient (low-energy SOMO) and the alkene is electron rich (high-energy HOMO), the SOMO/HOMO interaction is dominant, and there is net transfer of electron density from the alkene to the radical moiety.

As an illustration, consider the addition of cyclohexyl radicals to methyl acrylate. Secondary radicals are considered moderately electron rich, and the electron-withdrawing carbomethoxy substituent renders the alkene electron deficient. Cyclohexyl radicals were generated by $NaBH_4$ reduction of cyclohexylmercuric acetate, and relative rates of addition of this radical to methyl acrylate and a series of acrylates substituted at the alkene position α to the carbomethoxy group were measured. The correlation of log k_{rel} with polar substituent constants (σ or σ^-) is rather good and the reaction constant, $\rho > 3$. The positive ρ value indicates that a substantial fraction of negative charge accumulates on the α carbon in the transition state, so that EDGs such as methyl and methoxy slow the rate, while EWGs like cyano and chloro accelerate addition. Presumably, partial positive charge accumulates on the cyclohexyl moiety in the transition state.

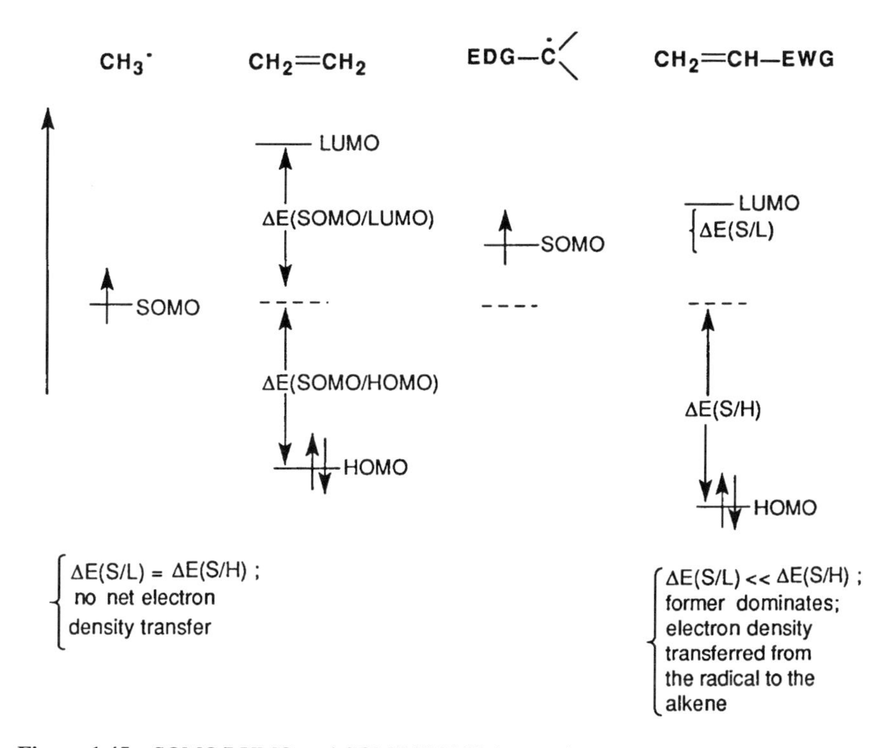

Figure 1.45 SOMO/LUMO and SOMO/HOMO interactions.

When substituents are present at the position β to the acrylate moiety, the dominant effect, as expected, is a steric effect, and log k_{rel} is found to correlate moderately well with the Taft steric parameter E_s. A *tert*-butyl group, for example, slows the rate of addition to the β carbon by a factor of 20,000, and this leads to a modest preference for addition at the α carbon. Very interestingly, the rate of addition to the α carbon is still quite enhanced in relation to the rate of addition to a monosubstituted alkene, even though the carbethoxy group would be expected to cause at least a small steric retardation toward addition at the α carbon. This is consistent with the expectation that a HOMO/LUMO interaction is operative whether the radical adds to either the α or β carbon. However, since the LUMO is more strongly concentrated on the β than the α carbon, the interaction has a greater magnitude when addition occurs at the carbon β to the conjugating substituent.

1.23 Radical Substituent Constants

A great deal of research has been dedicated to deriving a set of radical substituent constants analogous to the polar substituent constants of Hammett (σ) and Brown (σ^+). Ideally, such constants should reflect a purely radical stabilizing or destabilizing effect, as distinguished from polar and/or steric effects. The difficulty

Figure 1.46 Radical substituent constants.

inherent in this task is evident when one considers that the rates of many radical reactions correlate nicely with polar substituent constants and provide little evidence for a special stabilization effect on the radical character of the transition state. By carefully choosing a reaction system in which neither the product nor the transition state is significantly polar and in which differential steric effects are absent (Figure 1.46), a satisfactory set of constants, called $\sigma^{\bullet}$ (or σ_{rad}), has been derived.[54] These constants agree nicely with those derived from hyperfine splitting constants of the benzylic protons of *m*- and *p*-substituted benzyl radicals (called $\sigma_{\alpha}^{\bullet}$).[55] These splittings progressively decrease as the odd electron is delocalized more extensively onto the ring and the (*para*) substituent. In general, it is found that most *meta* substituents mildly destabilize a benzylic radical, and essentially all *para* substituents, whether donor or acceptor, exert significant stabilizing effects, as would be expected. The greatest stabilizing effects are exerted by NMe_2, NO_2, vinyl, phenyl, cyano, methylthio, and iodo (Table 1.2). It would appear that conjugative stabilization of the radical character of the transition state is operative for such groups as vinyl, phenyl, cyano, and carboalkoxy, while three-electron-type

Table 1.2 The Creary $\sigma^{\bullet}$ constants

Substituent	$\sigma^{\bullet}$
p-NMe_2	0.90
p-$CH{=}CH_2$	0.67
p-NO_2	0.57
p-C_6H_5	0.46
p-CN	0.46
p-SCH_3	0.43
p-I	0.41
p-CO_2Me	0.35
p-MeO	0.24
p-Br	0.13
p-CH_3	0.11
H	0.00

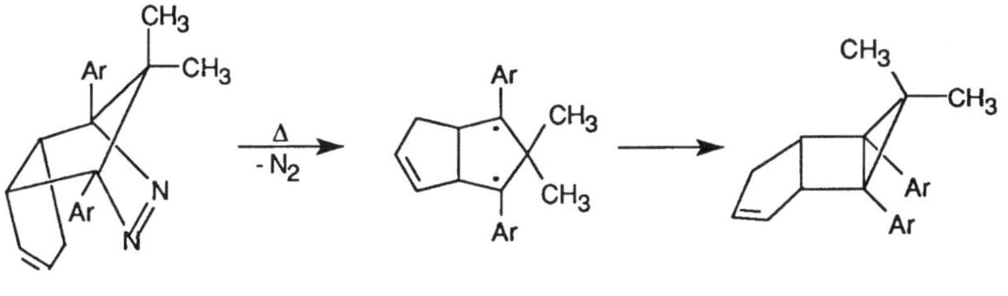

Ar = *m* or *p*-xC_6H_4-

Figure 1.47 Simultaneous polar and radical stabilization effects on a radical reaction.

bonding is presumably operative for the dimethylamino, methoxy, and methylthio groups. The relatively large $\sigma^\bullet$ values for iodo and methylthio suggest orbital overlap effects (valence-shell expansion).

A splendid example of the analysis of substituent effects on a radical reaction into polar and radical substituent effects has been provided (Figure 1.47).[56] Using Hammett σ values as polar substituent constants (σ_{pol}) and $\sigma^\bullet$ values as radical substituent constants (σ_{rad}), the rates of diradical formation were found not to correlate well with either σ or $\sigma^\bullet$, but instead to be correlated well by a two-parameter equation $\log k_{rel} = \rho_{rad}\sigma_{rad} + \rho_{pol}\sigma_{pol}$. The positive value of ρ_{pol} reflects differential stabilization of the reactant state by donor groups, based upon the $C^{+\delta}$–$N^{-\delta}$ dipoles, which are lost in the diradical. The nearly equal ρ values for polar ($\rho_{pol} = 0.48$) and radical ($\rho_{rad} = 0.50$) effects show that, even for a reaction that is not expected to develop any polar character in the transition state, polar effects on the reactant state can play an important role.

1.24 General Methods for the Generation of Specific Carbon Radicals

A wealth of synthetically convenient chemistry is available for delivering a carbon free radical site from a variety of functional groups, and new methods are constantly being developed. The tin hydride method mentioned earlier is especially prominent in the case of halogen and phenylthio functions, but carbon–halogen bonds can also be reduced by anion radicals, such as that of biphenyl, to generate radical sites.[57] Carboxyl groups can be converted to peroxide functions, but the recently developed Barton approach appears much more attractive.[58] In this approach, alkyl radicals are generated efficiently from N-acyloxythiopyridones via the decarboxylation of intermediate acyloxy radicals (Figure 1.48). The required substrates are prepared in a straightforward manner from the appropriate acyl chloride and N-hydroxythiopyridone. The extremely high reactivity of radicals toward sulfur (thiophilicity) is exploited in these reactions.

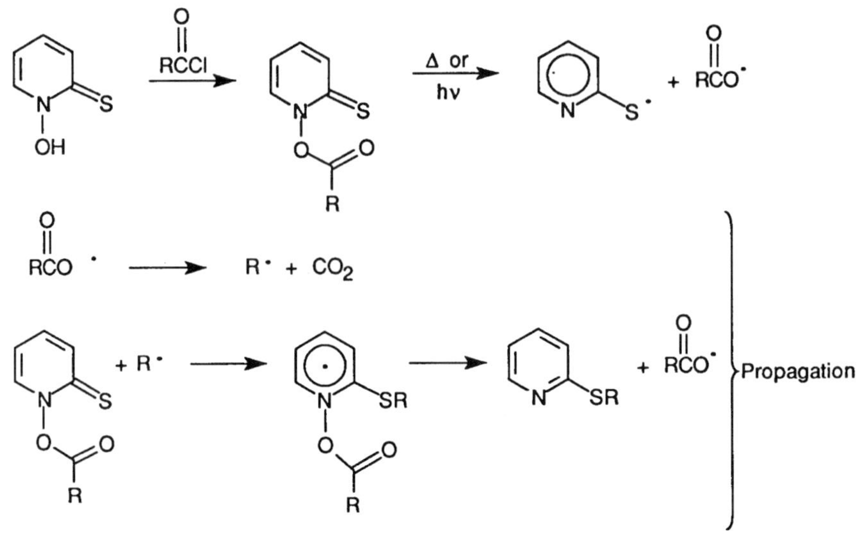

Figure 1.48 The Barton method for radical generation.

A classic and still attractive alternative method for radical generation from a carboxylic acid function is the Hunsdiecker reaction, which is thought to involve the chain decomposition of an intermediate acyl hypobromite to an alkyl bromide (Figure 1.49).[68] A double Hunsdiecker reaction has apparently even been used to generate a bromine-bridged radical (Figure 1.50).[59]

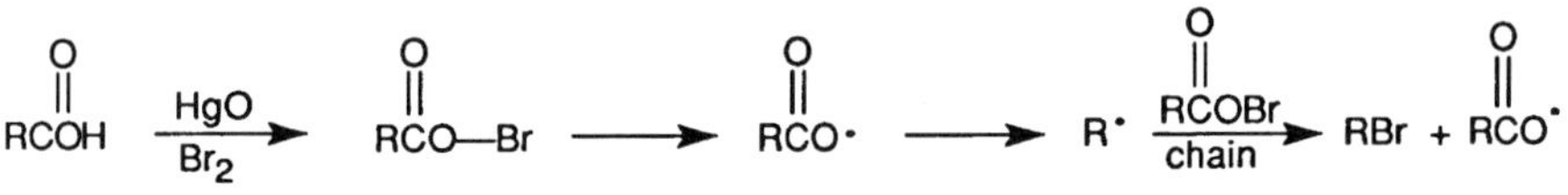

Figure 1.49 The Hunsdiecker reaction.

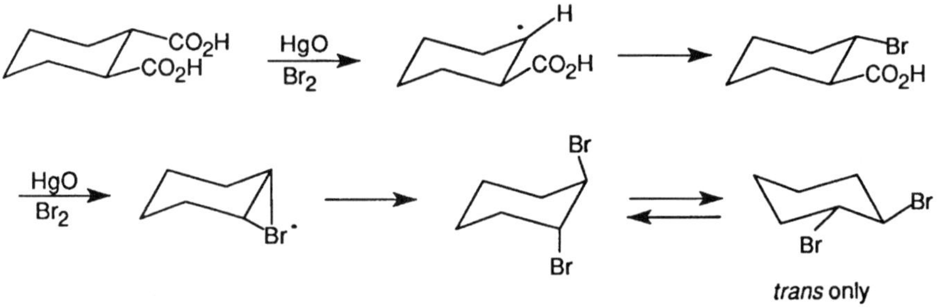

Figure 1.50 A double Hunsdiecker reaction.

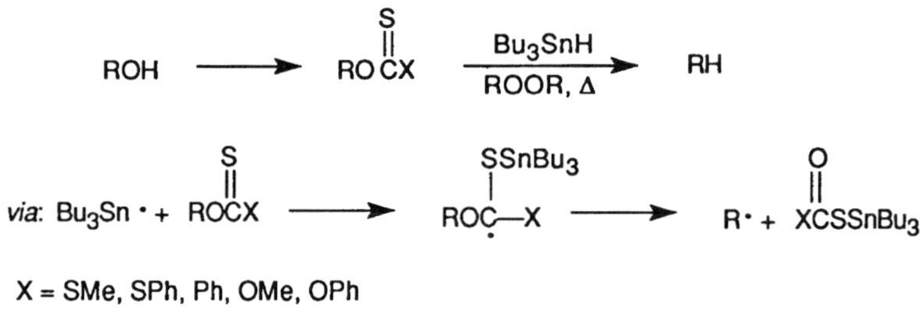

$$\mathrm{ROH} \longrightarrow \mathrm{ROC(=S)X} \xrightarrow[\mathrm{ROOR},\ \Delta]{\mathrm{Bu_3SnH}} \mathrm{RH}$$

via: $\mathrm{Bu_3Sn\cdot + ROC(=S)X \longrightarrow ROC^{\bullet}(SSnBu_3)\!-\!X \longrightarrow R\cdot + XC(=O)SSnBu_3}$

X = SMe, SPh, Ph, OMe, OPh

Figure 1.51 The Barton–McCombie reaction.

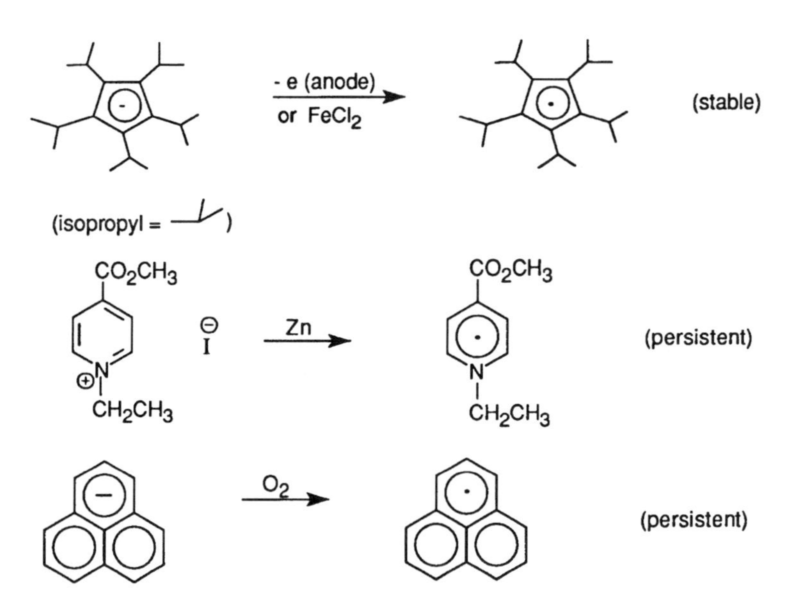

Figure 1.52 Anions and cations as radical precursors.

Radicals can also be generated easily from secondary alcohols by first converting them to thionoesters (Figure 1.51) and then treating the latter with tributyltin hydride under radical conditions (peroxide, Δ or hν; the Barton–McCombie reaction).[60]

The Grignard function can also serve as an efficient radical precursor via its reaction with dioxygen, which has been established as a radical chain process. Perhaps the simplest reaction for radical generation is oxidation/reduction of a stable anion/cation (Figure 1.52).[61,62]

References

General References

Bachmann, W. E. *Organic Chemistry*, 2nd ed., Vol. 1, H. Gilman, Ed., Wiley, New York, 1943, Chapter 6.

Waters, W. A. *The Chemistry of Free Radicals*, 2nd. ed., Oxford University Press, London, 1948.

Steacie, E. W. R. *Atomic and Free Radical Reactions*, 2nd ed., Reinhold, New York, 1954.

Walling, C. *Free Radicals in Solution*, Wiley, New York, 1957.

Pryor, W. A. *Free Radicals*, McGraw-Hill, New York, 1966.

Huyser, E. S. *Free-Radical Chain Reactions*, Wiley, New York, 1970.

Ingold, K. U.; Roberts, B. P. *Free-Radical Substitution Reactions*, Interscience, New York, 1971.

Kochi, J. K., ed. *Free Radicals*, Vols. I and II, Wiley, New York, 1973.

Borden, W. T., ed. *Diradicals*, Wiley, New York, 1982.

Wentrup, C. *Reactive Molecules*, Wiley, New York, 1984.

Leffler, J. E. *An Introduction to Free Radicals*, Wiley-Interscience, New York, 1993.

References and Footnotes

1. Bartlett, P. D.; Hiatt, R. R. *J. Am. Chem. Soc.* **1958**, *80*, 1399.
2. Hammond, G. S.; Wu, C.-H. S.; Trapp, D. D.; Warkentin, J.; Keys, R. T., *J. Am Chem Soc*. **1960**, *82*, 5394.
3. Fenton, H-J. H. *J. Chem. Soc.* **1894**, *65*, 899.
4. Walling, C. *Acc. Chem. Res.* **1975**, *8*, 125.
5. Sheldon, R. A.; Kochi, J. K. *J. Am. Chem. Soc.* **1970**, *92*, 4395.
6. Gomberg, M. *J. Am. Chem. Soc.* **1900**, *22*, 757.
7. Gomberg, M. *Ber. deutsch. chem. Ges.* **1900**, *33*, 3150.
8. Jacobson, P. *Ber. deutsch. chem. Ges.* **1905**, *38*, 196.
9. Lankamp. H.; Nauta, W. T.; MacLean, C. *Tetrahedron Lett.* **1968**, 249.
10. Marvel, C. S.; Kaplan, J. F.; Himel, C. M. *J. Am. Chem. Soc.* **1941**, *63*, 1892.
11. Lichtin, N. N.; Glazier, H. *J. Am. Chem. Soc.* **1951**, *73*, 5537.
12. Griller, D.; Ingold, K. U. *Acc. Chem. Res.* **1976**, *9*, 13–19.
13. Paneth, F.; Hofeditz, W. *Ber. deutsch. chem. Ges.* **1929**, *62*, 1335.
14. Kharasch, M. S.; Englemann, H.; Mayo, F. R. *J. Org. Chem.* **1937**, *2*, 288.
15. Hey, D. H.; Waters, W. A. *Chem. Revs.* **1937**, *21*, 169.
16. Flory, P. J. *J. Am. Chem. Soc.* **1937**, *59*, 241.
17. Walling, C. *Free Radicals in Solution*, Wiley, New York, 1957, p. 7.
18. Goldschmidt, S.; Renn, K. *Ber. deutsch. chem. Ges.*, **1922**, *55*, 628.
19. Coppinger, G. M. *J. Am. Chem. Soc.* **1957**, *79*, 501.
20. Hoffmann, A. K.; Henderson, A. T. *J. Am. Chem. Soc.* **1961**, *83*, 4671.
21. Briere, R.; Lemaire, H.; Rassat, A. *Bull. Soc. Chim. Fr.* **1965**, 3273.

22. Ziegler, K.; Schnell, B. *Ann.* **1925**, *445*, 266.

23. Kuhn, R.; Treschmann, H. *Monatsch.* **1964**, *95*, 547.

24. Ballester, M. *Advances in Physical-Organic Chem.* **1989**, *25*, 267.

25. Staab, H. A.; Rao, K. S.; Brunner, H. *Chem. Ber.* **1971**, *104*, 2634. This dimer has the normal structure in which the two radicals bond at thier benzylic positions.

26. Reichel, C. L.; McBride, J. M. *J. Am. Chem. Soc.* **1977**, *99*, 6758. The odd electron in the tripticyl radical is almost completely localized at the bridgehead position (C1).

27. Sogo, P. B.; Nakazi, M.; Calvin, M. *J. Chem. Phys.* **1957**, *26*, 1343.

28. For a relevant discussion of the phenalenyl system: Streitweiser, A. *Molecular Orbital* Theory, Wiley New York, 1961, p. 164. Note that the Hückel $4n + 2$ rule applies only to monocyclic systems.

29. Bartlett, P. D.; Funahashi, T. *J. Am. Chem. Soc.* **1962**, *84*, 2596. Also, about 70% of the radicals are efficient in initiating styrene polymerization. Bevington, J. C. *Trans. Faraday Soc.* **1955**, *51*, 1392.

30. Hammond, G. S.; Soffer, L. M. *J. Am. Chem. Soc.* **1950**, 72, 4711.

31. Root, K. S.; Hill, C. L.; Lawrence, L. M.; Whitesides, G. M. *J. Am. Chem. Soc.* **1989**, *111*, 5405.

32. Janzen, E. G. *Acc. Chem. Res.* **1971**, *4*, 31.

33. Wilt, J. W. in *Free Radicals*, Vol. I.; Kochi, J. K., Ed., Wiley, New York, 1973, 420.

34. Lal, D.; Griller, D.; Husband, S.; Ingold, K. U. *J. Am. Chem. Soc.* **1974**, *96*, 6355.

35. Kochi, J. K.; Krusic, P. J., *J. Am. Chem. Soc.* **1969**, *91*, 3940.

36. Chatgilialoglu, C.; Ingold, K. U.; Scaiano, J. C. *J. Am. Chem. Soc.* **1981**, *103*, 7739.

37. Corey, E. J.; Suggs, J. W. *J. Org. Chem.* **1975**, *40*, 2554.

38. Bowry, V. W.; Lusztyk, J.; Ingold, K. U. *J. Am. Chem. Soc.* **1991**, *113*, 5687.

39. Martin-Esker, A. A.; Johnson, C. C.; Horner, J. H.; Newcomb, M. *J. Am. Chem. Soc.* **1994**, *116*, 9174.

40. Newcomb, M. J.; Manek, M. B. *J. Am. Chem. Soc.* **1990**, *112*, 9662.

41. Ortiz de Montellano, P. R.; Stearns, R. A. *J. Am. Chem. Soc.* **1987**, *109*, 3415.

42. Newcomb, M.; Chestney, D. L. *J. Am. Chem. Soc.* **1994**, *116*, 9753.

43. Newcomb, M.; LeTadic-Biadatti, M-H.; Chestney, D. L. *J. Am. Chem. Soc.* **1995**, *117*, 12085.

44. Wilt, J. W. in *Free Radicals*, Vol. I; Kochi, J. K., Ed., Wiley, New York, 1973, 335.

45. Winstein, S.; Seubold, Jr., F. H. *J. Am. Chem. Soc.* **1947,** *69,* 2916. See also Wilt, J. W., in *Free Radicals*, Vol. I; Kochi, J. K., Ed., Wiley, New York, 1973, 346.

46. Wilt, J. W. in *Free Radicals*, Vol. I; Kochi, J. K., Ed., Wiley, New York, 1973 466.

47. Martin, J. C.; Bentrude, W. G. *Chem. Ind.* **1959**, *192*; Martin, J. C.; Bentrude, W. G. *J. Am. Chem. Soc.* **1960**, *84*, 1561.

48. Pearson, R. E.; Martin, J. C. *J. Am. Chem. Soc.* **1963**, *85*, 3142.

49. Russell, G. A.; Williamson, Jr., R. C., *J. Am. Chem. Soc.*, **1964**, *86*, 2357; Van Heldon, R.; Kooyman, E .C. *Rec. Trav. Chem.* **1964**, *73*, 269.

50. Russell, G. A. in *Free Radicals*, Vol. I, Kochi, J. K., Ed., Wiley, New York, 1973, 295.

51. Bartlett, P. D.; Rchardt, C. *J. Am. Chem. Soc.* **1960**, *82*, 1756.

52. Wong, M. W.; Pross, A.; Radom, L. *J. Am. Chem. Soc.* **1994**, *116*, 11938.

53. Giese, B. *Angew. Chem. Int. Ed. Engl.* **1983**, *22*, 753.

54. Creary, K. *J. Org. Chem.* **1980**, *45*, 280; Creary, K.; Mehrsheikh-Mohammadi, M. E.; McDonald, S. *J. Org. Chem.* **1987**, *52*, 3254; Timberlake, J. W.; Hodges, M. L. *Tetrahedron Lett.* **1970**, *11*, 4147; Jiang, X.-K.; Liu, W. W.-Z.; Wu, S.-H. *Tetrahedron* **1994**, *50*, 7503.

55. Dust, J. M.; Arnold, D. R. *J. Am. Chem. Soc.* **1983**, *105*, 1221 and 6531.

56. Nau, W. M.; Harrer, H. M.; Adam, W. *J. Am. Chem. Soc.* **1994**, *116*, 10972.

57. Sargent, G. D.; Cron, J. N.; Bank, S. *J. Am. Chem. Soc.* **1966**, *88*, 5363.

58. Barton, D. H. R.; Jaszberenyi, J. C.; Theodorakis, E. A. *J. Am. Chem. Soc.* **1992**, *114*, 5904.

59. Aplequist, D. E.; Werner, N. D. *J. Org. Chem.* **1963**, *28*, 48.

60. Barton, D. H. R.; Zard, S. Z. *Pure & Appl. Chem.* **1986**, *58*, 675.

61. Sitzmann, H.; Bock, H.; Boese, R.; Dezember, T.; Havlas, Z.; Kaim, W.; Moscherosch, M.; Zanathy, L. *J. Am. Chem. Soc.* **1993**, *115*, 12003.

62. Kosower, E. M.; Poziomek, E. J. *J. Am. Chem. Soc.* **1964**, *86*, 5515.

Exercises

1.1 The tropenyl radical is generated with surprising ease by simply heating the dimer **A** to 80°C. The experimentally measured bond dissociation energy, D, of the weak carbon–carbon bond in **A** is 35 kcal/mol. Assuming that a normal sp^3–sp^3 carbon–carbon bond has $D \approx 86$ kcal/mol (i.e., the same as that in ethane), calculate the experimental resonance stabilization energy of the tropenyl radical.

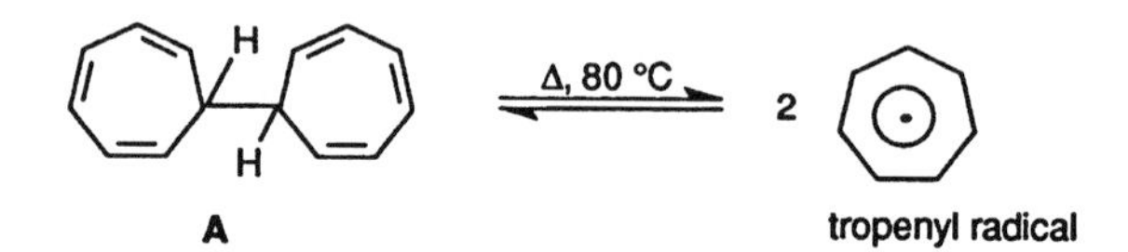

(Vincow, G.; Morrell, M. L.; Volland, W. V.; Dauben, H. J., Jr.; Hunter, F. R. *J. Am. Chem. Soc.* **1965**, *87*, 3527.)

1.2 Borinate radicals ($R_2BO^•$) have recently been observed to be substantially stabilized radicals. What type of radical stabilization is likely to be involved? Write resonance structures that correspond to the effect and then construct an MO diagram that illustrates it.

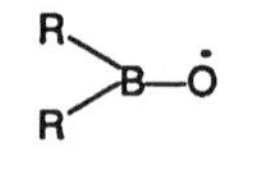

(Chung, T. C.; Janvikul, W.; Lu, H. L. *J. Am. Chem. Soc.* **1966**, *118*, 705.)

1.3 The bond dissociation energies of the methyl and tertiary carbon–hydrogen bonds of methylcubane are estimated to be, respectively, 100.5 and 107 kcal/mol.

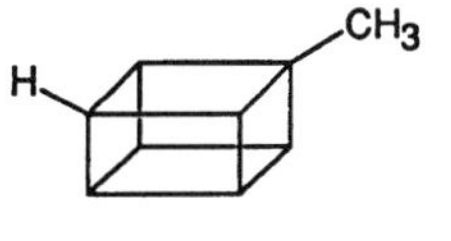

CH_3
H

D(C—H, 1°) ≈ 100.5 kcal/mol
D(C—H, 3°) ≈ 107 kcal/mol

(a) Provide a possible explanation for the weaker primary than tertiary bonds and in particular the unusually strong tertiary CH bond. (b) *Ab initio* SCF MO calculations indicate that abstraction of a tertiary hydrogen by a methoxy radical has an activation energy approximately 2.0 kcal/mol *lower* than for abstraction of a primary hydrogen. Construct transition-state models for these two abstraction reactions that include any polar effects expected to be operative and explain the predicted higher reactivity of the stronger tertiary carbon–hydrogen bonds. (Hrovat, D. A.; Borden, W. T. *J. Am. Chem. Soc.* **1994**, *116*, 6459.)

1.4 The reduction of 6-bromo-1-hexene by $Cr^{II}(en)_2$ (en = ethylenediamine) in 83%/vol DMF/H_2O at 25°C produces both 1-hexene (H) and methylcyclopentane (MC). The ratio of H/MC formed is linearly dependent upon the molar concentration of the $Cr^{II}(en)_2$ reducing agent, with slope 5×10^2. (a) Write a plausible mechanism for the reduction of 6-bromo-1-hexene by $Cr^{II}(en)_2$ that is consistent with the formation of these two products and with the linear dependence of their relative ratio on the concentration of the reducing agent. (b) Assuming that the rate constant for the clock reaction of the 5–hexenyl radical in the DMF/H_2O solvent mixture is the same as that in benzene ($1 \times 10^5\ s^{-1}$), calculate the rate constant for the reaction of this radical with $Cr^{II}(en)_2$. (Kochi, J. K.; Powers, J. W. *J. Am. Chem. Soc.* **1970**, *92*, 137.)

1.5 The metalloporphyrin-catalyzed epoxidation of alkenes has been studied extensively as a model for the epoxidation of alkenes by enzymes such as cytochrome P-450. One of the mechanisms that was initially considered plausible is the radical mechanism shown below. Using appropriate substrate probes, it has been possible to exclude the radical mechanism.

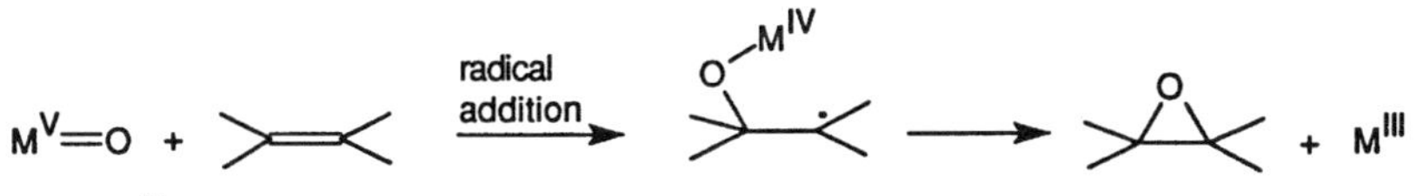

M^V=O + [alkene] —radical addition→ [O–M^{IV} radical adduct] ⟶ [epoxide] + M^{III}

(where $\underline{M}^V$ is a metalloporphyrin in which the metal has a formal oxidation state of +5)

(a) Suggest an appropriate alkene probe molecule for such a study, based upon the rapid probe reaction types discussed in the text. Contrast the results expected for a concerted reaction with those expected for the hypothetical radical mechanism. (b) In the hypothetical circumstance that the probe substrate you have chosen does, in fact, give evidence of having undergone the probe reaction, how could you exclude a carbocation mechanism? (Ostovic, D.; Bruice, T. C. *Acct. Chem. Res.* **1992**, *25*, 314.)

1.6 The rearrangement shown below, catalyzed by vitamin B12, was studied as a model for the methylmalonyl-CoA to succinyl-CoA carbon skeleton rearrangement in order to detect a possible cyclopropylcarbinyl radical intermediate.

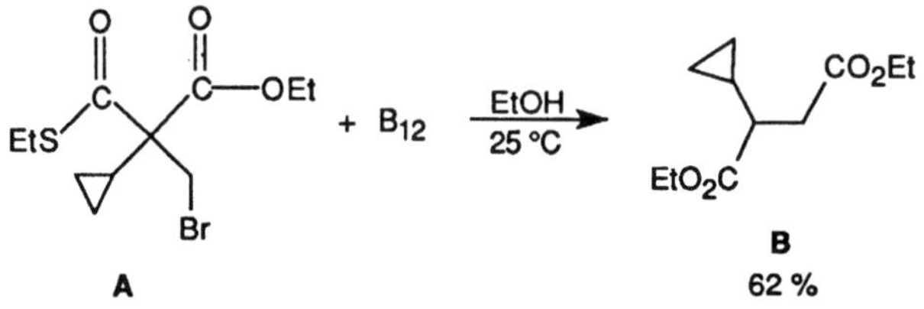

EtS(C=O)–C(cyclopropyl)(CH₂Br)–C(=O)–OEt + B_{12} $\xrightarrow[25\,°C]{EtOH}$ B

A   B 62 %

(a) Treatment of bromide **A** with tributyltin hydride yielded only **C**. No succinate ester (e.g., **B** or an analogous thiolester) was formed. Write a mechanism for the formation of **C** and show a hypothetical mechanism for rearrangement to a succinate ester. Although radical rearrangements are not common, explain why this rearrangement might be plausible.

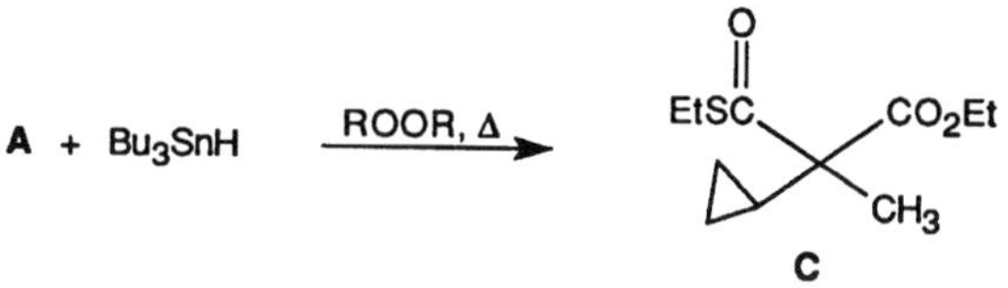

A + Bu_3SnH $\xrightarrow{ROOR,\ \Delta}$ EtSC(=O)–C(cyclopropyl)(CH₃)–CO₂Et

C

(b) The following reaction was also studied. Write a mechanism for the observed reaction, and explain what relevance it has to the mechanistic question being explored.

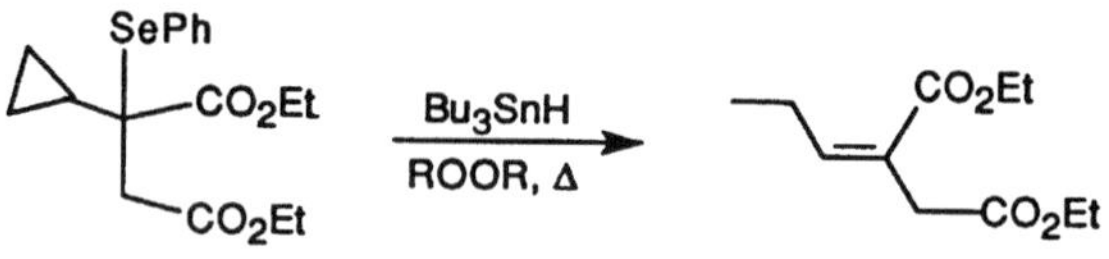

SePh, CO₂Et, CO₂Et substrate $\xrightarrow[ROOR,\ \Delta]{Bu_3SnH}$ CO₂Et, CO₂Et alkene product

(c) The probe reaction shown below has a rate constant $k = 1 \times 10^7\ s^{-1}$. What conclusions can you draw concerning the possibility of a radical mechanism for the conversion of **A** to **B** as catalyzed by vitamin B12?

cyclopropyl–$\dot{C}$HC(=O)OR ⟶ $\dot{C}H_2$–CH₂–CH=CH–CO₂R

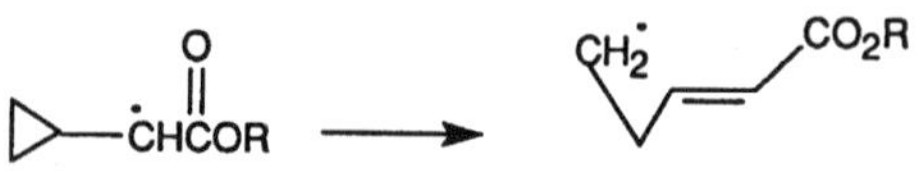

(He, M.; Dowd, P. *J. Am. Chem. Soc.* **1996**, *118*, 711.)

1.7 (a) Draw a dotted-line/partial-charge model for the transition state for the addition of cyclohexyl radicals to the beta carbon of a series of α-substituted methyl acrylate esters. Based upon this structural representation, predict the qualitative sequence of reactivities where the α substituent (Z) is H, CN, MeO, CO_2Et, and CH_3. Explain the basis for your rate sequence.

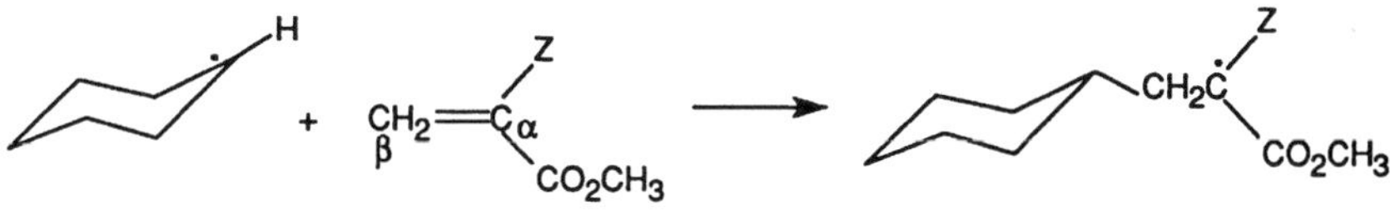

(b) Where the series of substituents H, CH_3, Et, isopropyl, and *tert*-butyl are present at the β position of a methyl acrylate ester, the relative rates of addition of cyclohexyl radicals to the β carbon correlate with E_s, the Taft steric parameter. However, more polar substituents such as Cl and CN deviate seriously from the log k_1 versus E_s plot. Specifically, the chloro derivative reacts 4.3 times as fast as would be expected based upon its steric parameter, and cyano reacts 92 times as fast. Explain the strongly accelerated rate for the β-cyano derivative using a dotted-line partial-charge transition state model and also using frontier orbital (FO) theory. (Note: Addition also occurs at the position α to the ester function, but this provides a regiochemically different product with a distinct rate constant.)

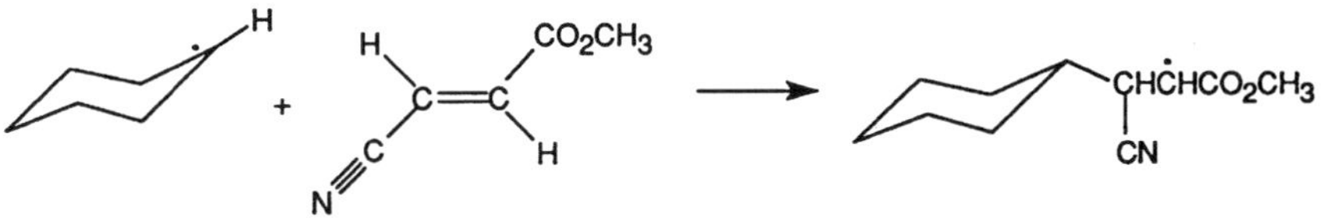

(c) Since a cyano group beta to the carboxyl group accelerates the rate of addition by a factor (92) that is not so much less than when it is present in the α position [310, see part (a) for the specific case where Z = CN versus Z = H], it has been said that polar effects have relatively little influence upon the regiochemistry of such radical additions. Rather, the regiochemistry is said to be mainly determined by steric effects, which retard addition at the alpha position to a substituent and have little or no effect on the rate of addition to a position β to a substituent. Discuss this interesting proposition critically, considering especially whether this conclusion can be extended to include the regiochemistry of additions to *monosubstituted* alkenes such as methyl acrylate or acrylonitrile, and styrene. (d) Based upon FO theory, should the substituent Z produce a greater acceleration when present at the α or β position of an acrylate ester? Explain qualitatively.

1.8 Write mechanisms for the following reactions and predict the major product or products. For each step that involves a radical intermediate, indicate what specific type of radical reaction is involved.

(a) 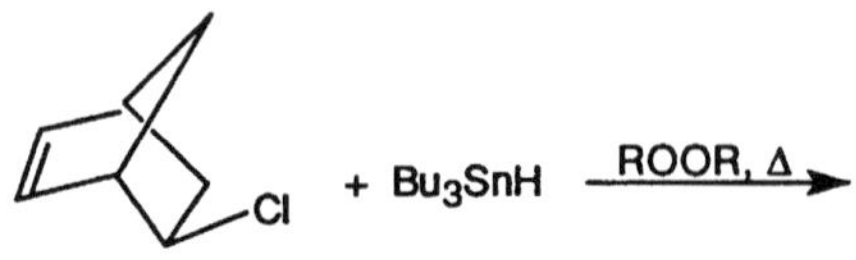 [bicyclic chloride structure] + Bu_3SnH $\xrightarrow{\text{ROOR, }\Delta}$

(b) $Ph_3\,CCH_2Cl$ + Bu_3SnH $\xrightarrow{\text{ROOR, }\Delta}$

(c) 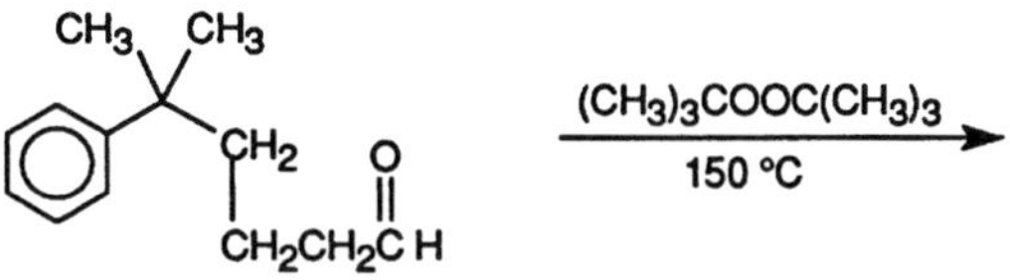 [structure: phenyl–C(CH_3)(CH_3)–$CH_2CH_2CH_2CHO$] $\xrightarrow[150\ ^\circ C]{(CH_3)_3COOC(CH_3)_3}$

(d) $(CH_2{=}CHCH_2CH_2CH_2CH_2\overset{O}{\overset{\|}{C}}O)_2$ $\xrightarrow[\text{S—H}]{\Delta}$

S—H = solvent having abstractable hydrogens

(e) 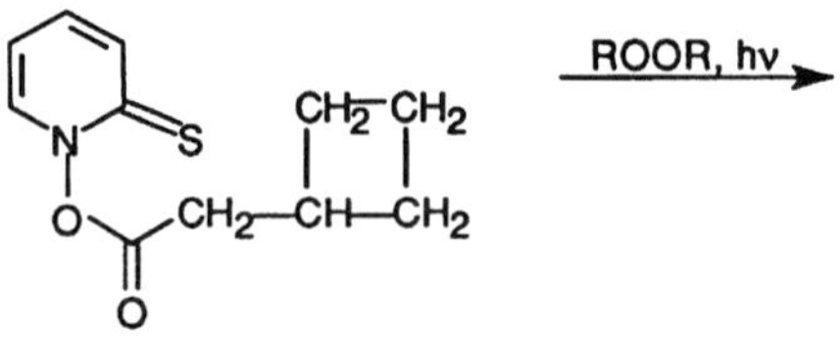 [structure: N-acyloxy pyridine-2-thione ester of cyclobutylacetic acid] $\xrightarrow{\text{ROOR, }h\nu}$

1.9 (a) The HMO (Hückel molecular orbital) energy levels for the allyl radical are given below. Count the number of π electrons in either canonical structure (two for a double bond, one for a radical center) and populate the HMOs progressively, beginning with the MO of lowest energy and using "up" and "down" arrows to represent electrons of α and β spin, respectively. Which MO is the SOMO?

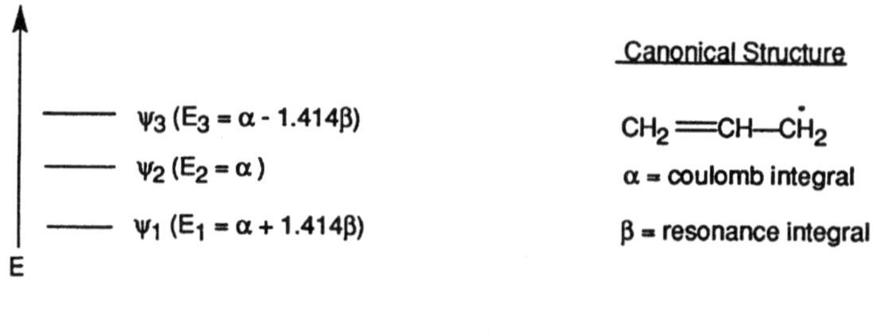

E

ψ_3 ($E_3 = \alpha - 1.414\beta$)

ψ_2 ($E_2 = \alpha$)

ψ_1 ($E_1 = \alpha + 1.414\beta$)

Canonical Structure

$CH_2{=}CH{-}\dot{C}H_2$

α = coulomb integral

β = resonance integral

(b) The coefficients of the allyl HMOs are given below. Since the odd electron density (ρ_i) at any carbon atom i is given, in this approximation, by $\rho_i = a^2_{\mathrm{SOMO,\,i}}$, that is, by the square of the coefficient of that carbon in the SOMO, calculate ρ_i for each carbon of the allyl radical and compare your results with those obtained from a conventional resonance treatment of the allyl radical.

$$\psi 1 = 0.500\ \phi 1 + 0.707\ \phi 2 + 0.500\ \phi 3$$

$$\psi 2 = 0.707\ \phi 1 - 0.707\ \phi 3$$

$$\psi 3 = 0.500\ \phi 1 - 0.707\ \phi 2 + 0.500\ \phi 3$$

(c) Given that the HMO total π-electron energy ($E\pi$) is the sum of orbital energies (see below), calculate E_π for the allyl radical.

$$E_\pi = \sum_{\mathrm{i}}^{\text{occupied MOs}} N_i E_i$$

where N_i = the orbital population (1 or 2)
E_i = the HMO energy of the occupied MOψ_i

(d) To obtain the calculated delocalization energy (DE), subtract from this the π-electron energy of the canonical structure, which includes two π electrons of ethene ($E_{\mathrm{BMO}} = \alpha + \beta$ for each electron) and the single electron of a carbon radical site ($E = \alpha$).

$$\mathrm{DE} = E_\pi(\text{allyl}) - E_\pi(\text{ethene}) - E\pi\ (\text{radical site})$$

CHAPTER

2

Radical Reactions

The reactions of radicals can be conveniently classified as *chain* or *non-chain* processes. Nonchain reactions can be further sub-classified as *cage* or *non-cage* reactions. Both chain and nonchain radical mechanisms are discussed in this chapter, the latter being considered first.

2.1 Nonchain Radical Reactions

Nonchain radical reactions most commonly involve radical pairs. Since radical pairs often undergo both cage and noncage reactions, the nonchain reactions of radicals are not always uniquely classified as belonging in the cage or noncage category (Figure 2.1). However, this section on nonchain reactions begins with an emphasis on cage reactions, and then proceeds to a discussion of predominantly noncage reactions. Finally, the reactions of singlet diradicals are discussed.

2.1.1 Reactions of Caged Radical Pairs

Homolysis by either thermal or photochemical means produces a pair of radicals that are in close proximity to each other and that are not separated by solvent molecules. Such a radical pair, enclosed in a solvent cavity or cage, is referred to as a *caged radical pair* or a *geminate radical pair.*[1] Symbolically, the caged nature of the radical pair is indicated by a long bar over the structures of both radicals. Assuming that the caged radical pair is generated in a singlet state, proximity and energy considerations make *geminate recombination* (cage recombination) a very

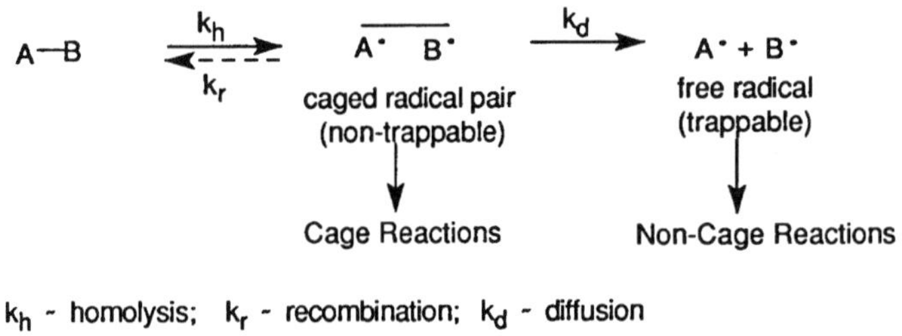

Figure 2.1 Cage and noncage radical reactions.

attractive option, and this radical coupling is typically very fast. However, escape of the individual radicals from the solvent cage to yield free radicals is usually at least equally facile, since this occurs at the rate at which the radicals diffuse through the solvent. That is to say, the solvent cage is by no means rigid or impenetrable. In some cases, intramolecular rearrangement, abstraction, addition, or fragmentation of one or both components of the radical pair is rapid enough to compete with diffusion and recombination. Fragmentation, especially, can sometimes occur in concert with the relevant homolysis and thus provide acceleration for this cleavage. Neighboring group assistance of homolysis has also been exemplified.

2.1.2 Cage Reactions of Peroxides

In the homolysis of dibenzoyl peroxide, essentially 100% of the benzoyloxy radicals can be trapped by I_2 or galvinoxyl.[2] That essentially all of the benzoyloxy radicals become free radicals does not necessarily imply that cage recombination is absent, however. Since the decarboxylation of benzoyloxy radicals is rather slow, essentially the only cage reaction available is recombination to give the starting peroxide, and this reaction would, of course, go undetected in routine trapping studies. Recombination is, however, detectable by isotopic labeling studies (Figure 2.2), and here it has been found that 4.7% of the benzoyloxy

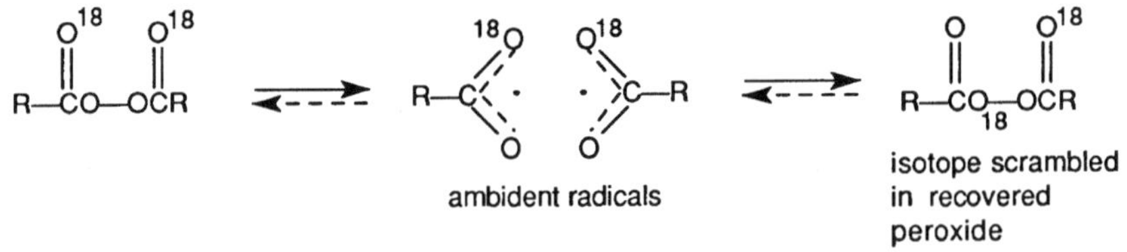

R	% Scrambling
Ph	4.7 (isoctane)
	18 (mineral oil)
CH_3	38 (isoctane)

Figure 2.2 Cage recombination in peroxide homolysis.

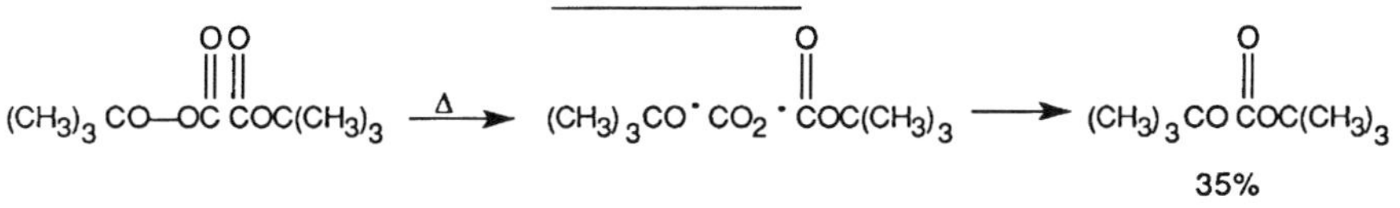

$$(CH_3)_3CO{-}OC(=O)C(=O)OC(CH_3)_3 \xrightarrow{\Delta} (CH_3)_3CO^{\cdot}\ CO_2\ {}^{\cdot}C(=O)OC(CH_3)_3 \longrightarrow (CH_3)_3CO\,C(=O)OC(CH_3)_3$$

35%

Figure 2.3 Cage recombination in multibond homolyses.

radicals recombine in isoctane, a relatively nonviscous solvent.[3] Cage recombination is enhanced in a more viscous solvent, such as mineral oil (18% recombination), since the diffusion rate is much lower in such a solvent. Recombination is even more extensive in the case of acetyl peroxide. The observation of scrambling of an ^{18}O label from the carbonyl oxygen to the peroxidic oxygen in these reactions supports the formulation of these reactions as simple (i.e., one-bond) homolyses. Concerted two- (or three-) bond cleavages resulting in fragmentation to one or two carbon dioxide molecules and one or two phenyl or methyl radicals would not be consistent with the regeneration of the starting peroxide and thus with scrambling in the recovered starting material. In the case of dibenzoyl peroxide, simple one-bond homolysis is also decisively demonstrated by the ability to trap 100% of the benzoyloxy radicals. In contrast to the slow decarboxylation of benzoyloxy radicals, acetoxy radicals decarboxylate rapidly enough to produce significant amounts of caged methyl radical/ methyl radical pairs, which produce ethane, even in the presence of traps.[4] Consequently, the efficiency of production of free (uncaged) radicals is less for acetyl peroxide than benzoyl peroxide.

Cage recombination is more easily detected when homolysis and fragmentation are concerted, since cage recombination in these instances yields stable products that are formed even in the presence of radical traps. The concerted decomposition of di-*tert*-butyl-monoperoxyoxalate in the presence of Galvinoxyl (Figure 2.3), for example, yields 50% of cage recombination products, including especially di-*tert*-butylcarbonate.[5]

2.1.3 Concerted versus Stepwise Decomposition

The distinction between concerted and stepwise homolysis/fragmentation of peroxides is straightforward when, as in the case of benzoyl peroxide decomposition, radical scavengers prevent fragmentation. In such a case, homolysis/ fragmentation is obviously stepwise. Equally, a concerted homolysis/fragmentation is strongly indicated by a major rate acceleration, as in the case of the *tert*-butyl perester of phenylacetic acid, which involves the concerted formation of benzyl radicals, carbon dioxide, and the *tert*-butoxy radical.[6] The presence of ^{18}O scrambling is indicative of a stepwise process, as noted above. In the specific case of the peroxy esters of arylacetic acids, the observation of a rate correlation with σ^+ is also a strong indication of a concerted two-bond cleavage, since this indicates the generation of positive charge on the benzylic carbon. Secondary kinetic deuterium isotope effects have also been used to make this interesting mechanistic distinction.

Finally, solvent viscosity effects are a potentially powerful tool for this mechanistic distinction. Since increased solvent viscosity increases the amount of cage recombination, the rate of a one-bond homolysis is decreased, because recombination regenerates the starting peroxide. In contrast, recombination generates a stable product in the case of a multiple bond cleavage, so that increasing the solvent viscosity has little or no effect on the decomposition rate.[7]

2.1.4 Cage Reactions: Azo Compounds

Trapping studies using galvinoxyl reveal that 34% of the tetramethylsuccinonitrile formed in the decomposition of AIBN in benzene at 60°C arises via cage recombination.[8] The relatively large extent of cage recombination, together with the large rate acceleration compared to an unsymmetrical azo analogue (Figure 2.4)[9] indicate that this decomposition is concerted. Secondary kinetic isotope effects suggest that the transition state is symmetrical, involving equal or comparable extents of cleavage of both C–N bonds.[10] Similar observation has been made for symmetrical α-phenylethylazo compounds.[11] On the other hand, unsymmetrical azo compounds appear to cleave in stepwise fashion.[12]

2.1.5 Formation of Grignard Reagents

This is probably the most familiar example of a nonchain radical mechanism in organic chemistry (Figure 2.5). It was previously noted that, in one instance, trapping of an intermediate alkyl radical by TEMPO has been achieved in high yield.[13] In this instance, at least, cage recombination must be minimal. This result

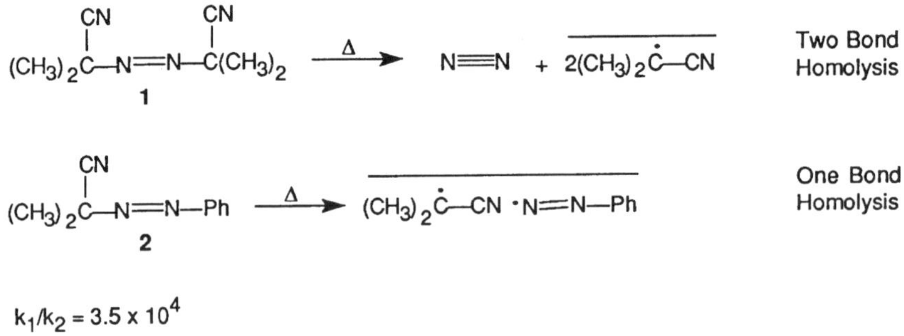

$$(CH_3)_2C(CN)\text{—N=N—}C(CN)(CH_3)_2\ (\mathbf{1}) \xrightarrow{\Delta} N{\equiv}N + \overline{2(CH_3)_2\dot{C}\text{—}CN}$$ Two Bond Homolysis

$$(CH_3)_2C(CN)\text{—N=N—Ph}\ (\mathbf{2}) \xrightarrow{\Delta} \overline{(CH_3)_2\dot{C}\text{—}CN\ \cdot N{=}N\text{—}Ph}$$ One Bond Homolysis

$k_1/k_2 = 3.5 \times 10^4$

Figure 2.4 Concerted versus stepwise decompositions of symmetrical and unsymmetrical azo compounds.

$$R\text{—}X + \dot{M}g \xrightarrow[\text{transfer}]{\text{electron}} \overline{R^{\cdot}\ \overset{\ominus}{X}\ \overset{\oplus}{\dot{M}g}} \xrightarrow{\text{diffusion}} R^{\cdot} + \overset{+}{\dot{M}g}\overset{\ominus}{X} \xrightarrow{\text{coupling}} R\text{—}\overset{\oplus}{Mg}\overset{\ominus}{X}$$

Figure 2.5 Nonchain radical mechanism for the formation of rignard regents.

seems somewhat surprising, but could be a consequence of the heterogeneity of the reaction; that is, the Mg(I) ion may be relatively unreactive because it initially is generated on the surface where it remains bonded to other magnesium atoms, as well as to the halide ion.

2.1.6 Reduction of Organomercury Derivatives

The oxymercuration of alkenes has been developed as a reaction of synthetic value for the net hydration of alkenes without the skeletal rearrangements encountered in acid-catalyzed hydration but with the same regiochemistry observed in the latter reaction (Figure 2.6).[14] The intermediate β-acetoxymercuri alcohols are reduced to simple alcohols, in this approach, by sodium borohydride/THF/water. The mechanism of the reduction of organomercurials remained uncertain for quite some time, but is now generally considered to proceed predominantly via a caged radical pair mechanism.[15] That the decomposition of the intermediate organomercuric hydride does not occur via a concerted mechanism is indicated by its non-stereospecificity (Figure 2.7). Since neither carbocation nor carbanion behavior is observed, the caged radical mechanism is indicated as probable.

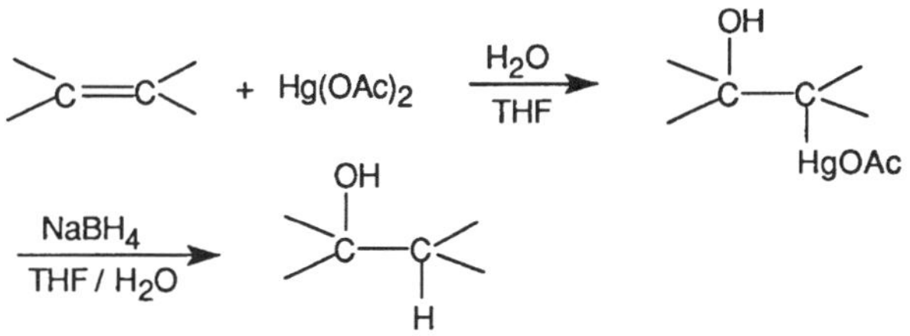

Figure 2.6 Hydration by oxymercuration/reduction.

2.1.7 The Kolbe Coupling Reaction

Although not currently considered to be as useful synthetically as it once was, this venerable reaction can still be useful for decarboxylative coupling (Figure 2.8).[16] Carboxylate salts are subjected to anodic oxidation (e.g., in methanol), generating acyloxy radicals, which rapidly decarboxylate, yielding alkyl radicals in high concentration at or near the anode. The alkyl radicals then undergo coupling, often rather efficiently. Benzoate salts do not undergo the reaction, however, in part because of the slow decarboxylation of benzoyloxy radicals. In some cases, certain radicals may undergo further oxidation to the corresponding carbocation, which then reacts with available nucleophiles (the abnormal Kolbe reaction).[17]

2.1.8 Homolytic Aromatic Substitution

The reactions of radicals, and particularly phenyl radicals, with aromatics has been studied extensively, in part because of the opportunity to compare homolytic with

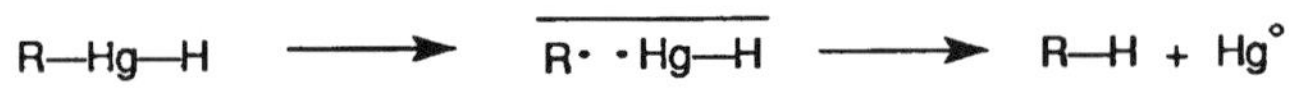

$$R{-}Hg{-}H \longrightarrow \overline{R\cdot \ \cdot Hg{-}H} \longrightarrow R{-}H + Hg^{\circ}$$

Non-stereospecific:

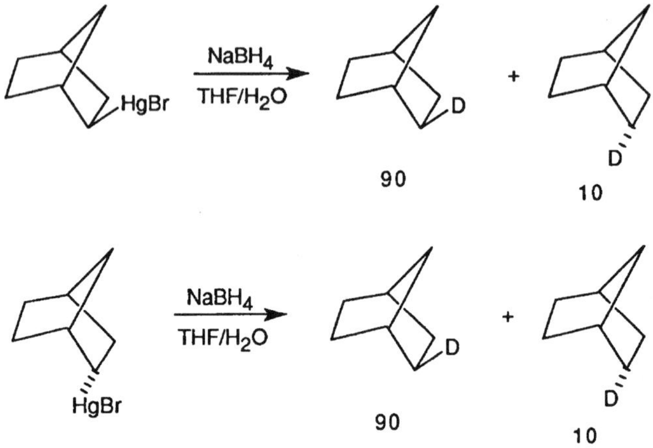

HgBr; $\xrightarrow[THF/H_2O]{NaBH_4}$ D + D; 90; 10

HgBr; $\xrightarrow[THF/H_2O]{NaBH_4}$ D + D; 90; 10

No Carbocation Rearrangements:

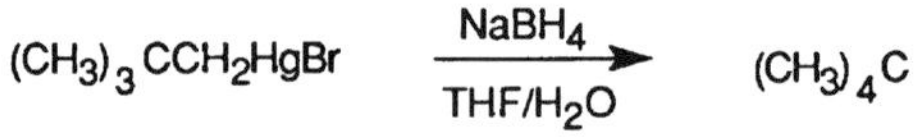

$$(CH_3)_3CCH_2HgBr \xrightarrow[THF/H_2O]{NaBH_4} (CH_3)_4C$$

No Abstraction of Solvent Protons by an Intermediate Carbanion:

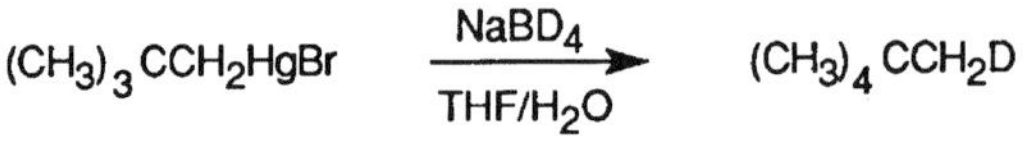

$$(CH_3)_3CCH_2HgBr \xrightarrow[THF/H_2O]{NaBD_4} (CH_3)_4CCH_2D$$

Figure 2.7 Radical cage mechanism for the decomposition of organomercuric hydrides.

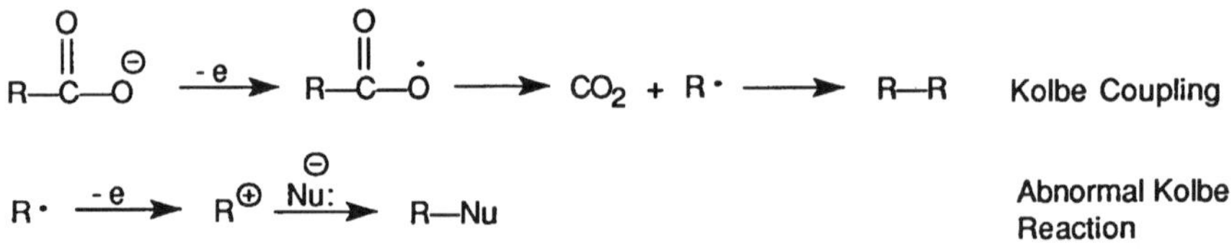

$$R{-}C(=O){-}O^{\ominus} \xrightarrow{-e} R{-}C(=O){-}\dot{O} \longrightarrow CO_2 + R\cdot \longrightarrow R{-}R$$ Kolbe Coupling

$$R\cdot \xrightarrow{-e} R^{\oplus} \xrightarrow{Nu:^{\ominus}} R{-}Nu$$ Abnormal Kolbe Reaction

Figure 2.8 The Kolbe and abnormal Kolbe reactions.

the more familiar electrophilic aromatic substitution. Phenyl radicals are often generated by decomposition of dibenzoyl peroxide, but many useful methods are available (Figure 2.9).[18] Incidentally, the circumstance that benzoyloxy radicals are relatively long lived with respect to decarboxylation does not seem to present a major problem in the dibenzoyl peroxide decompositions, since the resonance-stabilized benzoyloxy radicals add to benzene reversibly. Evidence for a radical intermediate analogous to the familiar arenium ions of electrophilic aromatic substitution is strong (Figure 2.10). This intermediate can and does undergo both coupling and disproportionation, but the yield of biphenyl, the substitution product, is sharply increased by the inclusion of dioxygen. The directive effects of substituents in homolytic phenylation are also of interest. Substituents with large $\sigma^{\bullet}$ values such as nitro modestly accelerate reaction at the *o*, *p* positions but have little effect on the rate at the *m* position. For the most part substituent effects are rather small (Figure 2.11).

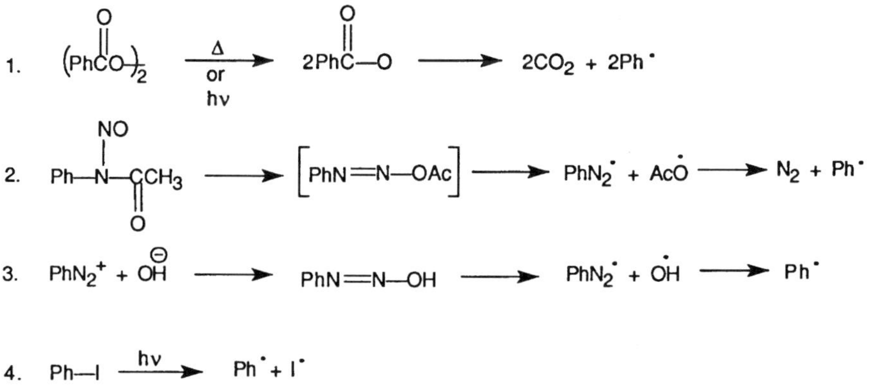

1. $(PhC(=O)O)_2 \xrightarrow[h\nu]{\Delta \text{ or}} 2PhC(=O)O^{\bullet} \longrightarrow 2CO_2 + 2Ph^{\bullet}$

2. $Ph{-}N(NO){-}C(=O)CH_3 \longrightarrow [PhN{=}N{-}OAc] \longrightarrow PhN_2^{\bullet} + AcO^{\bullet} \longrightarrow N_2 + Ph^{\bullet}$

3. $PhN_2^{+} + OH^{\ominus} \longrightarrow PhN{=}N{-}OH \longrightarrow PhN_2^{\bullet} + {}^{\bullet}OH \longrightarrow Ph^{\bullet}$

4. $Ph{-}I \xrightarrow{h\nu} Ph^{\bullet} + I^{\bullet}$

Figure 2.9 Methods for generating phenyl radicals.

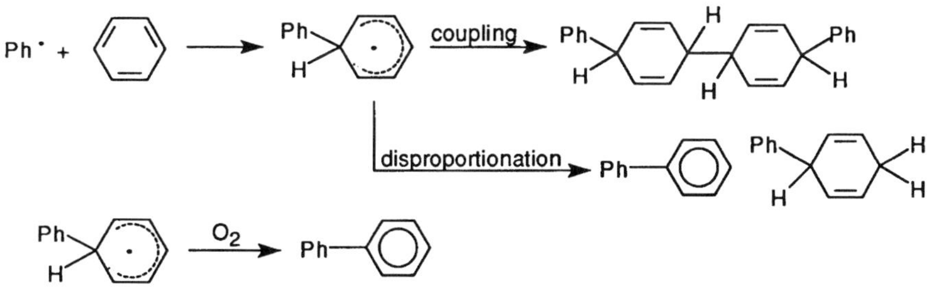

Ph$^{\bullet}$ + benzene ⟶ Ph(H)-cyclohexadienyl radical —coupling⟶ Ph(H)–C$_6$H$_5$(H)–(H)C$_6$H$_5$–Ph(H) dimer

Ph(H)-cyclohexadienyl radical —disproportionation⟶ Ph–C$_6$H$_5$ + Ph(H)C$_6$H$_5$(H)

Ph(H)-cyclohexadienyl radical $\xrightarrow{O_2}$ Ph–C$_6$H$_5$

Figure 2.10 Mechanism of homolytic phenylation of benzene.

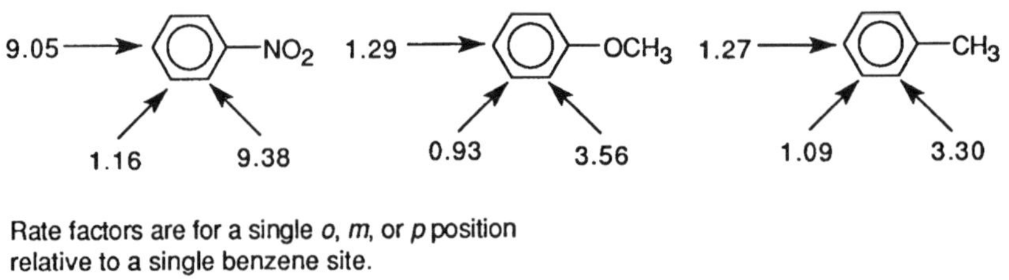

9.05 → C_6H_5–NO_2 (o 9.38, m 1.16) 1.29 → C_6H_5–OCH_3 (o 3.56, m 0.93) 1.27 → C_6H_5–CH_3 (o 3.30, m 1.09)

Rate factors are for a single *o*, *m*, or *p* position relative to a single benzene site.

Figure 2.11 Partial rate factors for the phenylation of arenes by acetylarylnitrosamines.

2.1.9 Diradicals

When, at least formally in terms of canonical structures, two radical moieties are present in the same molecule, the species is designated as a "diradical," as distinguished from a caged radical pair. Both singlet and triplet diradical states are, of course, possible, but singlet diradicals normally are produced by homolysis of singlet precursors. Triplet diradicals will be discussed separately, in Chapter 7. In the limit where the diradical moieties are widely separated and noninteracting, diradical behavior is presumed to approach closely that of caged radical pairs. Such diradicals are sometimes referred to as "double doublets." More interesting are the 1,3 and 1,4 diradicals, in which the radical sites are close enough to interact significantly. The interaction, which potentially involves both through space and through bond interactions, splits the two SOMOs into a HOMO and a LUMO (Figure 2.12). Since both electrons occupy the HOMO, the diradical character of the species becomes somewhat obscured, but is nevertheless reflected in its chemical

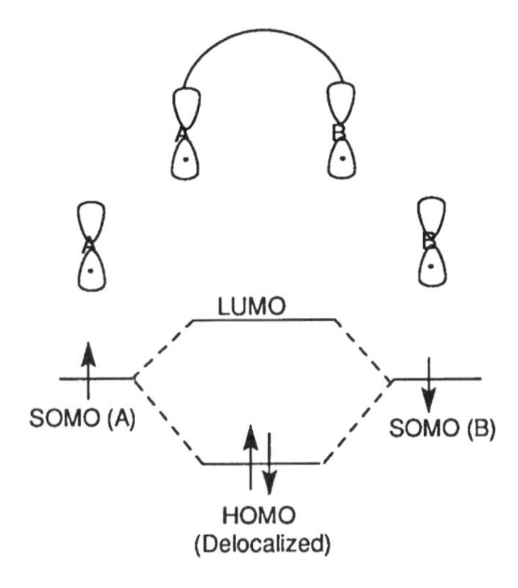

Figure 2.12 MO diagram of a singlet diradical in which the two radical sites are close enough to interact (directly or through bonds).

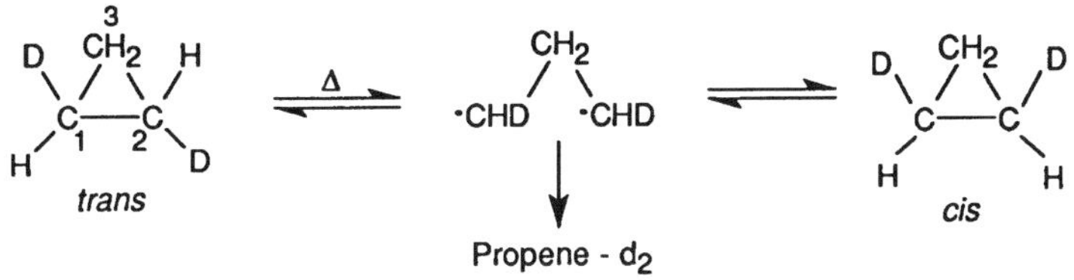

Figure 2.13 Thermolysis of cyclopropane.

behavior. An appropriate definition of diradicals in molecular orbital terms has been advanced: *A diradical is an atom or molecule in which two electrons occupy two degenerate or nearly degenerate orbitals.*[19]

Historically, trimethylene (or 1,3-propanediyl) was the first diradical to be postulated and investigated.[20] This species was invoked in the thermolysis of cyclopropane, which gives propene. It was subsequently found that *cis*- and *trans*-1,2-dideuterocyclopropane undergoes geometric isomerization in competition with the isomerization to propene (Figure 2.13).[21]

The proposed behavior of trimethylene therefore reflects both intramolecular coupling (to re-form the cyclopropane subsequent to rotational equilibration) and intramolecular disproportionation (formation of propene), both apropriate behaviors of a diradical. Incidentally, the C_1–C_3 and C_2–C_3 bonds of 1,2-dideuterocyclopropane are also cleaved, and these homolyses also result in *cis*/*trans* isomerization and rearrangement.

2.1.10 Trimethylenemethane Diradicals

Although intramolecular coupling of 1,3-diradicals is often too rapid to permit their interception by even very efficient radical scavengers, it should not be assumed that coupling is inevitable. The 2-isopropylidenecyclopentane-1,3-diyl radical is a classic example of a singlet diradical that can be trapped cleanly (Figure 2.14).[22] Cycloadditions to dienophiles (in this context *diylophiles*) are stereospecific, as befits a singlet species. If trapping agents are absent, intersystem crossing occurs, and the triplet diradicals dimerize.

2.1.11 1,4-Diradicals

Tetramethylene (1,4-butanediyl) diradicals can be formed by a wide variety of routes (Figure 2.15).[23] The formation of a common intermediate is supported by the obtaining of very similar product distributions from all these reactions. The main reaction channels are (1) fragmentation to two molecules of ethene; (2) cyclization to a cyclobutane; and (3) rotational equilibria that yield *cis*/*trans* cyclobutane mixtures in appropriately disubstituted precursors.

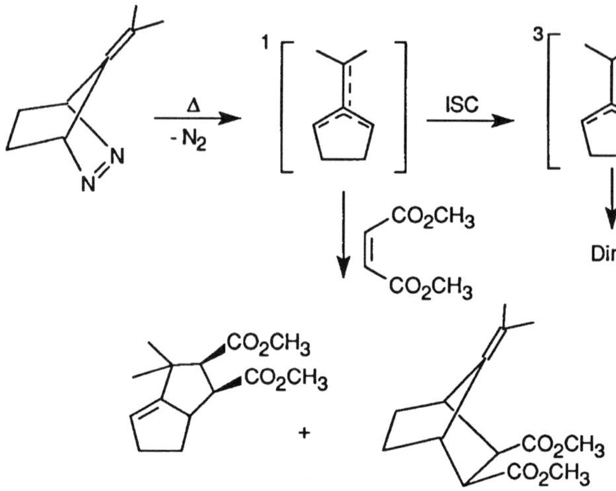

Figure 2.14 A trappable singlet trimethylenemethane diradical.

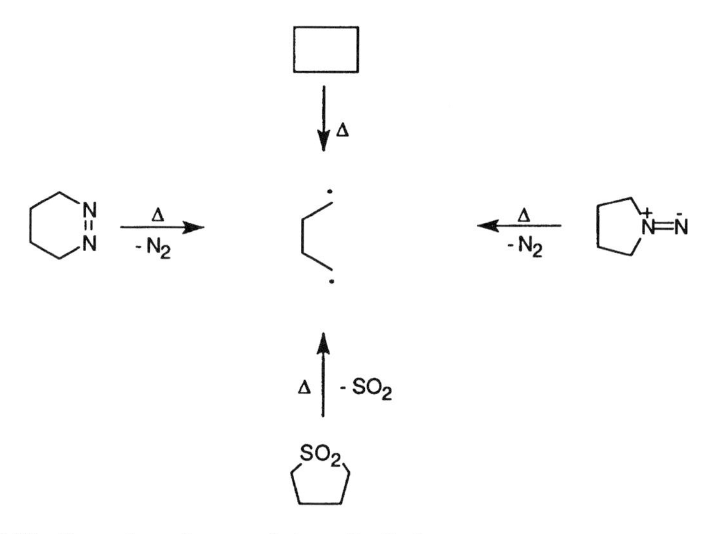

Figure 2.15 Generation of tetramethylene diradicals.

Theoretical studies also appear to support unambiguously the proposal that tetramethylene diradicals are true intermediates in these reactions. *Anti* and *gauche* rotamers are indicated, with the *anti* isomer representing the global minimum (Figure 2.16). Interestingly, the preferred geometry at the terminal carbons apears to be the one that permits hyperconjugative coupling with the C_2–C_3 bond and, importantly, through bond (specifically through the C_2–C_3 bond) interaction of the two radical sites. Direct (through space) interaction should be negligible in the *anti* rotamer.

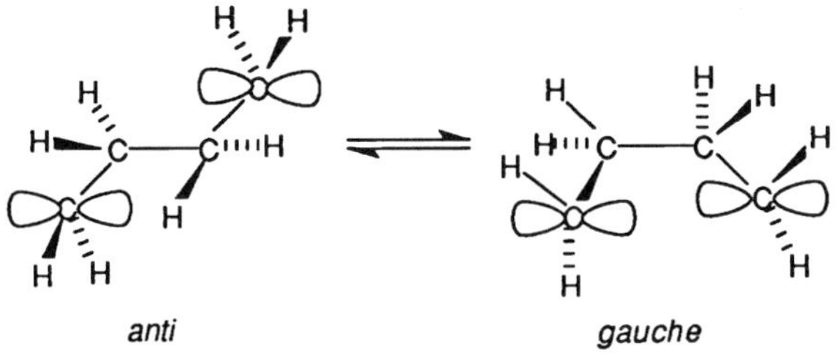

anti gauche

Figure 2.16 Rotamers of the tetramethylene diradical.

2.1.12 Diradical Cycloadditions and Cycloreversions

1,4-Diradicals are key reaction intermediates in many [2+2] cycloadditions and cycloreversions, since doubly suprafacial, concerted [2+2] cycloadditions are rare and are considered to be symmetry forbidden. Characteristically, these reactions are nonstereospecific, since rotational equilibrium in the diradical intermediate usually is competitive with cyclization (Figure 2.17).[24] It is plausible to assume, but not yet experimentally proved, that the reaction preferentially occurs via the more stable *anti* rotamer of the 1,4-diradical. Direct cyclization would not be possible from this rotamer, thus possibly allowing rotational equilibration to proceed essentially to completion.

2.1.13 Dehydroaromatic Diradicals

The thermal cycloaromatization of 1,6-dideutero-*cis*-3-hexene-1,5-diyne generates 1,4-dehydrobenzene, a 1,4-diradical, as a transient intermediate that can undergo retrocyclization to either the starting material or 3,4-dideutero-*cis*-3-hexene-1,5-diyne (Figure 2.18).[25] This reaction, often referred to as the Bergman reaction or the enediyne cyclization, apparently plays a major role in the potent antitumor activity of the enediyne class of antibiotics, including calicheamicin II, esperamicin A, dynemicin A, and neocarzinostatin.[26,27] The enediyne moiety in these antibiotics is contained in a strained ten-membered ring, which evidently greatly facilitates the

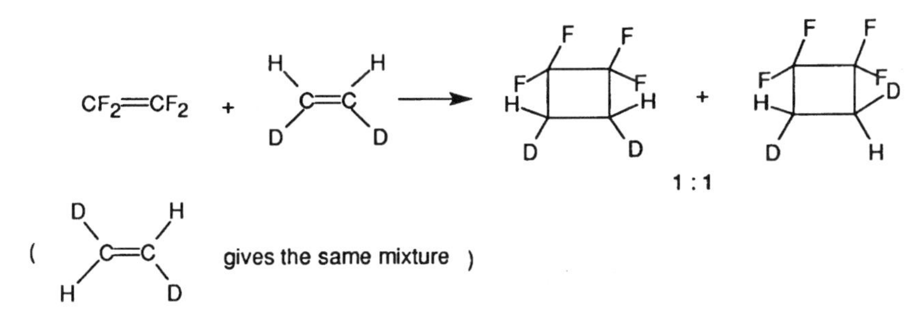

Figure 2.17 Diradical cycloaddition.

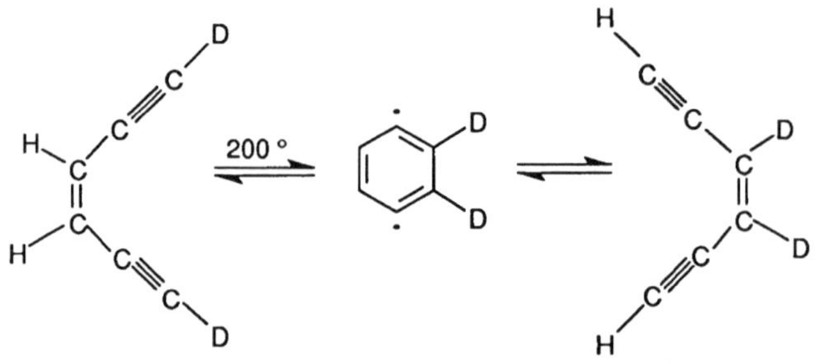

Figure 2.18 1,4–Dehydrobenzene.

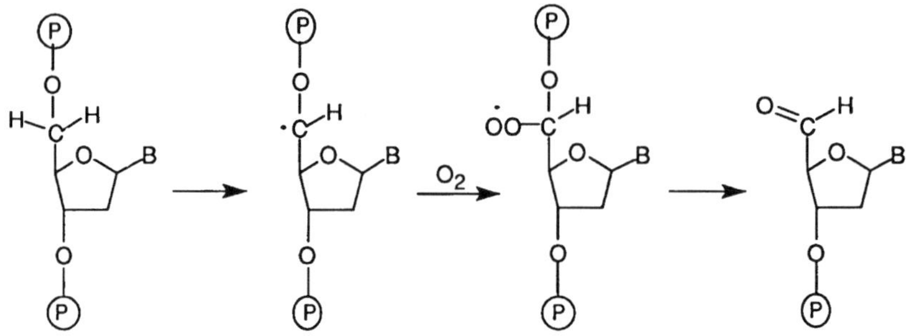

Figure 2.19 Hydrogen abstraction from DNA by a 1,4-diradical derived from an enediyne.

cyclization. The resulting 1,4-diradical apparently abstracts a primary hydrogen atom from the 5′-carbon of cellular DNA, thus initiating double-stranded cleavage of the DNA (Figure 2.19).

This novel type of cycloaromatization reaction has been extended and developed for synthetic purposes. An especially interesting application is the trapping of both radical centers by pendant alkene functions via 5-hexenyl radical type cyclizations (Figure 2.20).[28] In the example cited, 1,4-cyclohexadiene is included as a good hydrogen atom donor for trapping the cyclic radicals.

2.2 Radical Chain Reactions

The efficiency and synthetic value of free radical reactions is greatly enhanced by the existence of the radical chain mechanistic format. Several examples of such radical chain reactions have already been invoked in the previous chapter to illustrate the generation of specific organic radicals and to observe basic aspects of their structure and reactivity. The fundamental mechanistic aspects of radical chain reactions are specifically addressed in this chapter.

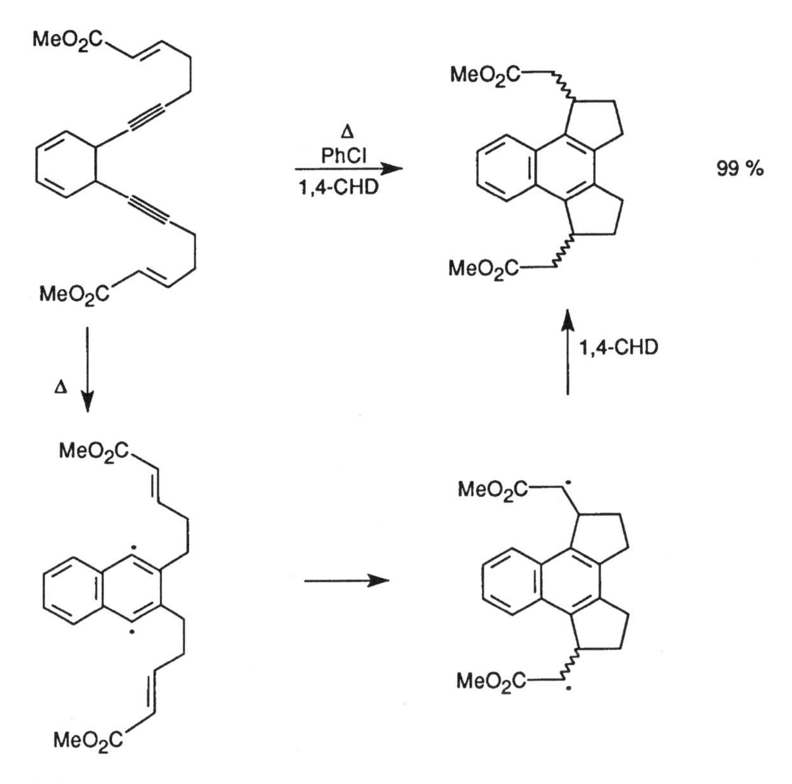

Figure 2.20 Trapping both radical sites of a 1,4-dehydrobenzene.

2.2.1 Homolytic Additions (Ad_H)

The radical chain addition of hydrogen bromide to alkenes, first observed in 1937 by Kharasch, is the classic example of this reaction type.[29] As is well known, the discovery of the reaction was facilitated by the circumstance that the regiochemistry of the reaction is opposite to that of the well-known electrophilic addition of hydrogen bromide (Figure 2.21).

The radical chain process was found to be favored by nonpolar solvents, light, and impure solvents (peroxide impurities), while the electrophilic reaction was favored by pure, polar solvents, the exclusion of light, and especially by the addition of small amounts of certain substances (inhibitors) like hydroquinone or phenols.

$$CH_3CH{=}CH_2 + HBr \xrightarrow[\Delta \text{ or } h\nu]{ROOR} CH_3CH_2CH_2Br$$

$$CH_3CH{=}CH_2 + HBr \xrightarrow{\text{electrophilic}} CH_3CH(Br)CH_3$$

Figure 2.21 Radical chain and electrophilic hydrobromination of propene.

These characteristics define the now-classic profile of radical chain reactions (Figure 2.22). In the case of hydrobromination of unsymmetrical alkenes, the regiochemistry can also be regarded as an extension of this mechanistic profile. The mechanism of the latter reaction is given in Figure 2.23.

The reaction consists of three phases, *initiation*, *propagation*, and *termination*. In the *initiation phase*, alkoxy radicals are first provided by the homolysis of the peroxidic bond of an *initiator*. These reactive radicals then abstract hydrogen atoms from hydrogen bromide molecules to provide bromine atoms. This indirect, two step, method for generating bromine atoms is necessary because the H–Br bond is too strong to be homolyzed under ordinary thermal conditions. The *propagation phase* consists of the iterated cycle of reactions that generate the macroscopically observable product (1-bromopropane). The first step of the propagation cycle

1. Require initiation by some source of free radicals (such as peroxides and heat or light).
2. Rates are not highly dependent on solvent polarity.
3. Reactions are suppressed by sub-stoichiometric amounts of inhibitors.

Figure 2.22 Profile of radical chain reactions.

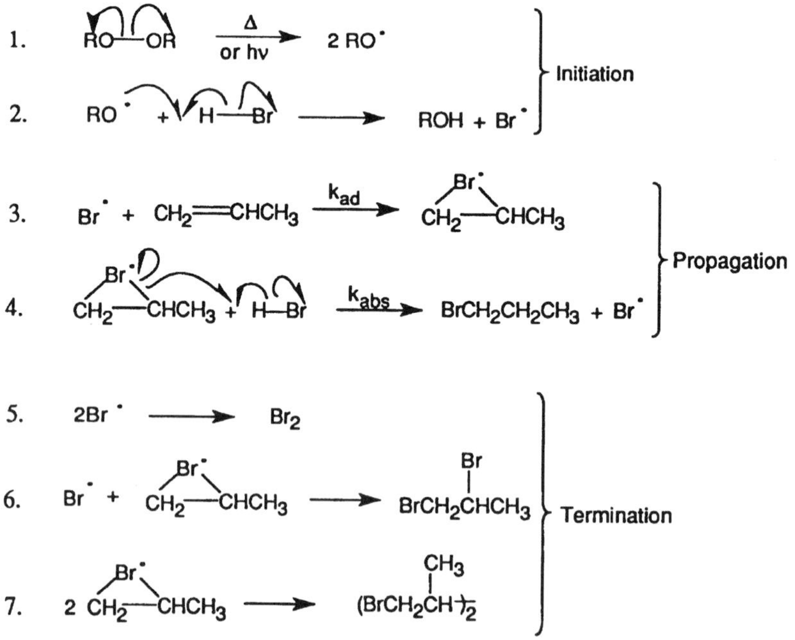

Figure 2.23 Radical chain hydrobromination of propene.

involves addition of bromine atoms to the π bond. It is now recognized that this addition does not yield a simple alkyl radical intermediate but rather a bromine-bridged radical. The second propagation step involves hydrogen abstraction from hydrogen bromide by this bridged radical. The final stage, *termination*, consists of one or more of the inevitable radical-consuming reactions (e.g., coupling).

The observed (anti-Markovnikov) regiochemistry of the reaction, which produces 1-bromopropane rather than 2-bromopropane, is that which would be expected for the formation of the more stable, secondary alkyl radical (Figure 2.24), based upon the 2° versus 1° alkyl radical character in the transition state for addition. It is likely that this transition state would also have significant polar (carbocation) character, which would also favor addition of $Br^{\bullet}$ to the less highly substituted carbon. The latter approach would also be favored by steric effects. Hydrogen abstraction by the 2° radical intermediate would then yield the anti-Markovnikov product, the 1-bromoalkane. The interpretation of the observed regiochemistry is complicated by the circumstance that the radical intermediate is bridged.[30] A resonance theoretical description of the bridged radical is useful in guiding an analysis (Figure 2.25). The three main canonical structures indicate that radical character (odd electron density) is delocalized over both of the alkene carbons and the bromine atom. In the case of an unsymmetric bridged radical (resulting from $Br^{\bullet}$ addition to an unsymmetrical alkene), structures **B** and **C** are not isoenergetic, and the radical character is greater on the secondary carbon because structure **B** is more stable than structure **C**. Consequently, it would appear reasonable that *hydrogen abstraction occurs preferentially at the site of greatest*

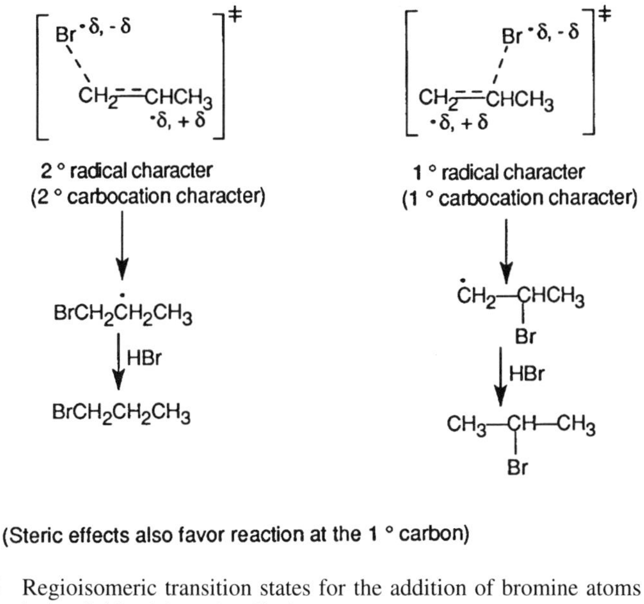

$[Br^{\cdot\delta,-\delta} \cdots CH_2{=}CHCH_3\ ^{\cdot\delta,+\delta}]^{\ddagger}$ $[CH_2{=}CHCH_3 \cdots Br^{\cdot\delta,-\delta}]^{\ddagger}$

2 ° radical character (2 ° carbocation character) | 1 ° radical character (1 ° carbocation character)

$BrCH_2\dot{C}H_2CH_3$ → (HBr) → $BrCH_2CH_2CH_3$ | $\dot{C}H_2{-}CH(Br)CH_3$ → (HBr) → $CH_3{-}CH(Br){-}CH_3$

(Steric effects also favor reaction at the 1 ° carbon)

Figure 2.24 Regioisomeric transition states for the addition of bromine atoms to propene via a hypothetical unbridged (open) radical.

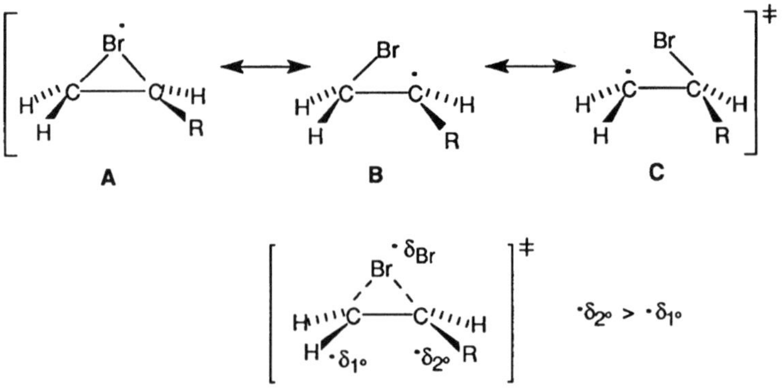

Figure 2.25 The bridged radical intermediate.

radical character. Also, if the actual bridged radical resembles **B** more than **C**, the primary C–Br bond is more completely "made" than the secondary C–Br bond. Therefore, both the odd electron density and the geometric structure of the bridged radical are unsymmetrical, and both should favor abstraction at the secondary carbon. Steric effects would, however, favor abstraction at the primary carbon.

2.2.2 Stereochemistry

The intermediacy of bromine-bridged radicals as opposed to "open" or classical β-bromoalkyl radicals in hydrogen bromide additions to simple alkenes is strongly supported by stereochemical studies of the DBr addition to *cis*- and *trans*-2-butene (Figure 2.26).[30] The hypothetical β-bromoalkyl radicals from *cis*- and *trans*-2-butene would be expected to interconvert rapidly, prior to deuterium abstraction,

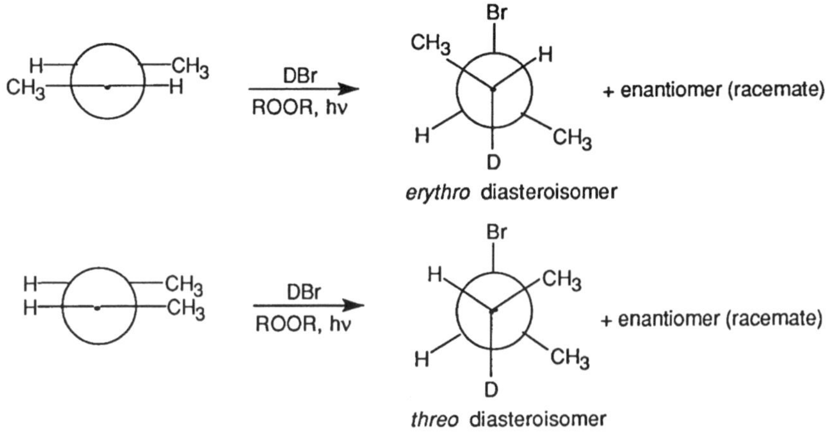

Figure 2.26 *Anti* stereospecificity of DBr addition to the 2-butenes.

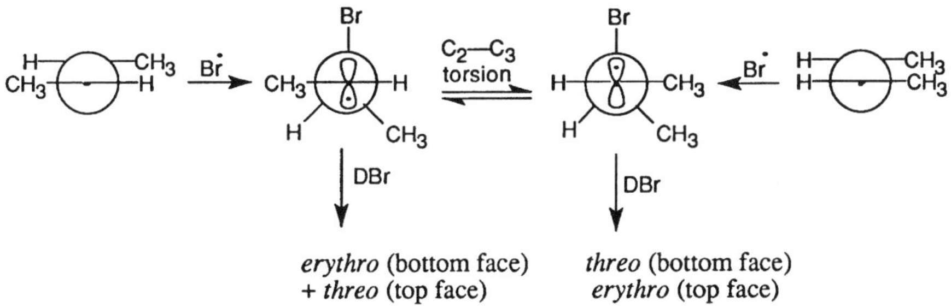

Figure 2.27 Nonstereospecific addition of DBr via a hypothetical open β-bromoalkyl radical.

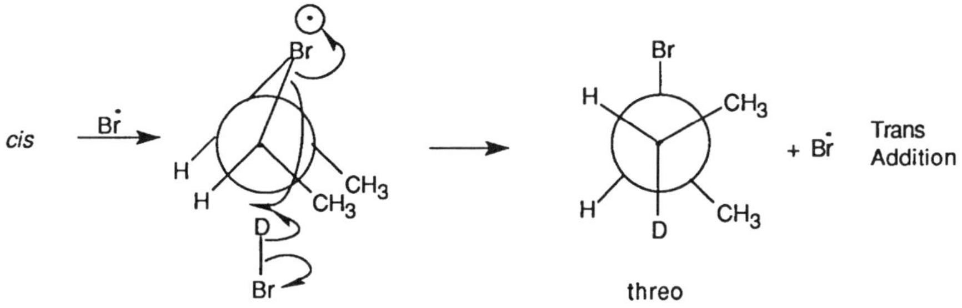

Figure 2.28 Stereospecific anti Opening of the bridged radical.

by rotation around the C–C bond, giving rise to nonstereospecific addition (Figure 2.27). Further, even in the unlikely absence of torsional equilibration, both of these intermediates should be able to abstract deuterium from either face of the trigonal plane of the radical center, giving a mixture of *erythro* and *threo* diastereoisomers from each of the alkenes. Bridging of bromine serves to block the *syn* (or top) face of the intermediate, leading to preferred abstraction at the bottom face and *anti* stereospecific addition (Figure 2.28). It is also noteworthy that the abstraction of deuterium from DBr should be substantially slower than the abstraction of hydrogen from HBr. Therefore, if the bridged radicals are able to resist thermal opening prior to the slower deuterium abstraction, they would clearly not have a sufficiently long lifetime to undergo opening in the case of hydrogen abstraction.

2.2.3 Energetics

Assuming that adequate initiation is provided, the essential requirement for the existence of an efficient radical chain reaction is that the rate of propagation must be much greater than that of termination. That is, the radical–molecule reactions of the propagation steps must be faster than the radical–radical reactions of termination. Since radical coupling reactions have rate constants that closely approach the diffusion rate, it is obviously impossible to envision radical/molecule reaction rate constants that are much larger. In general, the latter reactions will actually have significantly smaller rate constants. How, then, can propagation be dominant over termination? By means of concentration effects, of course. Whereas the concentration of alkene and HBr are normally in the range 0.1–1M, the steady state concentrations of the highly reactive radical species are minuscule. The rates (as opposed to rate constants) of bimolecular radical–radical (termination) reactions are therefore often relatively slow compared to the rates of radical–molecule (propagation) reactions. Nevertheless, in order to compete with termination reactions that are so efficient, propagation must be extremely fast. Since the propagation cycle consists of two consecutive steps, both propagation steps must be quite fast. Fortunately, the activation energies of radical–molecule reactions are often quite small, in part because of the intrinsic high reactivity of radicals but presumably also in part because solvent reorganization energies (solvent effects) are small and do not add much to the barrier.

For the purpose of discussion of the rates of radical–molecule reactions it is useful to define the term *excess activation enthalpy*, ΔH_{int}. As indicated in Figure 2.29, this is

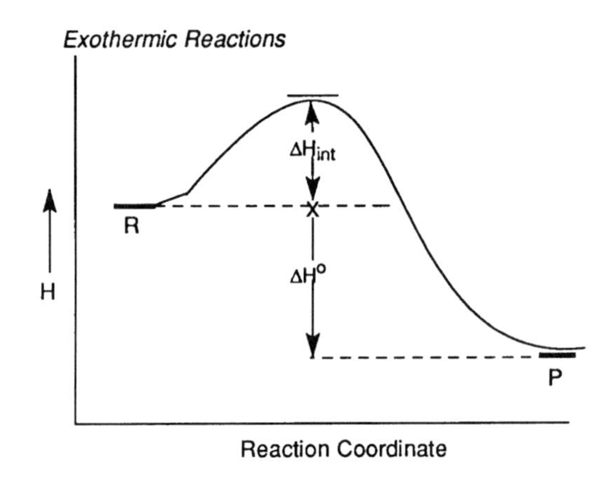

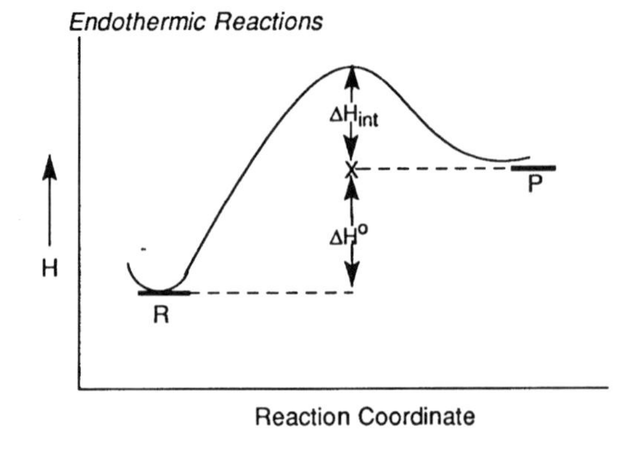

Figure 2.29 Excess activation enthalpies (ΔH_{int}).

$$\Delta H^o \approx \underset{\text{BROKEN}}{\overset{\text{BONDS}}{\Sigma D}} - \underset{\text{FORMED}}{\overset{\text{BONDS}}{\Sigma D}}$$

$$\Delta H_{ad} \approx D(C\overset{\pi}{=}C) - D(C\text{—}Br) = 65 - 70 = -5$$

$$\Delta H_{abs} \approx D(HBr) - D(C\text{—}H) = 87 - 94 = -7$$

Figure 2.30 Approximate enthalpy changes for the propagation steps of hydrobromination.

given by $\Delta H\ddagger$ for an exothermic reaction or by $\Delta H\ddagger - \Delta H^0$ for an endothermic reaction. It is therefore the portion of the activation energy that is not imposed by reaction endothermicity. Because of the high intrinsic reactivity of radicals, the excess activation enthalpies are usually rather small and especially so when polar effects stabilize the transition state (*vide infra*; the discussion of chlorination). Consequently, exothermic radical–molecule reactions are usually very fast unless steric hindrance is substantial. Even mildly endothermic reactions can be fast when excess activation enthalpies are small. It appears to be a useful empirical generalization that radical–molecule reactions will not be effective as propagation steps if $\Delta H^\circ \geq 10$ kcal/mol. This is, of course, not meant to imply that reactions with smaller endothermicities or even exothermic reactions will automatically succeed as propagation steps. Approximate enthalpy changes for the propagation steps of radical chain hydrobromination can be calculated from bond dissociation energies (Figure 2.30). Both the addition and propagation steps are exothermic, and both benefit from polar effects.

2.2.4 The Scope of Radical Chain Additions

The requirement that *both* propagation steps be very fast imposes an unusually stringent condition for successful radical chain additions, resulting in a relatively narrow scope for these reactions. Of the hydrohalogenation possibilities, only hydrobromination clearly meets this requirement. Hydroiodination is frustrated by an endothermic addition step [$D(C\text{-}I) \approx 54$, $D(C\overset{\pi}{=}C) \approx 65$], while hydrofluorination has a strongly endothermic abstraction step as a consequence of the very strong H–F bond. Hydrochlorination appears to be borderline, with an exothermic addition step and a mildly endothermic abstraction step, but can be accomplished in the presence of excess HCl.[31]

2.2.5 Thiol Additions

Though formally analogous to hydrobromination, this very facile reaction illustrates several new and interesting aspects of radical chain additions that are basic to a general understanding of the mechanistic type. In sharp contrast to hydrobromination, the addition of thiols to *cis*- and *trans*-2-butene is nonstereospecific and the recovered 2-butene is extensively geometrically isomerized (Figure 2.31).[32] The negative apparent activation energy (E_a), which is a composite activation energy, has its basis in a negative contribution to E_a from the $\Delta E°$ for the reversible and exothermic radical addition step, which is larger than the positive E_a contribution from the abstraction step. In effect, higher temperatures encourage the reversal of the addition step energetically (as well as entropically) and are unfavorable to addition. The extensive *cis/trans* isomerization observed in the recovered alkene indicates that not only is addition of thiyl radicals reversible, but that rotational equilibration in the radical intermediate is at least competitive with both deuterium abstraction and the elimination of thiyl radicals. This circumstance could be interpreted either in terms of an unbridged or a relatively weakly bridged intermediate. The latter interpretation proves to be the correct one, since the addition of DBr, a better chain transfer agent (i.e., hydrogen atom donor) than RSD to the RSD/2-butene reaction mixture results in a completely stereospecific thiol addition at –78°C.[33] The strategy of this experiment is to allow the radical intermediate less time to equilibrate rotationally by accelerating the rate of the hydrogen abstraction step. It is known, for example, that methyl radicals abstract hydrogen atoms from HBr at a rate about sixty times that of the analogous abstraction from butanethiol.[34] Thus when the chain transfer agent deuterium

1. ROOR ⟶ 2RO•
2. RO• + RSD ⟶ ROD + RS• } Initiation
3. RS• + *cis*-2-butene ⇌ bridged radical ⇌ (rotation) bridged radical ⇌ RS• + *trans*-2-butene
4a. bridged radical + RSD ⟶ adduct + RS•
4b. bridged radical + RSD ⟶ adduct + RS•
5. 2RS• ⟶ (termination) RSSR

Figure 2.31 Nonstereospecific addition of a deuterated thiol to *cis*-2-butene.

$$\left[\overset{\cdot\delta}{\underset{+\delta}{R}}\text{---H---}\overset{\cdot\delta}{\underset{-\delta}{Br}}\right]^{\ddagger} \qquad \left[\overset{\cdot\delta}{\underset{+\delta}{R}}\text{---H---}\overset{\cdot\delta}{\underset{-\delta}{SR'}}\right]^{\ddagger}$$

R = alkyl radical

Figure 2.32 Polar effects in hydrogen abstraction reactions.

bromide is provided and the temperature is lowered to –78°C (also to slow the torsional rate), *trans*-2-butene gives only the *erythro* thiol adduct, and *cis*-2-butene gives only the *threo* thiol adduct. An open (unbridged) radical should be able to abstract hydrogen from either trigonal face to give a mixture of *erytho/threo* diastereoisomers. The attentive reader will no doubt realize that the bromine atoms generated by chain transfer will also lead to competing radical chain addition of DBr to 2-butene, but the occurrence of this side reaction by no means precludes the determination of the stereochemistry of the thiol adduct. Incidentally, the additions of thiols to cycloalkenes, which cannot as readily undergo analogous rotational equilibration of the radical intermediate, tend to occur *trans* stereospecifically even without the addition of a good chain transfer agent.[35] In summary, the reversibility of the addition step of the thiol–alkene reaction can be viewed as the result of a hydrogen abstraction step which is not sufficiently fast to prevent reversal of the addition step.

The circumstance that HBr is a much better chain transfer agent than alkanethiols is interesting in view of the nearly equal bond dissociation energies of these two types of hydrogen donors (87 and 88 kcal/mol, respectively). It does appear that intrinsic activation energies of radical–molecule reactions are rather strongly affected by polar contributions to the transition states, and the transition state for abstraction of hydrogen from HBr presumably is more strongly stabilized by such contributions (Figure 2.32). Benzenethiol, of course, is a much better hydrogen donor than alkanethiols (D = 75 kcal/mol).

2.2.6 Vinyl Polymerization

In radical chain additions such as those of thiols and hydrogen bromide, the intermediate bridged radical normally abstracts a hydrogen atom to give the 1:1 alkene:thiol or alkene:HBr adduct. To a certain extent, however, addition to a second molecule of the alkene can occur competitively. The resulting radical can then abstract hydrogen to afford a 2:1 adduct, or it can continue to add a few more molecules of the alkene before hydrogen abstraction occurs. These "telomers" are typical byproducts of radical additions, and their proportions are minimized by a relatively fast hydrogen abstraction step (e.g., HBr) or by an excess of the addend reagent (*R*SH, HBr). When little or no hydrogen transfer agent is present and especially when the alkene is monosubstituted (i.e., of the vinyl monomer type), chain lengths for addition can be quite long, giving rise to

$$\begin{array}{lll}
1. & ROOR \xrightarrow{k_d} 2\,RO^{\bullet} & \\
 & \text{In = Initiator} & \\
2. & RO^{\bullet} + CH_2{=}CHS \xrightarrow{k_i} RO CH_2\dot{C}HS & \text{Initiation Stage} \\
 & \text{M = Monomer} \qquad\qquad M_1^{\bullet} & \\
3. & ROCH_2\dot{C}HS + CH_2{=}CHS \xrightarrow{k_p} ROCH_2CH(S)CH_2\dot{C}HS & \\
 & M_1^{\bullet} \qquad\qquad M \qquad\qquad\qquad M_2^{\bullet} & \text{Propagation Stage} \\
4. & M_2^{\bullet} + M \xrightarrow{k_p} M_3^{\bullet}\ (\text{etc.}) & \\
5. & 2M_n^{\bullet} \xrightarrow{k_t} M_n - M_n & \text{Termination}
\end{array}$$

S = H, CH_3, Ph, Cl, CN, CO_2Et, etc.

Figure 2.33 Vinyl polymerization.

addition polymers (Figure 2.33). Such vinyl polymerizations arguably represent both the simplest and the most important application of the radical chain reaction.

The kinetics of radical chain reactions have not yet been discussed, and the relatively simple kinetics of vinyl polymerization serves as an excellent introduction to radical chain kinetics.[36] We consider the addition polymerization of a monomer (M) initiated by a typical peroxidic initiator (In). The rate constant for the homolysis of the initiator is designated k_d (for decomposition) and that for the addition of alkoxy radicals ($RO^{\bullet}$) to the monomer to give a β-substituted alkyl radical ($M_1^{\bullet}$) as k_i (for initiation). These two reactions constitute the initiation stage of the radical chain process. The propagation phase consists of a series of reactions in which substituted alkyl radicals (such as $M_1^{\bullet}$) add to the monomer to give a new, longer-chain alkyl radical (such as $M_2^{\bullet}$) until, at length, termination occurs. All these closely analogous propagation reactions are assumed to have the same rate constant, k_p (for propagation). Termination (rate constant k_t) is assumed to occur by coupling or disproportionation of two long-chain alkyl radicals. The rate of monomer consumption is given in Eq. (1) (Figure 2.34) and simplified to Eq. (2), since for large-chain lengths the propagation rate is much greater than the rate of initiation. By applying the steady-state condition to $[M^{\bullet}]$ (Eq. 3), the expression for $[M^{\bullet}]$ in Eq. (4) results. Since the latter involves $[RO^{\bullet}]$, the steady-state assumption for $[RO^{\bullet}]$ is also invoked (Eq. 5) and solved for $[RO^{\bullet}]$ (Eq. 6). The fraction f that appears in these equations is traditional and represents the fraction of the $RO^{\bullet}$ radicals that are effective in initiating polymerization (i.e., the *initiator efficiency*). It will be evident that cage recombination can engender $f < 1$ for certain initiators. The expression for $[RO^{\bullet}]$ in Eq. (6) is then substituted into Eq. (4), giving the expression for $[M^{\bullet}]$ in Eq. (7). The latter is finally substituted into Eq. (2) to obtain the kinetic rate law of Eq. (8). This can be rewritten as in Eq. (9), where the rate of monomer consumption

1. $- d[M]/dt = k_i[RO^•][M] + k_p[M^•][M]$

2. $- d[M]/dt = k_p[M^•][M]$

3. $d[M^•]/dt = 0 = k_i[RO^•][M] - 2k_t[M•]^2$

4. $[M^•] = \{k_i[RO^•][M]/2k_t\}^{1/2}$

5. $d[RO^•]/dt = 0 = 2fk_d[In] - k_i[RO^•][M]$

6. $[RO^•] = 2fk_d[In]/k_i[M]$

7. $[M^•] = \{fk_d[In]/k_t\}^{1/2}$

8. $- d[M]/dt = k_p[M][fk_d[In]/k_t]^{1/2}$

9. $R_p = k_p[M][R_i/2k_t]^{1/2}$

R_p = Rate of Polymerization ; R_i = Rate of initiation

f = Initiator Efficiency

Figure 2.34 Kinetics of vinyl polymerization.

is equated to the rate of polymerization (R_p) and the expression $2fk_d$ [In] is equated to the rate of initiation (R_i). The latter generalizes the rate law to allow for other methods of initiation (i.e., photochemical). The resulting rate law is simple first order in monomer and one-half order in the initiator. It is important to keep in mind that the rate of any radical chain reaction depends upon at least three rate constants, k_p (propagation), $2fk_d$ (initiation), and k_t (termination). However, the rate of initiation can be controlled, leaving the reaction rate dependent mainly upon k_p and k_t. The one-half-order dependence upon the initiation and termination rates is dictated by the termination reaction, which is second order in $[M^•]$. Consequently, coupling and disproportionation, both of which are second order, are not kinetically distinguished.

Mechanistically, the involvement of simple (i.e., unbridged) alkyl free radicals as intermediates assures that these reactions occur nonstereospecifically, but the usual regiochemical preference observed in radical additions is found, and this provides for a predominantly head-to-tail polymer structure. The radical chain polymerization of a vinyl monomer does not, as the mechanism written previously implies, necessarily afford an unbranched polymer. This is especially true in the case of the simplest monomer, ethene, where the chain-propagating alkyl radicals are of the primary type and are thus especially highly reactive. Given the availability of a profusion of secondary carbon–hydrogen bonds nearby in the same molecule, it should not be surprising that intramolecular hydrogen atom abstraction occurs to form a more stable, secondary radical (Figure 2.35).[37] However, the secondary radical sites are still reactive enough to continue to propagate the chain from the new center, giving rise to a branched polymer structure with side chains of various

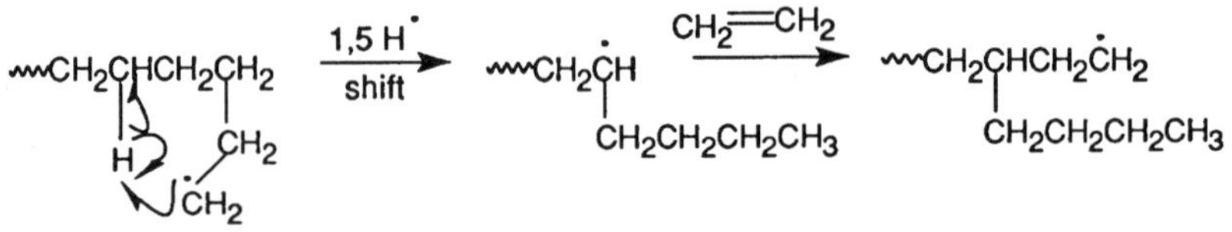

$$\sim CH_2CHCH_2CH_2(CH_2\dot{C}H_2)\ \xrightarrow[\text{shift}]{1,5\ H^\bullet}\ \sim CH_2\dot{C}H(CH_2CH_2CH_2CH_3)\ \xrightarrow{CH_2=CH_2}\ \sim CH_2CH(CH_2CH_2CH_2CH_3)CH_2\dot{C}H_2$$

Figure 2.35 The intramolecular mechanism for chain branching.

length, but butyl groups are especially prevalent. Intermolecular abstraction of hydrogen atoms from "dead" polymer chains by growing radical chains results in polymers having very long side chains. The chain branched poly(ethylene) produced by radical chain polymerization [also called low-density poly(ethylene)] stands in contrast to the linear poly(ethylene) produced by Ziegler–Natta polymerization, although both are produced in great quantity commercially.

2.2.7 More General Aspects of Radical Chain Reaction Kinetics

The kinetics of vinyl polymerization is simplified by the existence of a single propagation rate constant and a single termination reaction. More generally, however, radical chain reactions have two propagation steps with two distinct chain-carrying radicals and three possible, kinetically distinct, termination reactions. Although the kinetic rate law for the general case is rather complex, it simplifies nicely in a number of important special cases.[38] Consider, for example, the Kharasch reaction. The two propagation steps are addition of bromine atoms to the alkene (k_{ad}) and abstraction of hydrogen from hydrogen bromide by the bridged radical (k_{abs}). The possible termination reactions are the dimerizations of bromine atoms and of the alkyl radicals ($R^\bullet$) and the cross-coupling between bromine atoms and the bridged radicals. Although k_{ad} and k_{abs} are by no means necessarily equal, the overall rates of addition and abstraction must be equal, since every molecule of product formed requires cycling through both propagation steps. We therefore have:

$$k_{ad}[Br^\bullet][\text{alkene}] = k_{abs}[R^\bullet][HBr]$$

$$\frac{[Br^\bullet]}{[R^\bullet]} = \frac{k_{abs}[HBr]}{k_{ad}[\text{alkene}]}$$

If the ratio $[Br^\bullet]/[R^\bullet]$ is large, it follows that coupling will occur predominantly between bromine atoms. This requirement can be met if $k_{abs} \gg k_{ad}$ or when $[HBr] \gg [\text{alkene}]$, that is, by using a large excess of HBr. In this special case (coupling only between bromine atoms), a derivation analogous to that presented earlier for vinyl polymerization yields the rate law:

$$\text{Rate} = k_{ad}[\text{alkene}](R_i/2k_t)^{1/2}$$

Under these conditions, the addition of bromine atoms to the alkene is the formally rate-determining propagation step, and the reaction rate is independent of [HBr]. If, on the other hand, termination occurs only between alkyl radicals (excess alkene), the kinetics simplify to:

$$\text{Rate} = k_{abs}[\text{HBr}](R_i/2k_t)^{1/2}$$

and the abstraction step is formally rate determining.

2.2.8 Radical Cyclizations

The use of the 5-hexenyl radical cyclization as a mechanistically important radical probe reaction has already been discussed. This reaction and other related radical cyclizations have proved to be extremely versatile and efficient, and they are now generally considered to be one of the most useful types of radical chain reactions.[39,40] A priori, the 5-hexenyl radical could cyclize either to give the cyclopentylcarbinyl radical or the cyclohexyl radical. The latter radical is undoubtedly more thermodynamically stable than the former (2° versus 1° radical, less ring strain), but radical cyclization yields virtually exclusively the cyclopentylcarbinyl radical. The preference for the 5-*exo* cyclization mode over the 6-*endo* mode is usually considered to be the result of a combination of conformational and stereoelectronic effects. Presumably the preferred conformation of the 5-hexenyl radical is the extended (all *anti*) form. This requires two unfavorable *anti*-to-*gauche* conformational changes to generate the conformation required for 5-*exo* cyclization (Figure 2.36). The conformation appropriate for 6-*endo* cyclization requires a still further *anti*-to-*gauche* conversion and is therefore relatively less favorable. In terms of stereoelectronics, the 5-*exo* transition state appears to permit a more nearly linear approach of the $2p_z$ orbital at the radical site (C_1) to the $2p_z$ orbital at C_5 (and thus more efficient development of overlap) than is available in the approach of C_1 to C_6 in the 6-*endo* precursor. This preference for 5-*exo* cyclization is maintained even when the radical site is benzylic, although a

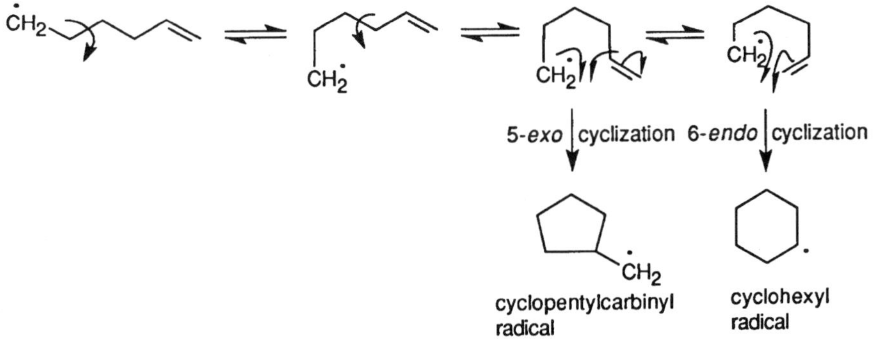

Figure 2.36 Conformational effects on the 5-*exo* versus 6-*endo* radical cyclization preference.

small amount of 6-*endo* product is formed in this case. When *two* strongly radical stabilizing substituents are present at C_1 and especially when a good hydrogen donor (e.g., Bu_3SnH) is *not* present, the 5-*exo* cyclization apparently becomes reversible, and the 6-*endo* mode is favored (Figure 2.37).

The cyclization of such stabilized 5-hexenyl radicals is rather convenient synthetically because the required radical can often be generated by hydrogen abstraction, rather than requiring an abstractable functional group such as bromine or iodine (Figure 2.38). This is feasible when the targeted C–H bond is the weakest one in the molecule. Note that an exceptionally good hydrogen donor is not used, the solvent serving as the donor.

The R_3SnH/RX method for radical generation has an undesirable feature, viz. the loss of a useful functionality (halogen) by reduction. Retention of the halogen function can be achieved when a variation of this basic method is employed (Figure 2.39). In this modification, the intermediate radicals are intercepted by iodine atom transfer from alkyl iodides (e.g., ethyl iodide), instead of hydrogen transfer. The alkyl radicals homolytically displace tributyltin radicals from the hexabutyldistannane reagent, which continue the chain by abstracting bromine atoms from the reactant bromide.[40]

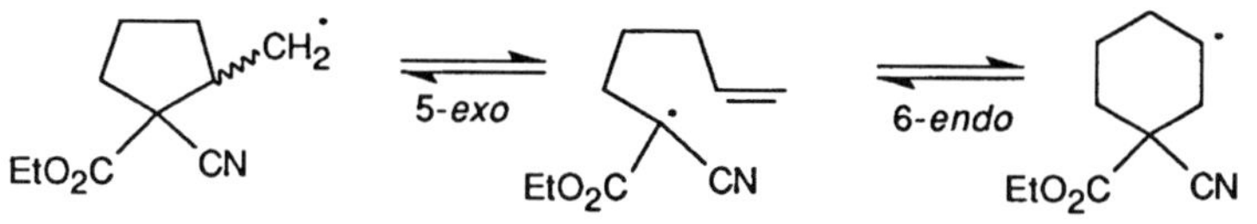

Figure 2.37 Thermodynamically controlled 6-*endo* cyclization.

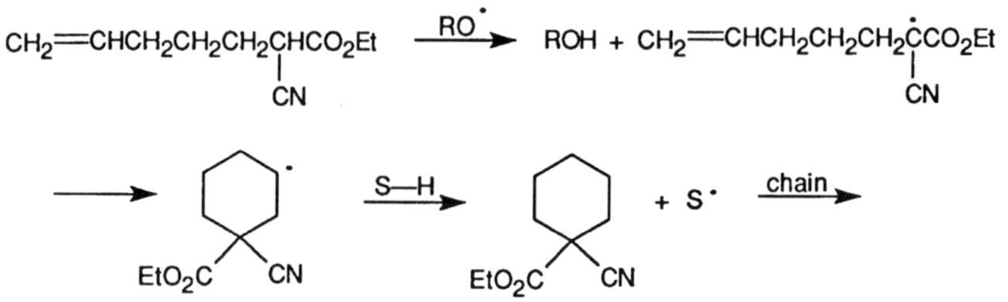

Figure 2.38 Generation of stabilized 5-hexenyl radicals by hydrogen abstraction.

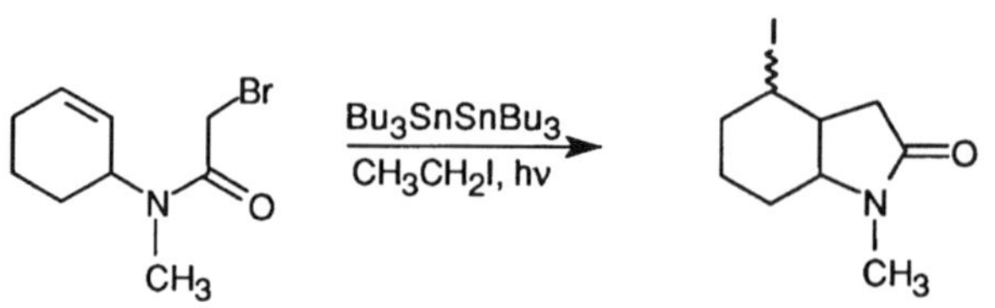

Figure 2.39 Cyclization with retention of a halogen functionality.

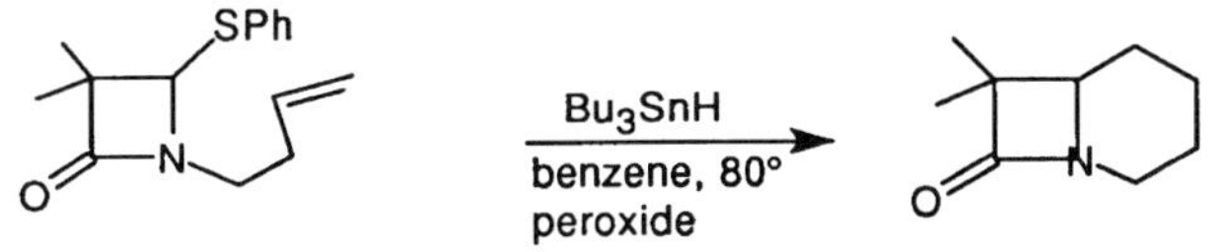

Figure 2.40 Generation of radicals by abstraction of phenylthio groups.

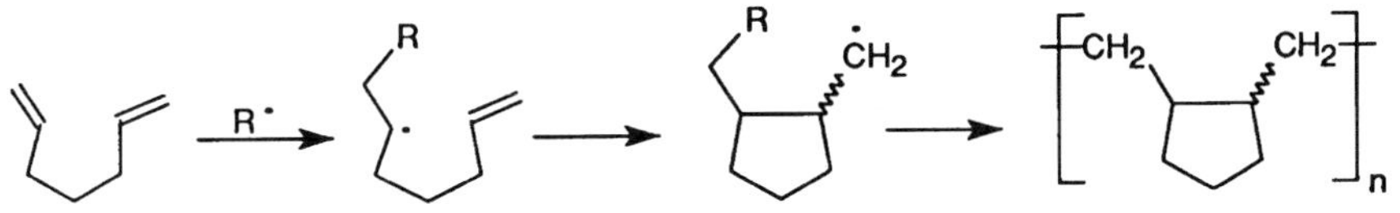

Figure 2.41 Cyclopolymerization.

It is also interesting to note that the phenylthio group is another effective homolytic leaving group for the purpose of generating radical sites (Figure 2.40).[40]

Radical cyclizations, and especially the 5-*exo* cyclization, serve as the basis for a novel polymerization mechanism known as "cyclopolymerization" (Figure 2.41).[41]

2.2.9 Homolytic Substitution (S_H)

The classic example of a radical chain substitution reaction is halogenation, but radical chain oxidation (autoxidation) rivals this reaction in importance. Other useful substitution reactions, including tributytin hydride reduction and aldehyde decarbonylation, have already been introduced in Chapter 1.

2.2.10 Chlorination

The chlorination of methane is a standard example of radical chain chlorination (Figure 2.42).[42,43] An external initiator (e.g., a peroxide) is unnecessary in this instance because the Cl–Cl bond is weak enough to supply initiating chlorine atoms at temperatures above 200°C. The propagation steps are both abstraction reactions. The abstraction of hydrogen atoms by chlorine atoms is 1 kcal/mol endothermic (Figure 2.43), but the excess activation energy is only 2.8 kcal/mol, making E_a = 3.8 kcal/mol. Once again, the very low excess activation energy would seem to reflect polar stabilization of the transition state. A further interesting comparison that supports the concept of polar stabilization of transition states for hydrogen abstraction is available (Figure 2.44).[44] In all three instances the reactions are aproximately thermoneutral ($\Delta H° \approx 0$), but the activation energies vary sharply in a manner consistent with the operation of polar effects.

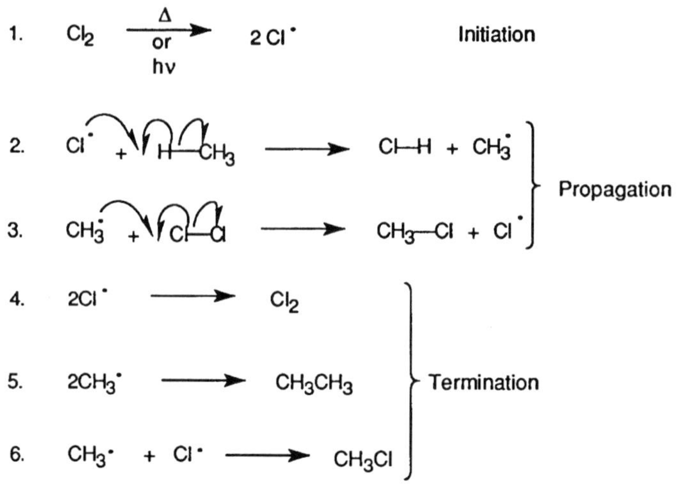

1. $Cl_2 \xrightarrow[\text{or } h\nu]{\Delta} 2\,Cl^{\bullet}$ Initiation

2. $Cl^{\bullet} + H{-}CH_3 \longrightarrow Cl{-}H + CH_3^{\bullet}$ Propagation

3. $CH_3^{\bullet} + Cl{-}Cl \longrightarrow CH_3{-}Cl + Cl^{\bullet}$ Propagation

4. $2Cl^{\bullet} \longrightarrow Cl_2$ Termination

5. $2CH_3^{\bullet} \longrightarrow CH_3CH_3$ Termination

6. $CH_3^{\bullet} + Cl^{\bullet} \longrightarrow CH_3Cl$ Termination

Figure 2.42 Radical chain mechanism for the chlorination of methane.

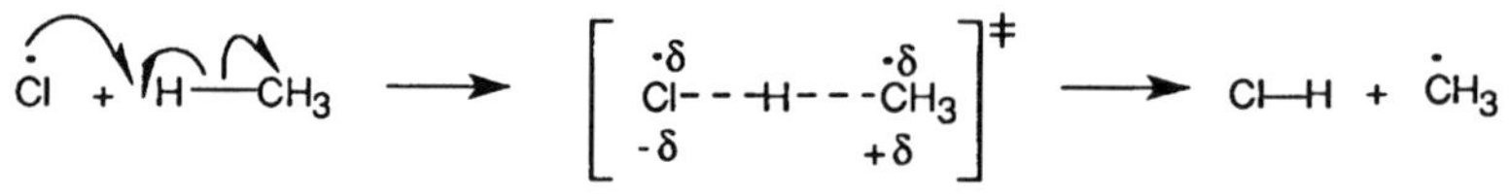

$$\dot{Cl} + H{-}CH_3 \longrightarrow \left[\overset{\bullet\delta}{\underset{-\delta}{Cl}}\cdots H \cdots \overset{\bullet\delta}{\underset{+\delta}{CH_3}} \right]^{\ddagger} \longrightarrow Cl{-}H + \dot{C}H_3$$

$\Delta H^{o} \approx D(CH_3{-}H) - D(H{-}Cl) = 104\text{-}103 = +1$ kcal/mol

$E_a = 3.8$ kcal/mol

$E_{int} = 2.8$ kcal/mol

Figure 2.43 Energetics of the hydrogen abstraction step.

$CH_3^{\bullet} + H{-}Cl \rightarrow CH_4 + Cl^{\bullet}$ $E_a = 3.8$ kcal/mol

$CH_3^{\bullet} + H{-}OCH_3 \rightarrow CH_4 + \dot{O}CH_3$ $E_a = 8$ kcal/mol

$CH_3^{\bullet} + H{-}CH_3 \rightarrow CH_4 + CH_3^{\bullet}$ $E_a = 14.5$ kcal/mol

Figure 2.44 Polar effects on activation energies for hydrogen transfer.

The second propagation step is highly exothermic, and, since favorable polar effects are also operative here, this also has an extremely low activation energy (< 1 kcal/mol; Figure 2.45).

Since primary, secondary, and tertiary alkane C–H bonds are all weaker than that of methane, the hydrogen abstraction step is exothermic and even more favorable in these cases. Chlorine atom abstraction, of course, remains exothermic for all the above C–H bond types. On the other hand, the abstraction of hydrogen atoms from *sp* or sp^2 C–H bonds is endothermic and therefore not as facile as addition to the π bonds of these systems.

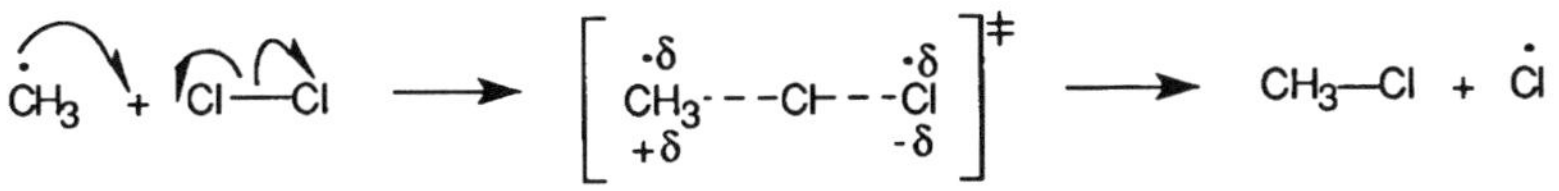

$$\dot{C}H_3 + Cl{-}Cl \longrightarrow \left[\overset{\cdot\delta}{\underset{+\delta}{CH_3}} \text{- - -} Cl \text{- - -} \overset{\cdot\delta}{\underset{-\delta}{Cl}} \right]^{\ddagger} \longrightarrow CH_3{-}Cl + \dot{C}l$$

$\Delta H^o \approx D(Cl{-}Cl) - D(CH_3{-}Cl) = 58\text{-}84 = -26$ kcal/mol

$E_a = E_{int} < 1$ kcal/mol

Figure 2.45 Energetics of the chlorine atom abstraction.

2.2.11 Bromination

The abstraction of hydrogen atoms from methane by bromine atoms is highly endothermic (Figure 2.46) and remains substantially endothermic for primary and even secondary C–H bonds.[42,43] However, abstraction of hydrogen from tertiary C–H bonds is only mildly endothermic and, of course, benefits from polar effects, which are likely to be enhanced by the tertiary nature of the carbocation character of the transition state. Not surprisingly, then, only tertiary C–H bonds are efficiently brominated among alkane C–H bonds. Benzylic and allylic C–H bonds are also weak enough to be efficiently abstracted by bromine atoms. This, of course, contrasts sharply with the low selectivity of chlorine atoms in hydrogen abstractions. The contrasting selectivities of chlorine and bromine atoms are also often discussed in terms of the Hammond–Leffler postulate, according to which the more endothermic abstraction by $Br^{\bullet}$ has a transition state with greater product (alkyl radical) character than that for exothermic abstraction by Cl• (Figure 2.47). Consequently, radical stabilization effects (3° > 2° > 1° > CH_3) are expressed more fully in the former.

$$Br\bullet + CH_4 \longrightarrow HBr + CH_3\bullet \qquad \Delta H = +17 \text{ kcal/mol}$$

$$Br\bullet + CH_3CH_3 \longrightarrow HBr + CH_3CH_2\bullet \qquad \Delta H = +11 \text{ kcal/mol}$$

$$Br\bullet + (CH_3)_2CH{-}H \longrightarrow HBr + (CH_3)_2CH\bullet \qquad \Delta H = +7 \text{ kcal/mol}$$

$$Br\bullet + (CH_3)_3C{-}H \longrightarrow HBr + (CH_3)_3C\bullet \qquad \Delta H = +4 \text{ kcal/mol}$$

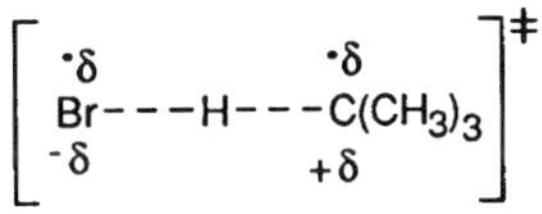

$$\left[\overset{\cdot\delta}{\underset{-\delta}{Br}} \text{- - -} H \text{- - -} \overset{\cdot\delta}{\underset{+\delta}{C(CH_3)_3}} \right]^{\ddagger}$$

Figure 2.46 Abstraction of hydrogen by bromine atoms.

$$\left[\overset{\cdot\delta}{Cl} \text{- - -} H \text{- - -} \overset{\cdot\delta}{R} \right]^{\ddagger} \qquad \left[\overset{\cdot\delta}{Br} \text{- - -} H \text{- - -} \overset{\cdot\delta}{R} \right]^{\ddagger}$$

Little radical character on R — Much radical character on R

Figure 2.47 Relative selectivity of $Cl^{\bullet}$ and $Br^{\bullet}$ in hydrogen abstraction.

2.2.12 Bromination by N-Bromosuccinimide

The traditional laboratory reagent for the selective bromination of allylic, benzylic, and tertiary carbon–hydrogen bonds has long been N-bromosuccinimide (NBS).[45] The radical chain nature of the reaction is clear from its susceptibility to inhibition and the requirement of initiation (e.g., by dibenzoyl peroxide in refluxing carbon tetrachloride).[46] The propagation cycle originally assumed and accepted for quite some time involved hydrogen abstraction by the succinimidyl radical, followed by the abstraction of a bromine atom from NBS by the intermediate alkyl radical (Figure 2.48).[47]

A very attractive alternative mechanism subsequently emerged that involves bromination by molecular bromine generated in a low steady-state concentration provided by the reaction of NBS with hydrogen bromide as it is produced in the propagation cycle of this reaction (Figure 2.49).[48] This sequence requires a small amount of HBr, of unspecified origin, for initiation. The revised mechanism is supported by kinetic studies of the bromination of *m*- and *p*-substituted toluenes by both NBS and by $Br_2/h\nu$ at low concentration. In both cases the relative rate data are correlated by the Hammett–Brown equation (i.e., using the σ^+ substituent constant) with $\rho = -1.36$, suggesting that hydrogen abstraction is effected by the same species in both reactions, viz. $Br^\bullet$.[49] The remote possibility that the succinimidyl radical and $Br^\bullet$ coincidentally exert precisely the same polar effect in benzylic hydrogen atom abstraction was rendered unlikely by studying the bromination of the same

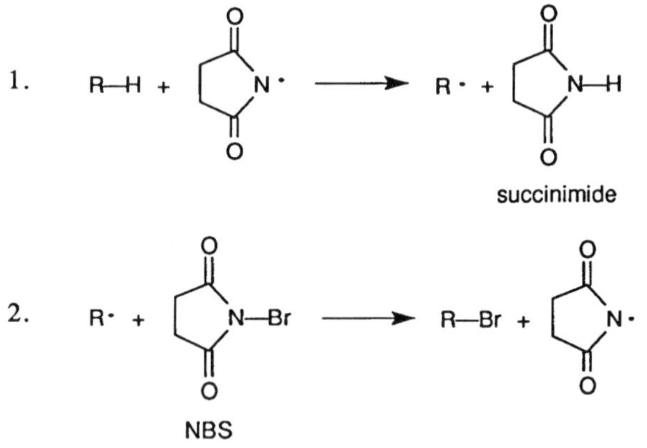

1. R—H + (succinimidyl radical, N·) ⟶ R· + (N—H) succinimide

2. R· + (N—Br) NBS ⟶ R—Br + (N·)

Figure 2.48 The original mechanism proposed for NBS bromination.

1. $R\cdot + Br_2 \longrightarrow R\text{—}Br + Br\cdot$

2. $Br\cdot + R\text{—}H \longrightarrow R\cdot + HBr$

3. $HBr + NBS \longrightarrow Br_2 + \text{succinimide}$

Figure 2.49 A revised mechanism of NBS bromination of benzylic and allylic substrates.

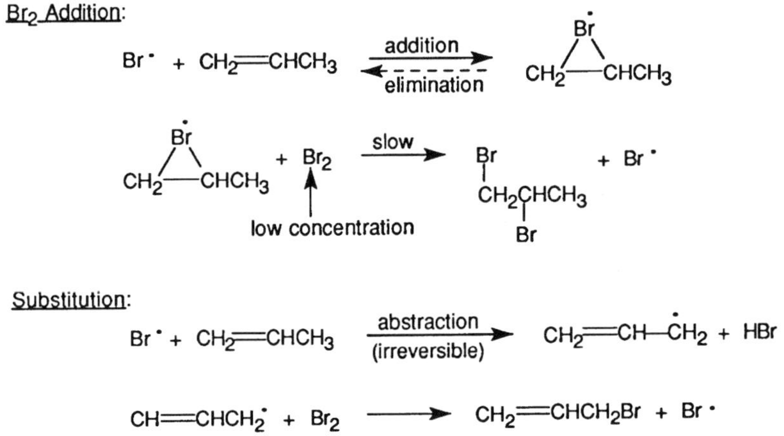

Br$_2$ Addition:

$$Br^{\bullet} + CH_2{=}CHCH_3 \underset{\text{elimination}}{\overset{\text{addition}}{\rightleftharpoons}} CH_2\overset{\overset{\bullet}{Br}}{\triangle}CHCH_3$$

$$CH_2\overset{\overset{\bullet}{Br}}{\triangle}CHCH_3 + Br_2 \xrightarrow{\text{slow}} CH_2(Br)CH(Br)CH_3 + Br^{\bullet}$$

low concentration

Substitution:

$$Br^{\bullet} + CH_2{=}CHCH_3 \xrightarrow[\text{(irreversible)}]{\text{abstraction}} CH_2{=}CH{-}\dot{C}H_2 + HBr$$

$$CH{=}CHCH_2^{\bullet} + Br_2 \longrightarrow CH_2{=}CHCH_2Br + Br^{\bullet}$$

Figure 2.50 Competition between addition and abstraction in the reaction of Br_2 with Alkenes.

substrates with tetrafluoro and also tetramethyl-NBS. Precisely the same value of ρ as for bromination by Br_2 and by NBS was again observed.[49] For the purposes of bromination of reactive (benzylic and allylic) substrates, NBS can be regarded as a synthetically convenient source of molecular bromine at low concentration. In the context of *allylic* bromination, this concentration aspect is potentially an important factor in circumvention of radical chain addition of bromine to the alkene π bond. In these reactions, the addition of $Br^{\bullet}$ to the π bond is competitive with allylic hydrogen abstraction, but at very low $[Br_2]$, the abstraction step of the radical chain addition (which involves abstraction of bromine atoms from Br_2) is slow (Figure 2.50). The addition of $Br^{\bullet}$ is therefore reversible under these conditions, and irreversible hydrogen abstraction by $Br^{\bullet}$ is the eventual result.

2.2.13 Autoxidation

The oxidation of organic molecules by atmospheric oxygen can often occur spontaneously and is therefore termed *autoxidation*. The autoxidation of diethyl ether, in which potentially explosive peroxides are produced in this very common solvent, offers a classic example of the potentially troublesome nature of this ubiquitous reaction. On the other hand, controlled autoxidation of substrates such as cumene is extremely useful. Cumene peroxide, the product of cumene autoxidation, is a commercial precursor for both phenol and acetone. An even more general application for radical chain oxidation is in the combustion of hydrocarbon fuels.

The autoxidation of cumene has received especially detailed attention. The reaction is usually studied at temperatures from 30 to 70°C so that the product peroxide is thermally stable.[50] Under these conditions a radical initiator must be provided. When carried out at higher temperatures, the peroxidic product acts as initiator and provides an autocatalytic process. The reaction mechanism (Figure

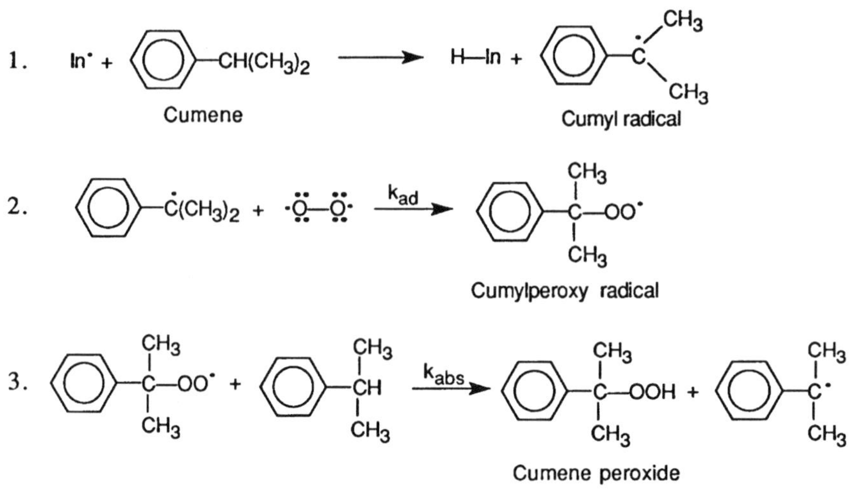

Figure 2.51 Cumene autoxidation mechanism.

2.51) involves addition of cumyl radicals to dioxygen to give cumylperoxy radicals in the first propagation step (Step 2 of Figure 2.51). Dioxygen (itself a triplet) is an excellent radical trap, so that this reaction closely approaches diffusion control even for the resonance-stabilized cumyl radical. The second propagation step, abstraction of hydrogen from cumene by the cumylperoxy radical, is the rate-controlling propagation step. Consequently the observed kinetic rate law at all but very low oxygen concentrations is:

$$\frac{-d[O_2]}{dt} = \frac{-d[\text{cumene}]}{dt} = \frac{d[R\text{OOH}]}{dt} = k_{\text{abs}}[\text{cumene}](R_i/2k_t)^{1/2}$$

Somewhat surprisingly for oxygen-centered radicals, peroxy radicals are relatively highly selective in regard to hydrogen abstraction, the result being that alkanes, for example, are not readily oxidized under these mild conditions. This selectivity is easily rationalized in terms of the relatively low dissociation energy of the O–H bond of peroxides [$D(R$OO–H) = 88 kcal/mol].[51] The weakness of this bond compared to the corresponding O–H bond of alcohols at least in part must reflect three-electron bonding (and delocalization) in the peroxy radical ($ROO^\bullet$). The substantial electrophilic character of peroxy radicals is reflected in the ρ value (−0.6) observed in the autoxidation of substituted toluenes.[52]

The kinetic rate law cited for cumene autoxidation indicates that termination involves cumylperoxy radicals and is bimolecular. This coupling has been shown to involve the formation of the tetroxide, which, at the usual autoxidation temperatures, decomposes to dioxygen and caged cumyloxy radicals (Figure 2.52).[52] The preponderance of these undergo diffusive separation and continue to provide initiation. Only about 10% of the cumyloxy radicals undergo cage

$$1.\quad 2RO\dot{O} \longrightarrow ROOOOR \text{ (tetroxide)}$$

$$2.\quad ROOOOR \longrightarrow O_2 + \overline{R\dot{O}\dot{O}R} \xrightarrow[10\%]{\text{coupling}} ROOR \text{ (stable) Cumene peroxide}$$

$$3.\quad \overline{R\dot{O}\dot{O}R} \xrightarrow{90\%} 2RO^{\bullet} \text{ (initiator radicals)}$$

Figure 2.52 Termination in cumene autoxidation.

recombination, and these alone represent the termination stage. Consequently, the inefficiency of termination contributes to a relatively high rate of cumene autoxidation. In this connection, Russell has made an extremely important point concerning relative rates of radical chain reactions.[53] It has been emphasized previously that the rate of a radical chain reaction depends not only on the rate of the slowest propagation step, but also upon the rates of iniation and termination. Although the initiation rate is controllable, the termination rate is not. The hydrocarbon tetralin is observed to undergo autoxidation at a rate about three times that of cumene. This is considered to reflect the greater ease of hydrogen atom abstraction from tetralin than from cumene in the respective rate-determining propagation steps. However, Russell observed that a small amount of added tetralin sharply *decreases* the overall oxidation rate in the cumene autoxidation. This has been confirmed to result from the much more efficient termination between tetralinperoxy radicals (Figure 2.53), which is possible when an α-hydrogen is present (as in a secondary hydroperoxide) but not in its absence (tertiary hydroperoxide). From this perspective, the only slightly faster autoxidation of tetralin than cumene conceals more dramatic circumstances, that is, that propagation is much faster for tetralin, but termination is also much faster. The importance of *inefficient termination* in providing more efficient radical chain reactions must therefore not be overlooked.

The facility of the autoxidation of ethers and other oxygenated organic molecules can be understood on the basis of weakened C–H bonds and the polar effect of

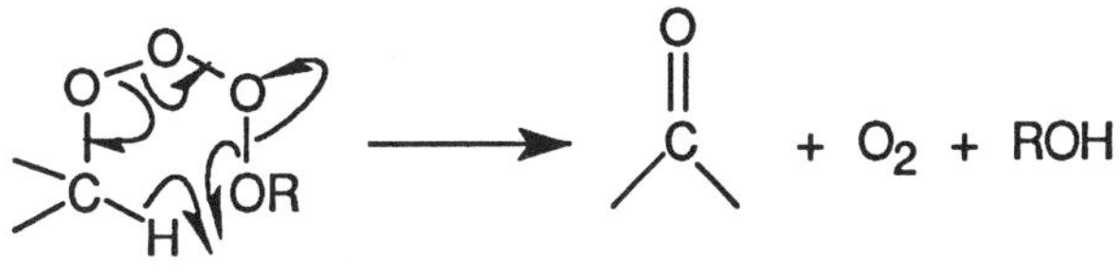

Tetralin tetroxide ⟶ tetralone + O_2 + ROH

Figure 2.53 Efficient termination of tetralinperoxy radicals.

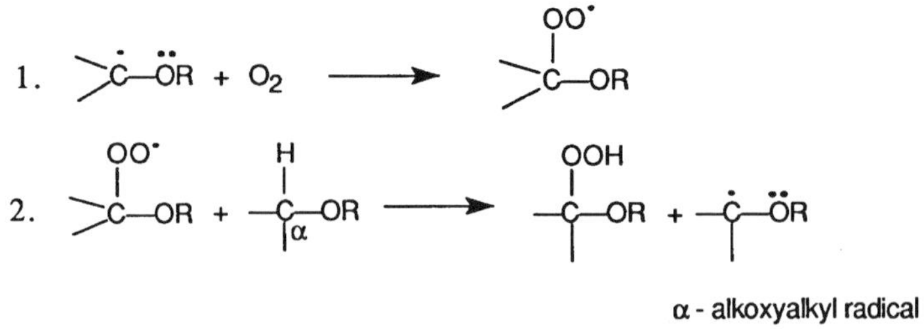

1. $>\dot{C}{-}\ddot{O}R + O_2 \longrightarrow >C(OO^{\cdot}){-}OR$

2. $>C(OO^{\cdot}){-}OR + {-}C(H)(\alpha){-}OR \longrightarrow {-}C(OOH){-}OR + {-}\dot{C}{-}\ddot{O}R$

α - alkoxyalkyl radical

Figure 2.54 Autoxidation of ethers.

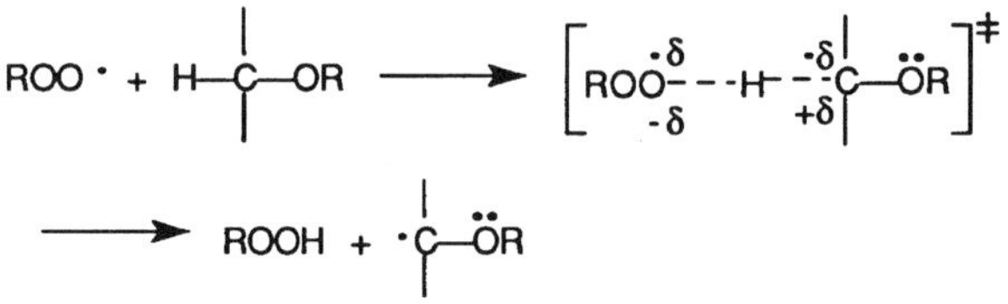

$$ROO\cdot + H{-}C{-}OR \longrightarrow \left[ROO^{\cdot\delta}_{-\delta}{-}{-}{-}H{-}{-}{}^{\cdot\delta}_{+\delta}C{-}\ddot{O}R\right]^{\ddagger} \longrightarrow ROOH + \cdot C{-}\ddot{O}R$$

Figure 2.55 Polar effect in the abstraction of an α-hydrogen from ether.

peroxy radicals in the hydrogen abstraction step. The α C–H bond is of course the weakest, presumably as a result of three-electron-bonding stabilization of the resulting α-alkoxyalkyl radical (Figure 2.54). The electrophilic nature of peroxy radicals engenders some carbocation character in the transition state for hydrogen abstraction, and this character is strongly stabilized by the electron-donating alkoxy function (Figure 2.55).[54]

2.2.14 Inhibition

The inhibition of radical chain reactions by hydroquinones, phenols, anilines, and a variety of other molecules implies that the chain-carrying radicals are rather efficiently intercepted by these molecules and that the new radical produced by this

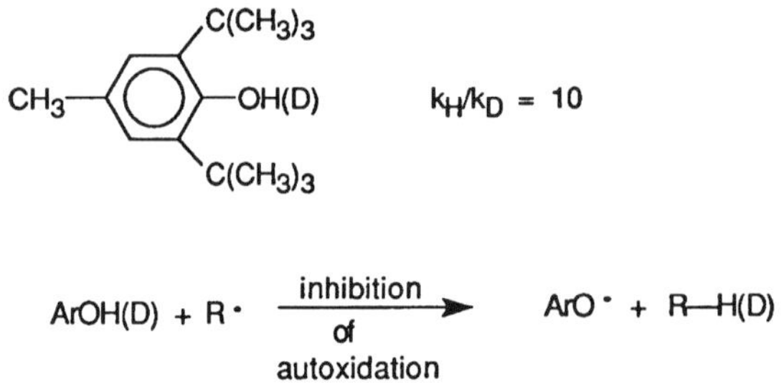

$CH_3{-}C_6H_2(C(CH_3)_3)_2{-}OH(D)$ $\quad k_H/k_D = 10$

$$ArOH(D) + R\cdot \xrightarrow[\text{autoxidation}]{\text{inhibition of}} ArO\cdot + R{-}H(D)$$

Figure 2.56 Isotope effects in inhibition by phenols.

radical–molecule reaction is not sufficiently reactive itself to initiate new chains. The mechanism of inhibition by hydroquinone and phenols, for example, is presumed to involve hydrogen abstraction from the phenolic oxygen, producing a resonance-stabilized phenoxy radical, which eventually undergoes coupling. This mechanism is supported by isotope effect studies that reveal that O-deuterated phenols are as much as ten times less efficient inhibitors than their protic counterparts (Figure 2.56).[55]

2.2.15 Hydrogen Abstraction from Electronegative Atoms by Electronegative Radicals

The activation energies for hydrogen abstraction from phenolic hydroxyl groups by peroxy radicals, the reaction type involved in inhibition of oxidation, are remarkably small. An activation energy of E_a = 0.5 kcal/mol has, for instance, been reported for the abstraction of hydrogen from 2,6-di(*tert*-butyl)phenol by *tert*-butylperoxy radicals (Figure 2.57).[56] Since both the reactant peroxy radical and the product aryloxy radical are substantially resonance stabilized, the abstraction is close to thermoneutral (i.e., $\Delta H° \approx 0$). Further, the transition state is not subject to a polar stabilization effect of the type previously discussed ($X^{+\delta}$ . . . H . . . $Y^{-\delta}$). The minimal activation energy therefore initially appears somewhat surprising. However, it is now clear that hydrogen abstractions of the type *X* . . . H . . . *Y*, where *X* and *Y* are both electronegative atoms such as halogen, oxygen, or nitrogen, have unusually low activation energies in comparison to cases where *X* and *Y* are both hydrogen, alkyl, or trialkylsilyl (Table 2.1).

$$(CH_3)_3COO^{\bullet} + \text{HO–C}_6\text{H}_3(C(CH_3)_3)_2 \xrightarrow{E_a = 0.5\ \text{kcal/mol}} (CH_3)_3COOH + {}^{\bullet}\text{O–C}_6\text{H}_3(C(CH_3)_3)_2$$

Figure 2.57 Abstraction of phenolic hydrogens by peroxy radicals.

Table 2.1 Activation Energies of *X* . . . H . . . *Y* Hydrogen Abstractions

X = *Y*	E_a (kcal/mol)	*D*(*X* – *H*) (kcal/mol)
$(CH_3)_3CO^{\bullet}$	2.6	105.1
$Cl^{\bullet}$	4.8	103.2
$RCH_2S^{\bullet}$	5.2	87.8
$H^{\bullet}$	9.6	104.2
$CH_3^{\bullet}$	14.5	104.9

$$\left[\begin{array}{ccc} \cdot\delta & & \cdot\delta \\ \text{RO}{-}{-} & {-}\text{H}{-} & {-}{-}\text{OR} \\ -\delta & +\delta & -\delta \end{array} \right]^{\ddagger}$$

Figure 2.58 A proposed polar stabilization effect on hydrogen abstraction transition states having X = Y = electronegative atom.

It is also evident that the effect is not correlated with bond dissociation energies. It does appear logical, however, to asociate the activation-energy-lowering effect of two terminal electronegative atoms with a slightly different kind of polar effect (Figure 2.58).

References

1. Franck, J.; Rabinowitch, E. *Trans. Faraday Soc.* **1934**, *30*, 120.
2. Hammond, G. S.; Soffer, L. M. *J. Am. Chem. Soc.* **1951**, *72*, 4711; Shine, H. J.; Waters, J.; Hoffman, D. *J. Am. Chem. Soc.* **1963**, *85*, 3613.
3. Martin, J. C.; Hargis, J. H. *J. Am. Chem. Soc.* **1969**, *91*, 5399.
4. Rembaum, A.; Swarez, M. *J. Am. Chem. Soc.* **1955**, *77*, 3486.
5. Bartlett, P. D.; Gontarev, B. A.; Sakurai, H. *J. Am. Chem. Soc.* **1962**, *84*, 3101.
6. Bartlett, P. D.; Rüchardt, C. *J. Am. Chem. Soc.* **1960**, *82*, 1756.
7. Pryor, W. A.; Smith, K. J. *J. Am. Chem. Soc.* **1970**, *92*, 5403; Neuman, R. C.; Lockyer, G. D., Jr.; Amrich, M. J. *Tetrahedron Lett.* **1972**, 1221.
8. Bartlett, P. D.; Funahashi, T. *J. Am. Chem. Soc.* **1962**, *84*, 2596.
9. Brooks, V. W.; Dainton, F. S.; Ivin, K. J. *Trans. Faraday Soc.* **1965**, *61*, 1437; Koenig, T. in *Free Radicals,* Vol. I, Kochi, J. K.; Ed., Wiley, New York, 1973, p. 143.
10. Seltzer, S. *J. Am. Chem. Soc.* **1961**, *83*, 2625; Seltzer, S.; Hamilton, E., Jr. *J. Am. Chem. Soc.* **1966**, *88*, 3775.
11. Cohen, S. G.; Wang, C. H. *J. Am. Chem. Soc.* **1955**, *77*, 2457 and 3628.
12. Seltzer, S.; Dunne, F. T. *J. Am. Chem. Soc.* **1965**, *87*, 2628.
13. Chapter 1, p. ?
14. Larock, R.C. *Solvation/Demercuration Reactions in Organic Synthesis*; Springer, New York, 1986.
15. Whitesides, G. M.; San Filippo, J., Jr. *J. Am. Chem. Soc.* **1970**, *92*, 6611.
16. Schfer, H. *Top. Curr. Chem.* **1990**, *152*, 91.
17. Corey, E. J.; Bauld, N. L.; LaLonde, R. T.; Casanova, J., Jr.; Kaiser, E. T. *J. Am. Chem. Soc.* **1960**, *82*, 2645.
18. Perkins, M. J. in *Free Radicals,* Vol. II, Kochi, J. K., Ed., Wiley, New York, 1973, p. 231.
19. Salem, L.; Rowland, C. *Angew. Chem. Int. Ed. Engl.* **1972**, *11*, 92.
20. Chambers, T. S.; Kistiakowsky, G. B. *J. Am. Chem. Soc.* **1934**, *56*, 399.
21. Rabinovitch, B. S.; Schlag, E. W.; Wiberg, K. B. *J. Chem. Phys.* **1958**, *28*, 504; Bergman, R. G. in *Free Radicals,* Vol. I, Kochi, J. K., Ed., Wiley, New York, 1982, p. 191.

22. Berson, J. A. *Acc. Chem. Res.* **1978**, *11*, 446; Berson, J. A. in *Diradicals*, Borden, W. T., Ed., Wiley, New York, 1982, p. 151.

23. Dervan, P. B.; Dougherty, D. A. in *Diradicals*, Borden, W. T., Ed., Wiley, New York, 1982, p. 107.

24. Bartlett, P. D.; Cohen, G. M.; Elliott, S. P.; Hummel, K.; Minns, R. A.; Sharts, G. M.; Fukunaga, J. Y. *J. Am. Chem. Soc.* **1972**, *94*, 2899; Lowry, J. H.; Richardson, K. S. *Mechanism and Theory in Organic Chemistry*, Third Edition, Harper and Row, New York, 1987, p. 907.

25. Bergman, R. G. *Acc. Chem. Res.* **1973**, *6*, 25.

26. Lee, M. D.; Ellestad, G. A.; Borders, D. B. *Acc. Chem. Res.* **1991**, *24*, 235.

27. Nicolau, K. C.; Daio, W.-M. *Angew. Chem. Int. Ed. Engl.* **1991**, *30*, 1387.

28. Grissom, J. W.; Calkins, T. L.; Egan, M. *J. Am. Chem. Soc.* **1993**, *115*, 11744.

29. Kharasch, M. S.; Engelmann, H.; Mayo, F. R. *J. Org. Chem.* **1937**, *2*, 288.

30. Skell, P. S.; Shea, K. J. in *Free Radicals,* Vol. I, Kochi, J. K., Ed., Wiley, New York, 1982, p. 809; Maj, S. P.; Symons, M. C. R.; Trousson, P. M. R. *J. Chem. Soc. Chem. Commun.* **1984**, 561.

31. Raley, J. H.; Rust, F. F.; Vaughn, W. E. *J. Am. Chem. Soc.* **1948**, *70*, 2767; Abell, P. I. in *Free Radicals,* Vol. I, Kochi, J. K., Ed., Wiley, New York, 1982, p. 70.

32. Abell, P. I. in *Free Radicals,* Vol. I, Kochi, J. K., Ed., Wiley, New York, 1982, p. 80.

33. Neuriter, N. P.; Bordwell, F. G. *J. Am. Chem. Soc.* **1960**, *82*, 5354.

34. Ingold, K. U. in *Free Radicals*, Vol. I, Kochi, J. K., Ed., Wiley, New York, 1982, p. 80.

35. Lebel, N.; DeBoer, A. *J. Am. Chem. Soc.* **1967**, *89*, 2784.

36. Stevens, M. P. *Polymer Chemistry*, Oxford University Press, New York, 1990, p. 199.

37. Reference 23, p. 205.

38. Walling, C. *Free Radicals in Solution,* John Wiley & Sons, New York, 1957, p. 243.

39. Wilt, J. W. in *Free Radicals,* Vol. I, Kochi, J. K., Ed., Wiley, New York, 1982, p. 191.

40. Motherwell, W. B.; Crick, D. *Free Radical Chain Reactions in Organic Synthesis*, Academic Press, New York, 1992.

41. Odian, G. *Principles of Polymerization*, Third Edition, John Wiley & Sons, New York, 1991, p. 512.

42. Walling, C. *Free Radicals in Solution*, John Wiley & Sons, New York, 1953, p. 352.

43. Poutsma, M. L. in *Free Radicals,* Vol. II, Kochi, J. K., Ed., Wiley, New York, 1973, p. 159.

44. Benson, S. W.; DeMore, W. B. *Ann. Rev. Phys. Chem.* **1965**, *16*, 397; Ingold, K. U. in *Free Radicals*, Vol. I, Kochi, J. K. Ed., John Wiley & Sons, New York, 1973, p. 69.

45. Ziegler, K.; Späth, A.; Schaaf, E.; Schumann, W.; Winkelmann, E. *Ann.* **1942**, *551*, 80.

46. Poutsma, M. L. in *Free Radicals*, Vol. II., Kochi, J. K., Ed., Wiley, New York, 1973, p. 420–433.

47. Bloomfield, G. F. *J. Chem. Soc.* **1944,** 114.

48. Adam, J.; Gosselain, P. A.; Goldfinger, P. *Nature* **1953**, *171*, 704.

49. Pearson, R. E.; Martin, J. C. *J. Am. Chem. Soc.* **1963**, *85*, 354 and 3142.

50. Howard, J. in *Free Radicals*, Vol. II, Kochi, J. K., Ed., John Wiley and Sons, New York, 1973, p. 4.

51. Mahoney, L. R.; DaRooge, M. A. *J. Am. Chem. Soc.* **1970**, *92*, 4063.

52. Howard, J. A.; Ingold, K. U.; Symonds, M. *Can. J. Chem.* **1968**, *46*, 1017.

53. Russell, G. A. *J. Am. Chem. Soc.* **1957**, *79*, 3871; Blanchard, H. J. *J. Am. Chem. Soc.* **1959**, *81*, 4548; Bartlett, P. D.; Traylor, T. G. *J. Am. Chem. Soc.* **1963**, *85*, 2407.

54. Reference 47, p. 23

55. Howard, J.; Ingold, K. U. *Can. J. Chem.* **1962**, *40*, 1851; Ref. 50, p. 43.

56. Zavitsas, A. A.; Chatgilialoglu, C. *J. Am. Chem. Soc.* **1995**, *117*, 10645.

Exercises

2.1 The photolysis of the 11β-nitrite ester of corticosterone generates an aldoxime that can be hydrolyzed to aldosterone acetate. Write a plausible radical mechanism for this conversion and indicate how you could determine whether the reaction is of the chain or nonchain type.

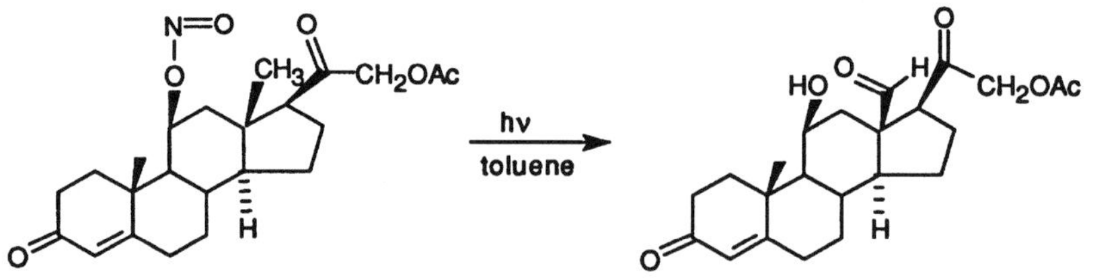

(Barton, D. H. R.; Beaton, J. M. *J. Am. Chem. Soc.* **1961**, *83*, 4083.)

2.2 The trityl radical reacts with diaroylperoxides in a nonchain, second-order reaction, giving a trityl ester as one of the products. The ρ value for a series of symmetrically disubstituted diaroylperoxides is ρ = +1.45. Classify the reaction by its mechanistic symbol and construct a dotted line/partial charge/partial radical character model for the transition state. Would you expect log k_{rel} to correlate with σ or σ^-?

$$Ph_3C\cdot + ArC(=O)OOC(=O)Ar \longrightarrow ArC(=O)OCPh_3$$

(Suehiro, T.; Kanoya, A.; Hara, H.; Nakahama, T.; Omori, M.; Komori, T. *Bull. Chem. Soc. Japan* **1967**, *40*, 668.)

2.3 The thermal, nonchain decomposition of a peroxide (P) should be a first-order process; that is, a plot of log[P] versus *t* should be linear with slope –*k*, where *k* is the first-order rate constant. However, for many peroxides the plot is curved, such that the slope decreases with time. Further, the initial slope increases as [P] increases. Linear plots, if they are obtainable at all, can be observed only at quite low peroxide concentrations. (a) Using the reaction of the preceding problem as a model, write a mechanism for the induced, radical chain decomposition of a peroxide in an essentially unreactive solvent. (b) Induced decompositions in ethereal solvents are often especially facile. Write a mechanism for the induced decomposition of a peroxide in diethyl ether that explains the formation of ethoxyethyl aroate esters in high yields. Point out any steps in which polar effects facilitate the reaction and provide TS models for these steps.

$$\mathrm{ArC(=O)OOC(=O)Ar} \xrightarrow{\mathrm{Et_2O},\ \Delta} \mathrm{ArC(=O)OCH(OEt)CH_3}\quad 90\%$$

(c) The decomposition of diaroylperoxides having an ^{18}O label in the carbonyl oxygen was studied in diethyl ether. What interesting mechanistic question(s) could this study resolve? Explain and illustrate the results anticipated for each mechanism under consideration.

$$\mathrm{ArC(={}^{18}O)OOC(={}^{18}O)Ar} \xrightarrow{\mathrm{Et_2O},\ \Delta}$$

(Cass, W. E. *J. Am. Chem. Soc.* **1947**, *69*, 500; Drew, E. H.; Martin, J. C. *Chem. Ind. (London)* **1959**, 925; Denney, D. B.; Feig, G. *J. Am. Chem. Soc.* **1959**, *81*, 5322.)

2.4 (a) Trialkylboranes are spontaneously oxidized by dioxygen in a radical chain process. Provide a plausible propagation cycle for this reaction. Indicate which propagation step is expected to be the slower one and explain why.

$$\mathrm{R_3B + O_2 \longrightarrow ROOBR_2}$$

(b) The reaction above involves a homolytic displacement on boron and the cleavage of a boron–carbon bond that is not especially weak. Explain why S_H reactions in general on boron are expected to be much faster than on carbon, nitrogen, or oxygen. Explain further why an S_H reaction involving an oxygen radical in particular should be more facile on boron than on these other atoms. (Ingold, K. U. *Chem. Commun.* **1969**, *911*; Davies, A. G.; Ingold, K. U.; Roberts, B. P.; Tudor, R. *J. Am. Chem. Soc. (B)* **1971**, 698.)

2.5 Discuss the bearing of the following hydrogen–deuterium secondary kinetic isotope effect data on the mechanism of thermal decomposition of these azo compounds.

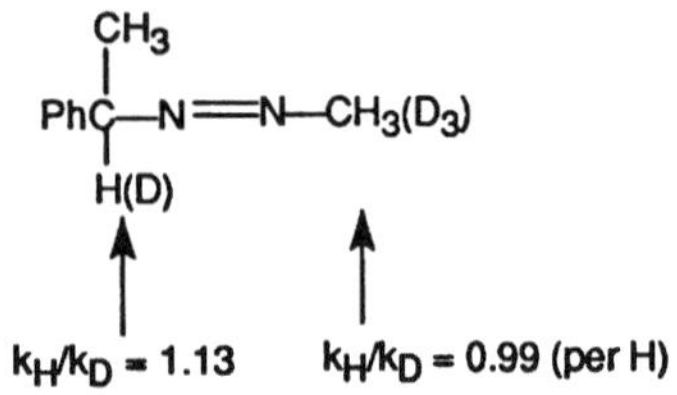

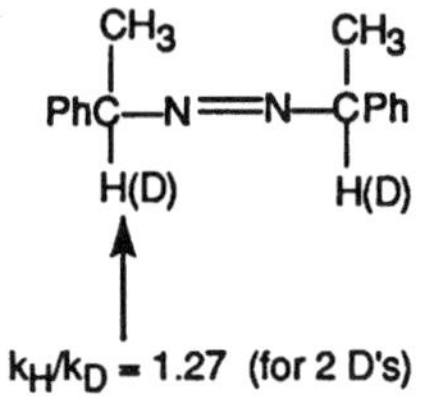

$PhC(CH_3)(H(D))—N=N—CH_3(D_3)$

$k_H/k_D = 1.13$ $k_H/k_D = 0.99$ (per H)

$PhC(CH_3)(H(D))—N=N—C(CH_3)(H(D))Ph$

$k_H/k_D = 1.27$ (for 2 D's)

(Seltzer, S.; Dunne, F. T. *J. Am. Chem. Soc.* **1965**, *87*, 2628.)

2.6 Provide a detailed mechanistic explanation for the observation that *cis*-4-(*tert*-butyl)cyclohexyl bromide reacts 15 times as fast as the *trans* isomer with the Br_2 under radical chain conditions. Your explanation should also explain the formation of a single dibromide from the *cis* isomer and a complex mixture of dibromides from the *trans* isomer.

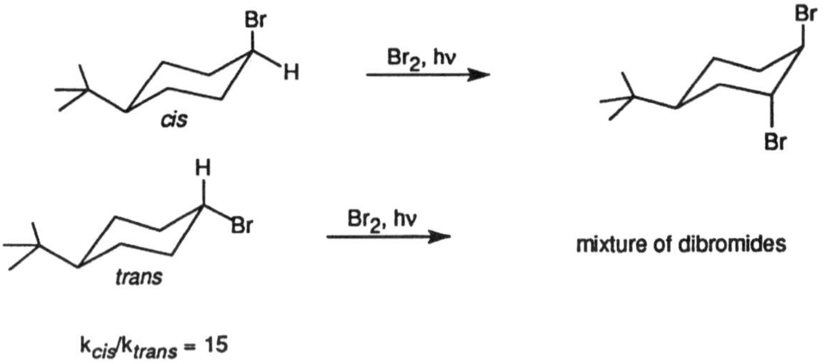

Br, H, *cis* $\xrightarrow{Br_2,\ h\nu}$ Br, Br

H, Br, *trans* $\xrightarrow{Br_2,\ h\nu}$ mixture of dibromides

$k_{cis}/k_{trans} = 15$

(Skell, P. S.; Readio, P. D. *J. Am. Chem. Soc.* **1964**, *86*, 3334.)

2.7 Analyze the propagation cycle for the addition of thiols to simple alkenes and predict the most probable rate law for these additions based upon your prediction of the rate-determining propagation step.

RSH + $>C=C<$ $\longrightarrow$ RS—C—C—H

2.8 The Grignard reagent of 5–bromo-1–hexene is readily prepared in the usual way. The product contains at most 5% of the cyclopentenylmethyl Grignard. It is known that the radical intermediates in Grignard-forming reactions are efficiently trapped by radical scavengers such as TEMPO. The 5-hexenyl radical, which is considered to be an intermediate in the conversion of 5-bromo-1-hexene to its Grignard reagent, has an internal trap–the cyclization to the cyclopentylmethyl radical.

(a) Rationalize the observation of only a small amount of cyclopentylmethyl Grignard in this reaction and reconcile this observation with the efficient trapping of such radicals by TEMPO.

Mg, Et_2O

Br → MgBr + MgBr

> 95% ≤ 5%

(b) The reaction of 5–hexenylmagnesium bromide with dioxygen yields a mixture of oxygenated products that is up to 80% cyclized, depending on the concentration of O_2. Write a mechanism for this reaction that is consistent with the finding of extensive cyclization.

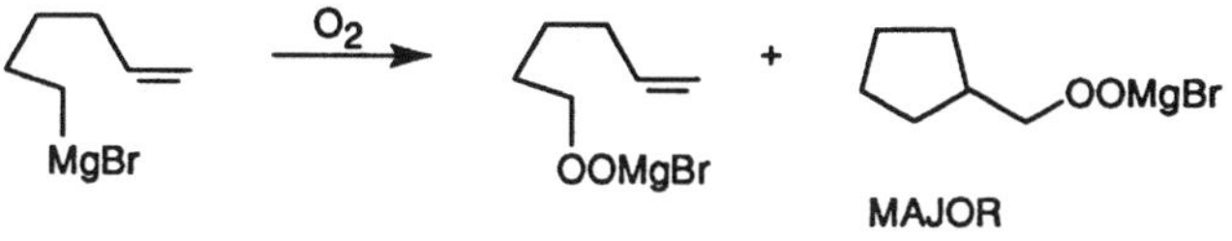

O_2

MgBr → OOMgBr + OOMgBr

MAJOR

(c) Given the triplet ground state of O_2, rationalize the circumstance that nucleophilic addition to oxygen is not observed.

$$R\text{–Mg}X + O_2 \rightarrow RO\text{OMg}X$$

(Walling, C.; Cioffair, A. *J. Am. Chem. Soc.* **1970**, *92*, 6609; Lamb, R. C.; Ayers, P. W.; Toney, M. K.; Garst, J. F. *J. Am. Chem. Soc.* **1966**, *88*, 4261.)

2.9 Dimethylphosphine adds to alkenes via a radical chain mechanism to give excellent yields of 1 : 1 adducts:

$(CH_3)_2PH$ + ⟶ $(CH_3)_2P$–C–C–H

(a) Write the most plausible propagation steps for the addition to ethene. (b) Analyze the approximate energetics (ΔH^0) of each propagation step using the bond dissociation energies provided and considering polar effects on the excess activation energy.

$D(\text{P–C}) = 62\,\text{kcal/mol}$

$D(\text{P–H}) = 77$

$D(\text{C–H}) = 98$

$D(\text{C}\overset{\pi}{-}\text{C}) = 63$

$$\Delta H^0 = \underset{\text{Broken}}{\overset{\text{Bonds}}{\Sigma D}} - \underset{\text{Formed}}{\overset{\text{Bonds}}{\Sigma D}}$$

(c) Which step is likely to be the slower of the two propagation steps? In that case, what kinetic rate law is expected? (d) When an excess of the alkene is used, the yield of the 1:1 adduct is diminished and byproducts are formed, but when an excess of phosphine is used, the yield of the 1:1 adduct approaches 100%. Reconcile these observations with your analysis in (c) and indicate the likely structures of the byproducts. (e) When the addition to *cis*-2-butene is carried out to less than 100% conversion, what is the expected stereochemistry of the recovered alkene? Explain. How might this be affected by the concentration of the phosphine? (Fields, R.; Haszeldine, R. N.; Wood, N. F. *J. Chem. Soc. (C)* **1970**, 1370, 744, and 197.)

2.10 The polymerization of ethene under radical conditions generates polymers that, besides the butyl and other short side chains produced by "back biting," have long, polymeric side chains. Propose a mechanism for the formation of these polymeric side chains.

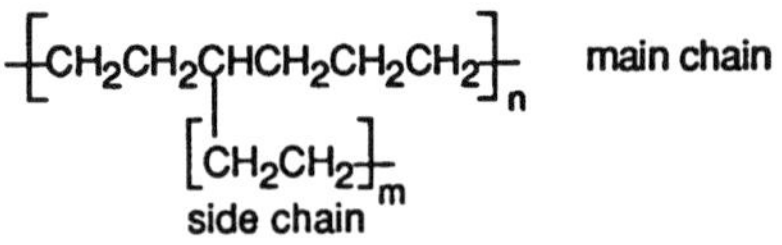

$$\left[CH_2CH_2\underset{\underset{\text{side chain}}{\left[CH_2CH_2\right]_m}}{\underset{|}{C}H}CH_2CH_2CH_2\right]_n \quad \text{main chain}$$

2.11 Write plausible radical chain mechanisms for the following transformations:

(a) 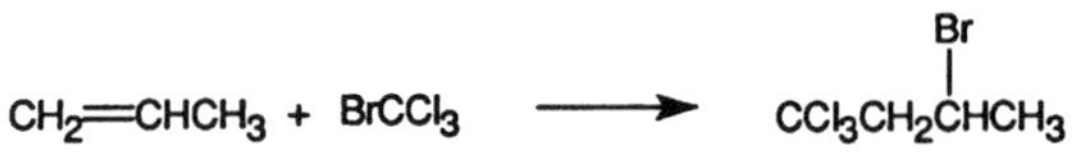 $CH_2{=}CHCH_3 + BrCCl_3 \longrightarrow CCl_3CH_2\overset{\overset{Br}{|}}{C}HCH_3$

(b) 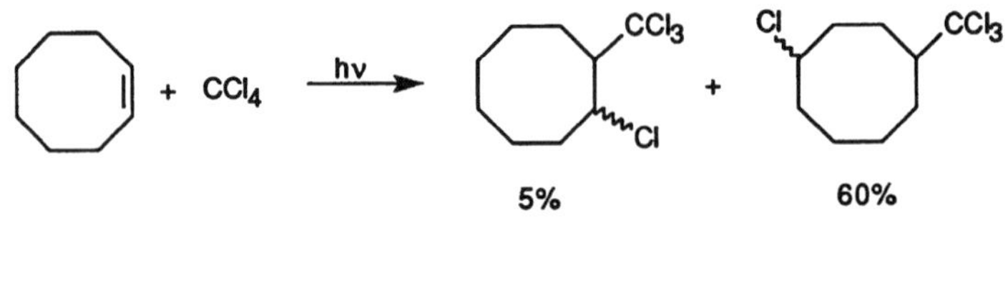 cyclooctene + CCl_4 $\xrightarrow{h\nu}$ 1-chloro-2-(trichloromethyl)cyclooctane (5%) + 1-chloro-4-(trichloromethyl)cyclooctane (60%)

5% 60%

(c) 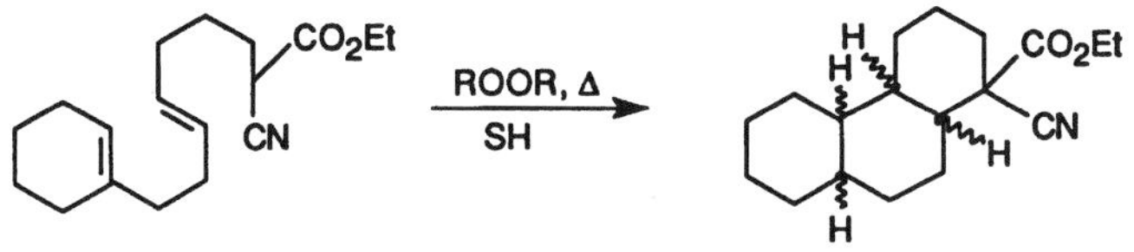 CO_2Et CN $\xrightarrow[\text{SH}]{\text{ROOR, }\Delta}$ H H CO_2Et CN H H

(SH = solvent having abstractable hydrogens.)

(d) 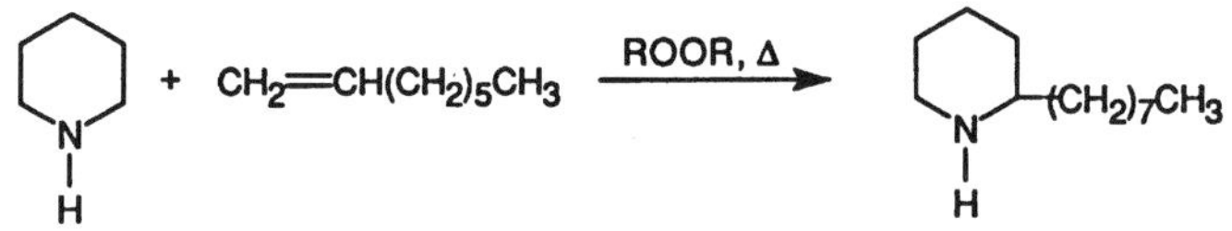 N–H + $CH_2{=}CH(CH_2)_5CH_3$ $\xrightarrow{\text{ROOR, }\Delta}$ N–H $(CH_2)_7CH_3$

(e) 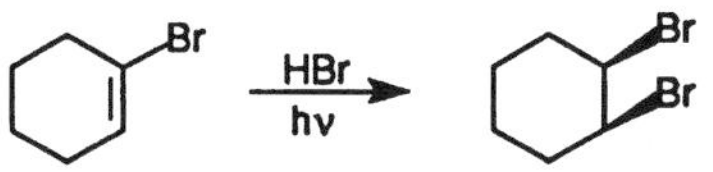 Br $\xrightarrow[h\nu]{\text{HBr}}$ Br Br

2.12 The *exo* and *endo* hydrogens at the 5-position of 1,4-diphenylbicyclo-[2.1.0]pentane equilibrate thermally in a process having $\Delta H^{\ddagger} = 12.2$ kcal/mol. This substance reacts with molecular oxygen at room temperature in a second-order process (first order in $[O_2]$ and first order in the bicyclopentane derivative) to give the endoperoxide shown. Write mechanisms for the "bridge flip" rotation and for the reaction with dioxygen that are consistent with these data.

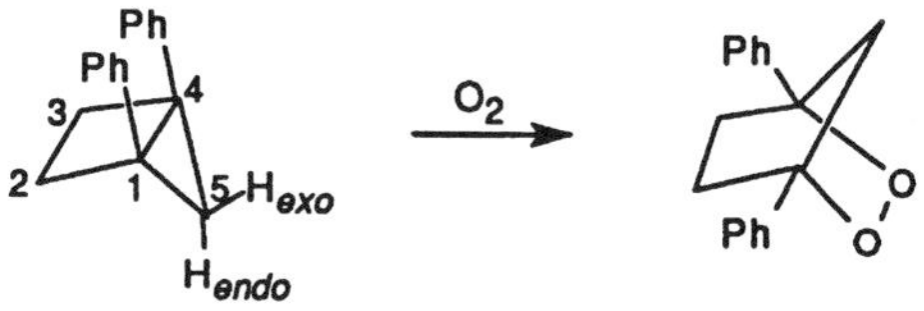

Ph Ph 3 4 2 1 5 H_{exo} H_{endo} $\xrightarrow{O_2}$ Ph Ph O O

(Coms, F. D.; Dougherty, D. A. *J. Am. Chem. Soc.* **1989**, *111*, 6894; Adam, W.; Platsch, H.; Wirz, J.; *ibid.*, 6896.)

CHAPTER

3

The Characterization of Radicals and Radical Pairs by ESR and CIDNP

3.1 Electron Spin Resonance

Undoubtedly the most definitive characterization of free radicals is provided by electron spin resonance (ESR) spectroscopy. This powerful spectroscopic method is capable not only of conclusively identifying specific free radicals, but also of providing detailed information on the spin-density distribution in the radical. The method is generally applicable to persistent radicals in solution and to radicals of all types in frozen matrices.

The basis for ESR spectroscopy is the existence of the two previously mentioned spin states of the electron and their energetic distinction in the presence of an external magnetic field. The α spin state (by convention this corresponds to the spin quantum number $m_s = +\frac{1}{2}$) has its spin vector approximately aligned with the external magnetic field (actually at a 35° 15′ angle from the direction of the external field). The β spin state has its spin vector approximately opposed to the external field. Since a spinning, *negatively* charged entity has a magnetic moment vector directed oppositely to its spin vector, *the β state of an electron is the state that has its magnetic moment vector most nearly in alignment with the field and that is therefore the more stable state* (Figure 3.1).

Radiation of the appropriate frequency can then induce transitions between the two states, and since the thermal Boltzmann distribution of states favors the more stable β state, there is a net absorption of radiation when the Einstein condition ($\Delta E = h\nu = g\beta H$) is met. The frequency of the absorbed radiation (which turns out to be in the microwave region) is characteristic for transitions between the spin states of unpaired electrons, whether free or present in molecules (e.g., as free radicals).

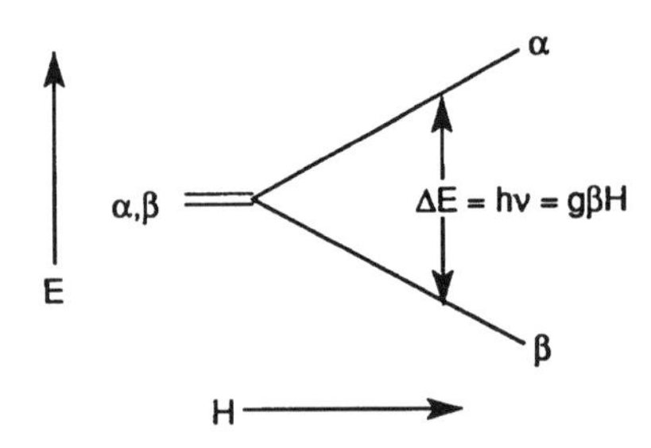

Figure 3.1 The ESR experiment. H is the external magnetic field; $\beta = eh/2mc$ (Bohr magneton); g = Landé g factor (~2.00 for free radicals).

Different free radicals actually absorb at very slightly different frequencies (at constant field) as a result of slightly different g values, but these "chemical shifts" are small and are of somewhat limited use as a means of identifying specific free radicals.

3.1.1 The Hydrogen Atom

The ESR spectrum of this prototype free radical consists of two absorptions separated by 50.7 mT (mT = millitesla; equivalent to 507 gauss).[1] The simulated ESR spectrum shown in Figure 3.2 illustrates the convention that it is the derivative of the absorption (not the absorption itself, in contrast to NMR) that is recorded in ESR spectroscopy. The splitting of the absorption into a doublet is, of course, the result of interaction between the magnetic moment of the electron and that of the proton, a phenomenon known as electron–nuclear hyperfine splitting (hfs). The magnitude of the doublet separation is defined as the hyperfine splitting constant, a = 50.7 mT. The existence of two, and only two, absorptions in the ESR spectrum of the hydrogen atom arises from the circumstance that there are now four possible spin (magnetic) states in the two-spin system corresponding to possible electron spin states α_e and β_e and proton spin states α_p and β_p. These spin states are depicted at the right of Figure 3.3, where the two solid arrows represent the two ESR

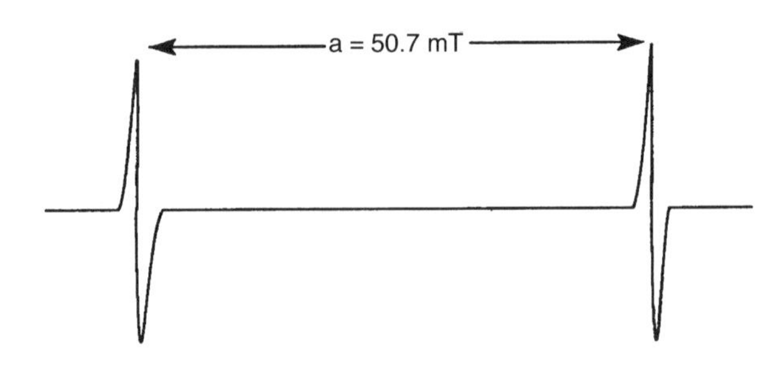

Figure 3.2 ESR spectrum of the hydrogen atom.

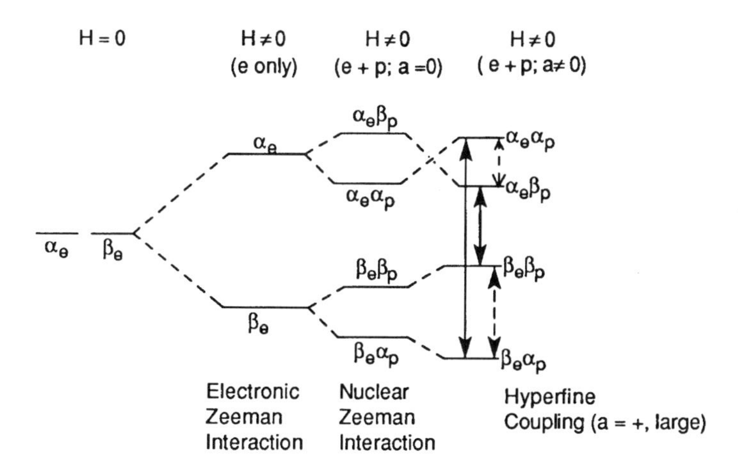

Figure 3.3 Electron–nuclear spin states of the hydrogen atom. *e*, electron; *p*, proton; *a*, hyperfine coupling interaction constant between electron and proton; $H = 0$, no magnetic field; $H \neq 0$, presence of magnetic field; $a = 0$, neglecting the hyperfine interaction; $a \neq 0$, including the hyperfine interaction.

(microwave) transitions and the two dashed arrows represent the two transitions in the NMR (radio-frequency) spectrum. The ESR transitions, of course, invert only the electron spin (e.g., $\beta_e\alpha_p \rightarrow \alpha_e\alpha_p$). The energetic ordering of these spins states can be seen by considering each of the three relevant magnetic interactions in a stepwise fashion. First, the interaction between the electron magnetic moment and the applied field (the electronic Zeeman interaction) distinguishes the electron spin states, making β_e the ground state. Second, the magnetic interaction between the proton and the applied field (nuclear Zeeman interaction) distinguishes the two spin states of the proton and makes the α_p state the one of lower energy (the proton is a spinning *positive* charge). Consequently, the $\beta_e\alpha_p$ state is lower in energy than the $\beta_e\beta_p$ state, etc. Finally, the proton–electron hyperfine interaction is considered. Recall that the α_p and β_e states both have their magnetic moments aligned with the external magnetic field and therefore with each other. In the simplest case (corresponding to a positive hyperfine splitting constant), the interaction of the α_p and β_e magnetic moments is a stabilizing one, so that the $\beta_e\alpha_p$ spin state is the ground spin state. Among the upper electron states (α_e) the $\alpha_e\beta_p$ state has the electron and proton moments aligned and experiences a hyperfine stabilization, while the $\alpha_e\alpha_p$ state is destabilized. Depending upon the magnitude of the splitting, the energies of these two states may or may not cross (the depiction is for a large hyperfine splitting). The four composite spin states exist even when there is no hyperfine interaction, of course, but the two ESR absorptions (as well as the NMR absorptions) are degenerate (i.e., occur at the same frequency) in that case. The existence of a hyperfine interaction is therefore the necessary and sufficient condition for the splitting of the ESR signal into a doublet. The interaction of the magnetic moment of the electron with that of the proton, incidentally, can only occur through that small fraction of the electron density which resides *at the nucleus* [i.e., through ψ^2(SOMO) evaluated at $r = 0$, where r is the electron–nuclear distance]. This so-called *Fermi contact interaction* is available only when the unpaired electron density occupies an atomic orbital of the s type, since all other AOs have nodes at the nucleus. Because the hydrogen atom has unit unpaired electron density in a 1s orbital, the hyperfine splitting is especially large.

3.1.2 The Methyl Radical; α Hyperfine Splittings

The ESR spectrum of the methyl radical (examined in a frozen matrix) consists of an evenly spaced quartet (1:3:3:1, i.e., binomial intensity distribution) with $a = -2.30\,\text{mT}$ (Figure 3.4).[1] By analogy to NMR specroscopy, the $n + 1$ rule for splittings by three equivalent protons predicts just such a quartet. Similarly, since ^{12}C is nonmagnetic, no carbon splitting is observed. The hyperfine splitting is only *ca.* 1/20 that in the hydrogen atom, since the unpaired electron occupies a carbon atomic orbital.

Assuming an sp^2 hybridization state, this AO is a $2p_z$ AO, for which the trigonal plane containing carbon and the three hydrogen atoms is a nodal plane. The unpaired electron can therefore not be delocalized onto these hydrogens. The interesting question therefore arises as to how net electron spin density is transmitted to the hydrogen nuclei.[2] To find an answer to this basic question, let us assume for concreteness that the unpaired electron (in the $2p_z$ AO, which is the SOMO) has a β spin (the ground spin state). Consider now the interaction of

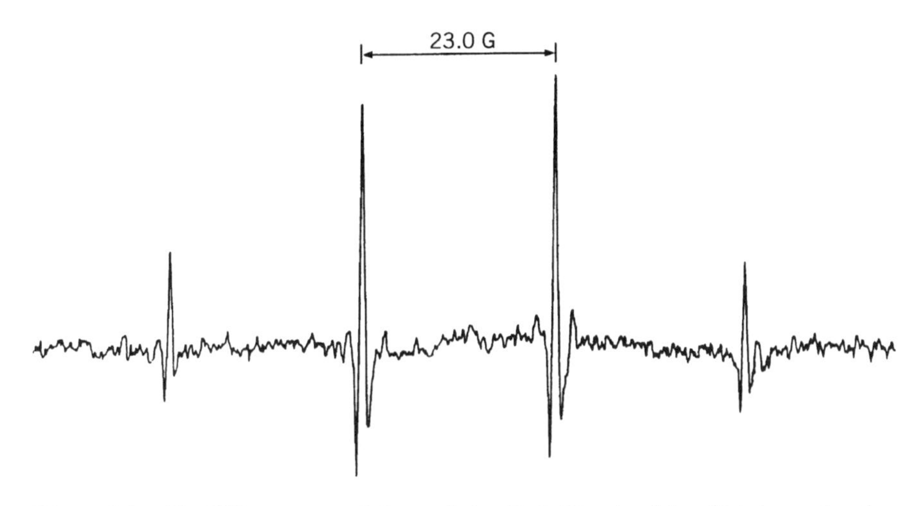

Figure 3.4 The ESR spectrum of the methyl radical. The signal is split into n + 1 = 4 lines by three equivalent protons. The line intensities are in the familiar 1:3:3:1 ratio. Reprinted with the permission of Professor R. W. Fessenden and R. H. Schuler and the American Institute of Physics.

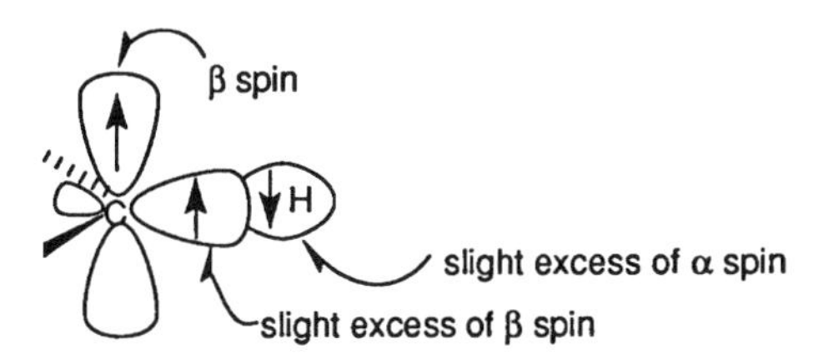

Figure 3.5 The spin-polarization mechanism for α splitting.

this electron with the two electrons in any one of the carbon–hydrogen bonds (Figure 3.5). Recall (e.g., the Pauli Exclusion Principle) that electrons of the same spin are less repulsive than electrons of the opposite spin, because of exchange stabilization. Consequently, the β electron of the C–H bond has a slightly higher electron density in the carbon sp^2 AO component of the bond, and the α electron has a slightly greater density in the hydrogen 1*s* orbital. This mechanism is called *spin polarization* and amounts to an induction of an unsymmetrical spin distribution in the C–H bonds by the unpaired electron as a result of differential electron repulsions. Interestingly, the spin density at hydrogen (α) is opposite to that of the molecule as a whole (β). It therefore is designated as "negative spin density" and corresponds to a negative value of the hyperfine constant, *a*. The existence of negative spin on hydrogen in the methyl radical is conclusively supported by both theory and experiment, but it should be noted that the sign of the splitting constant is not usually evident from routine solution ESR spectra. This foregoing discussion suggests that a distinction should perhaps be drawn between *odd electron density* (considered to be the spin density in the SOMO—a purely theoretical quantity) and net *spin density*.

The hyperfine interaction between electron and nuclear spins via bonds (including long-range bonding interactions) is called isotropic hyperfine interaction and is the only net hyperfine interaction observable in fluid solution. The direct (through-space) interaction is important in frozen matrices where the anisotropic, through-space interactions are not averaged out by rapid motions of the radical relative to the field.

3.1.3 The Ethyl Radical; β Hyperfine Splittings

As might be expected, the ESR signal of this radical is split into a quartet of triplets, that is, 12 lines (Figure 3.6).[1] This corresponds to $a_\alpha = -2.24$ mT (2H) and $a_\beta = +2.69$ mT (3H). The α hyperfine splitting is again negative and of similar magnitude to that of the methyl radical. The hyperfine splittings from the β hydrogens are positive and of roughly the same magnitude as the α splittings. The β splittings arise from delocalization of the odd electron (i.e., of the SOMO) onto the methyl hydrogens by hyperconjugative interaction of the C–H bonds with the 2*p* AO on carbon (Figure 3.7). The spin density on these hydrogens is therefore positive, that is, the same as the overall spin of the radical.

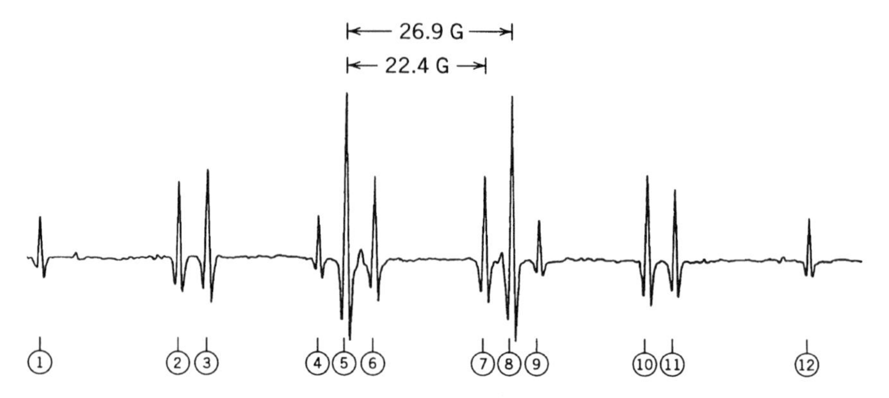

Figure 3.6 The ESR spectrum of the ethyl radical. The beta protons split the signal into a quartet, each line of which is further split into a triplet by two equivalent alpha protons (12 lines in all). Reprinted with the permission of Professor R. W. Fessenden and R. H. Schuler and the American Institute of Physics.

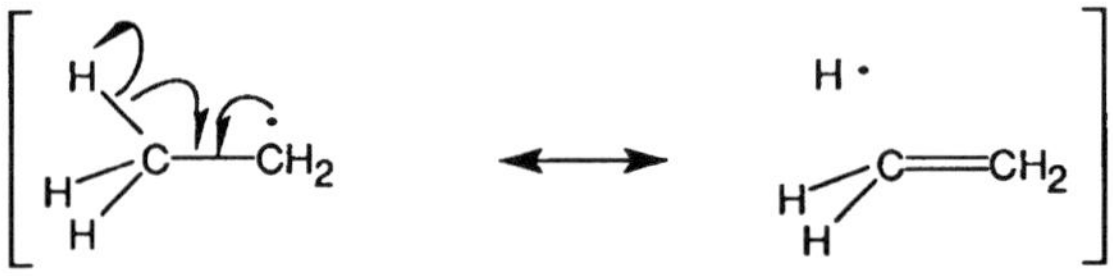

Figure 3.7 The hyperconjugative mechanism for β hyperfine splittings.

3.1.4 The Allyl Radical; The McConnell Equation

The allyl radical has a delocalized π SOMO that has a node at C_2. The odd electron densities at carbon are $\rho_1 = \rho_3 = 0.5$ and $\rho_2 = 0$. These odd electron populations occupy carbon $2p_z$ AOs, but as in the case of the methyl radical, they can spin polarize the α C–H bonds, giving rise to negative spin densities and thus to negative splittings from the four hydrogens α to C_1 and C_3 (Figure 3.8).[3,4] Of these, H_4 and

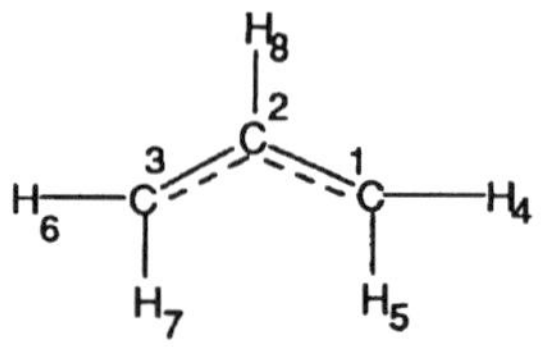

$a_4 = a_6 = -1.48$ (2H)
$a_5 = a_7 = -1.39$ (2H)
$a_8 = +0.41$ (1H)

Figure 3.8 Odd electron densities and splitting constants in the allyl radical.

$$a_{\alpha,i} = Q\rho_i$$

where: $Q \approx -2.3$ mT; ρ_i is the odd electron density at C_i; and $a_{\alpha,i}$ is the hyperfine splitting from (α) protons attached to C_i.

Figure 3.9 McConnell equation for α hyperfine splittings.

H_6 are symmetry equivalent, as are H_5 and H_7, but all four have very similar magnitudes. The splittings are smaller than for the methyl radical because the spin density at carbon (C_1 and C_3) is only one-half that in the methyl radical, so that the net spin induced on the attached protons and thus the hyperfine splittings are about one-half those in the methyl radical. The quantitative relationship between the odd electron density (ρ_i) in a $2p_z$ AO on the ith carbon atom and the hyperfine splitting constant ($a_{\alpha,i}$) of an α hydrogen attached to that carbon is called the McConnell equation (Figure 3.9).[2]

Such a relationship is useful when odd electron densities *at carbon* are symmetry determined or when they are available from Hückel MO theory. When net spin densities on *hydrogen* are directly calculated, as in any MO calculation that includes electron repulsions (e.g., SCF MO calculations) and in which formally "paired" electrons of different spin are permitted to occupy different spatial orbitals (the UHF method), a more appropriate and more precise relationship is $a = 507\rho_H$. This recognizes that the splitting constant from a hydrogen nucleus is rigorously proportional to the net spin density on hydrogen.

It may initially appear somewhat surprising that, in spite of the fact that the odd electron density at C_2 (i.e., ρ_2) is zero, a substantial doublet splitting (about one-third of the magnitude of the triplet splittings) arises from H_8 (i.e., the proton attached to C_2). Moreover, this splitting constant is *positive*, in contrast to those of the other α hydrogens. It is a reliable general observation that protons attached to carbons that are at nodal positions in the SOMO of a radical give rise to significant positive splittings. Once again, spin polarization is involved (Figure 3.10). Assume that the odd electron (in the SOMO) has β spin. The paired electrons in ψ_1 (the bonding MO) are spin polarized by the odd electron's β spin, with the result that the α electron tends to avoid C_1 and C_3 and the β electron density tends to increase at C_1 and C_3. Negative (i.e., α) electron density at C_2 therefore arises from the filled MO. Since there is no positive (β) spin density at C_2 in the SOMO, the net spin density at C_2 is negative. Note also that the positive spin density at C_1,$C_3 > 1$ as a result of the positive increment from ψ_1. This explains why the splittings of H_4–H_7 are somewhat greater than one-half the methyl radical splitting. When the negative (α) spin density at C_2 spin polarizes the C_2–H bond, a second spin inversion results, and the negative spin at C_2 is converted to positive spin at H_8. Actually, when the unrestricted Hartree–Fock (UHF) method is employed, as is required for radicals, α and β electrons are not forced into an MO (for example, ψ_1) that requires the same spatial distribution of both electrons, but a separate and distinctly different MO is

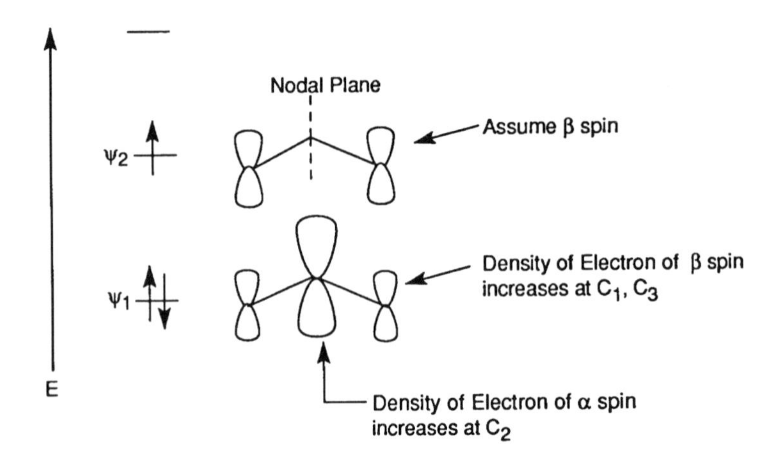

1. β,β repulsions are smaller than β,α repulsions because of exchange stabilization of electrons of the same spin.
2. The β electrons of ψ_1 tend to accumulate slightly at C_1, C_3; the α electrons avoid C_1, C_3, and accumulate at C_2.
3. Negative (α) spin in the $2p_z$ orbital at C_2 gives rise to β (i.e., positive) spin at H_8 via spin polarization of the C_2–H bond.

Figure 3.10 The origin of negative spin density at C_2 of the allyl radical.

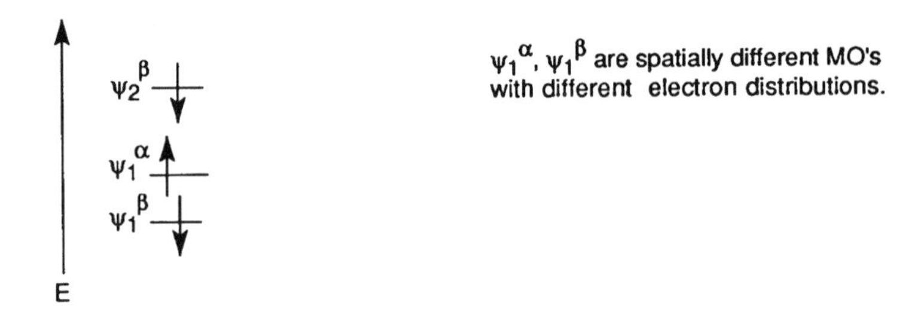

Figure 3.11 Unrestricted spin orbitals in the allyl radical.

provided for each electron (Figure 3.11). The MO designated $\psi_1{}^{\alpha}$, because of the greater repulsion between α and β electrons, has a higher energy than $\psi_1{}^{\beta}$, and $\psi_1{}^{\beta}$ has a higher density at C_1,C_3 than does $\psi_1{}^{\alpha}$. Conversely, $\psi_1{}^{\alpha}$ has a higher density (coefficient) at C_2 than does $\psi_1{}^{\beta}$.

3.1.5 Cyclic Delocalized Radicals

It is appropriate to re-emphasize that SCF MO calculations using the UHF method are capable of generating rather accurate values of spin densities on hydrogen and thus of predicting hyperfine splittings from the simple proportionality to the hydrogen atom splitting. Nevertheless, in simple cases where odd electron densities

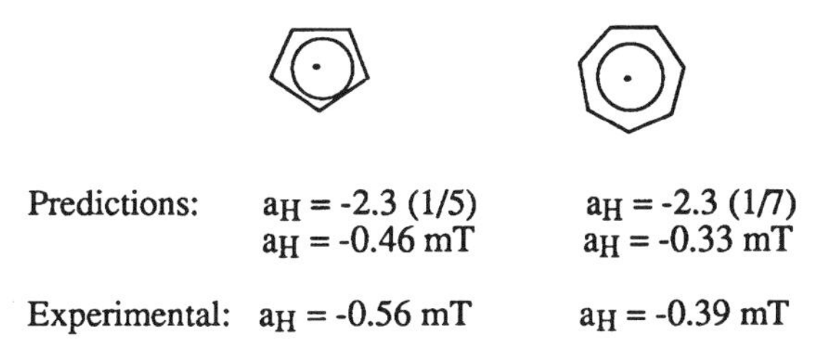

Figure 3.12 Experimental hyperfine splittings of delocalized radicals and predictions from the McConnell equation.

on carbon are symmetry determined, the McConnell relationship is useful in providing approximate predictions of α hyperfine splittings. The cyclopentadienyl radical ($\rho = \frac{1}{5}$) and the cycloheptatrienyl radical ($\rho = \frac{1}{7}$) are exemplary (Figure 3.12).[5]

3.1.6 Conformational Dependence of β Hyperfine Splitings

β Hyperfine splittings are extremely useful in the conformational analysis of radicals since they are highly sensitive to the alignment of the C_β–H bond with the ($2p_z$) SOMO. A second McConnell equation expresses the variation of a_β with the time-averaged value of $\cos^2\theta$, where θ is the dihedral angle between the relevant C_β–H bond and the $2p_z$ SOMO (Figure 3.13).[6] When θ = 90°, the C_β–H bond lies in the trigonal plane of the SOMO, and there is no hyperconjugative delocalization of the odd electron (Figure 3.14). The very small net spin density on the protons in this case, which is represented by the B_0 term, arises from spin polarization of the

$$a_\beta = B_0 + B_2 \langle \cos^2\theta \rangle \approx B_2 \langle \cos^2\theta \rangle$$

$$\text{where: } B_2 \approx +5.4\,\text{mT}$$

Figure 3.13 McConnell equation for β hyperfine splittings.

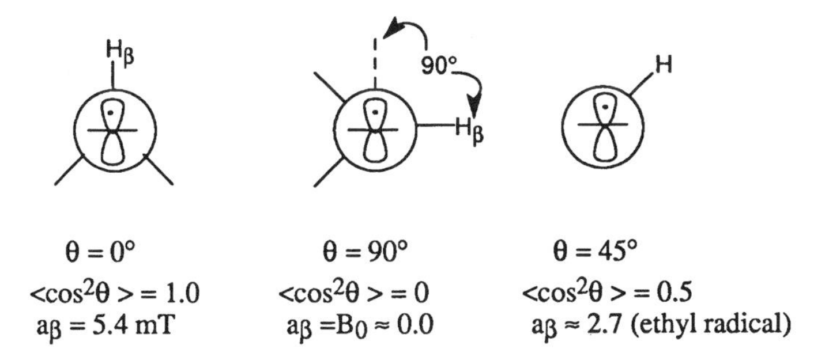

Figure 3.14 Conformational dependence of *a*β.

C_α–C_β bond, followed by a subsequent spin polarization of the C_β–H bond. In any case, this term is quite small and is often neglected. The value of the B_2 parameter cited (5.4 mT) is derived from the ethyl radical, in which a_β = 2.7 mT and $<\cos^2\theta> = 0.50$ (the methyl group is freely rotating and has $<\theta> = 45°$). In delocalized radicals, a_β is also dependent upon the spin density ρ_i at the radical site, if it is other than 1.0:

$$a_\beta = (B_0 + B_2 <\cos^2\theta>)\,\rho_i \approx (B_2 <\cos^2\theta>)\,\rho_i$$

The beta hyperfine splittings of β-substituted ethyl radicals ($XCH_2CH_2^{\bullet}$) therefore provide fundamental insights into the structures of these radicals.[7] The value of a_β for X = F (+2.34 mT) is close to that for the ethyl radical (2.69 mT), indicating a situation close to free rotation. When $X = Me_3Si$, however, the a_β value decreases to 1.77, indicating the carbon–silicon bond rather strongly prefers alignment with the SOMO. This arrangement would maximize hyperconjugative overlap with the C–Si bond, which is relatively weak and therefore especially amenable to hyperconjugation. In this conformation θ for the C–H bonds is 60° and $a_\beta \approx \frac{1}{4} B_2$ = 1.35 mT. For $X = Me_3Sn$, there is a slight further decrease in a_β to a = +1.58 mT. Finally, when X = Cl, a_β again decreases rather sharply, to +1.02 mT. It would appear, minimally, that Cl is locked into the θ = 0° conformation even more strongly than Si or Sn, but it is not at all evident that the C–Cl bond should be more amenable to hyperconjugative interaction than the much weaker C–Si and C–Sn bonds. Further, a_β has dropped well below the value expected for θ = 60° (i.e., 1.35 mT). Since the α and β protons are not equivalent (note the different hyperfine splittings; a_α = −2.18 mT) in this radical, a symmetrically bridged or even a rapidly equilibrating structure is clearly ruled out. The data seem to be most compatible with a weakly (i.e., unsymmetrically) bridged structure (Figure 3.15) in which a_β is further diminished by the deformation (widening) of the C–C–H_β angle occasioned by the bridging of chlorine. Finally, it is noted that a_α is still in the range expected for an sp^2-hybridized radical center, so that strong bridging is not indicated.

3.1.7 Reinforcement/Interference Effects upon β Hyperfine Splittings

The McConnell equation for β hyperfine splittings makes explicit that a_β is linearly dependent upon ρ_i, the spin density at the interacting radical site (C_α). Therefore, a_β is proportional to $a^2_{SOMO,\alpha}$, the square of the coefficient of the SOMO at C_α. An extremely interesting and dramatic efect, known as the Whiffen effect, results when the C–H bond interacts simultaneously with two radical sites (i,j) that are both

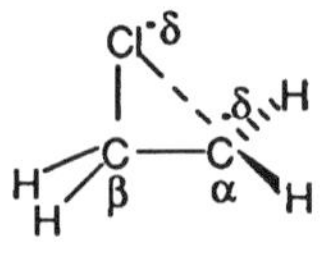

Figure 3.15 The unsymmetrically bridged β-chloroethyl radical.

a_β = 4.77mT

a_β =0.44mT

Figure 3.16 The Whiffen effect.

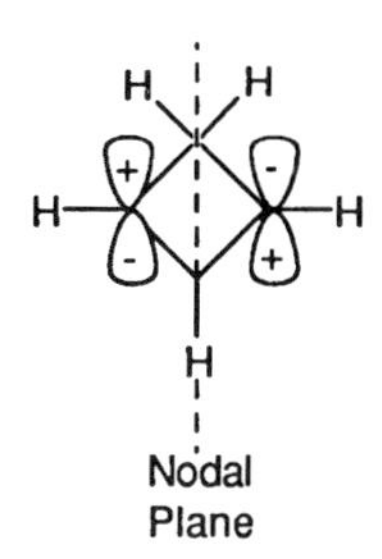

Figure 3.17 The antisymmetric SOMO of the cyclobutenyl radical.

contained in the same conjugated system.[8] A classic instance is provided by the sharply contrasting splittings of the β protons in the cyclohexadienyl and cyclobutenyl radicals (Figure 3.16). The splitting in the former is more than an order of magnitude greater than the latter. A key observation is that the SOMO of the allylic radical moiety of the cyclobutenyl radical is antisymmetric and has a nodal plane passing through the methylene group (including both hydrogens; Figure 3.17). As a result, the principal part of the electron spin, which is in the SOMO, is not delocalized at all over the methylene group. In contrast, the pentadienyl-like SOMO of the cyclohexadienyl radical is symmetric with respect to an analogous plane of symmetry, and the SOMO is delocalized extensively (via hyperconjugation) onto the methylene hydrogens. Obviously, the effect has major implications for cyclic radicals and ion radicals. Fortunately, a McConnell-type approach can still be maintained by noting that, rigorously, a_β is proportional to $(a_{i,\mathrm{SOMO}} + a_{j,\mathrm{SOMO}})^2$ and *not* to $a^2_{i,\mathrm{SOMO}} + a^2_{j,\mathrm{SOMO}}$ (i.e., $\rho_i + \rho_j$). In cases where $a_i = -a_j$, a_β vanishes (antisymmetric MO). Where $a_i = a_j$, the splitting is enhanced by a factor of two over the classical expectation of proportionality to $\rho_i + \rho_j$ (Figure 3.18).

3.1.8 Long Range Hyperfine Splittings

In close analogy to NMR spectroscopy, proton splittings of the α (analogous to germinal couplings) and β (analogous to vicinal couplings) type are by far the largest contributors to ESR spectra. Small splittings from more remote atoms can sometimes be observed, however, in well-resolved ESR spectra. Since these long range splittings are also quite geometry sensitive, they can be suprisingly large in

Classical

$a_\beta \propto a_i^2 + a_j^2 = \rho_i + \rho_j$

assuming $\rho_i = \rho_j$

$a_\beta \propto 2\rho_i$

Whiffen/Nonclassical

$a_\beta \propto (a_i + a_j)^2$

assuming $a_i = a_j$

$a_\beta \propto 4\, a_i^2 = 4\, \rho_i$

Modified McConnell Equation

$$a_\beta = Bo + B_2 < \cos^2\theta > (a_i + a_j)^2$$

Figure 3.18 ESR β hyperfine splittings enhanced by the Whiffen effect.

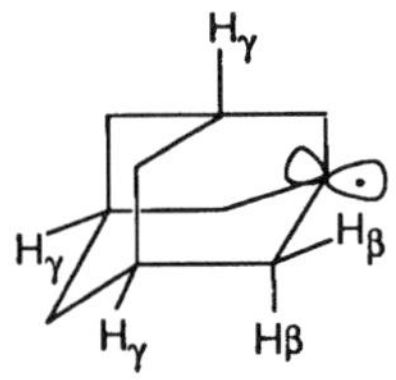

$a_\gamma = 0.47$ mT (3H)
$a_\beta = 0.66$ mT (6H)

Figure 3.19 Long range hyperfine splittings in the 1-adamantyl radical.

rigid and appropriately constituted molecules. An interesting case in point is the 1-adamantyl radical (Figure 3.19), in which the γ hydrogens have a splitting not much smaller than those of the β hydrogens.[9]

The latter splitting is relatively small for a β hyperfine splitting, in part because of the unfavorable dihedral angle for hyperconjugative interaction ($\theta = 60°$ and $<\cos^2\theta> = \frac{1}{4}$) and in part because of the sp^3 hybridization state of this radical. On the other hand, the γ splitting is enhanced by direct overlap of the back lobe of the SOMO with the back lobe of the C_γ–H bond and possibly also by W-plan through bond interaction.

3.1.9 Nitrogen Splittings

Nitrogen atoms have an important role in the stabilization of several of the radicals discussed in this chapter, including 2-cyano-2-propyl, DPPH, and the nitroxyl radicals. When there is net spin density on nitrogen and a Fermi contact interaction exists,

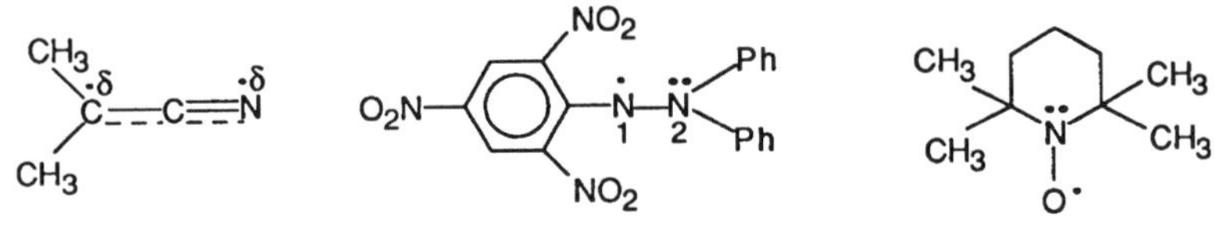

$a_\beta = 2.03$ mT (6H)

$a_N = 0.33$

(a septet of triplets)

$a_{N1} = 0.94$ mT

$a_{N2} = 0.78$ mT

$a_N = 1.59$ mT

Figure 3.20 Nitrogen hyperfine splittings.

N engenders a triplet splitting of the ESR signal. This is a very characteristic 1:1:1 intensity ratio triplet, as distinct from the 1:2:1 triplet splittings that arise from splitting by two equivalent protons. The nitrogen triplet arises because N has a spin of 1, with three spin states corresponding to $m = -1, 0, +1$. These states are statistically equally probable, resulting in an equal intensity triplet. A few pertinent examples of nitrogen hyperfine splittings are listed in Figure 3.20. Interestingly, the two non-equivalent nitrogen atoms of DPPH have splitting constants that are not very different, an observation that tends to support the concept of three-electron bonding.[10]

3.1.10 ^{13}C Hyperfine Splittings; The Hybridization of the Radical Site

As we have previously seen in the case of the methyl radical, an unpaired electron in a $2p_z$ orbital on carbon has no Fermi contact interaction with its nucleus and therefore makes no direct contibution to a hyperfine interaction. Indirectly, however, the unpaired electron does induce asymmetry in the spin-density distribution in the C_α–H bonds, producing a net negative spin density in the hydrogen 1*s* orbital and a corresponding net positive spin density in the carbon sp^2 orbital component of the C_α–H bond. The latter spin density, via the 33% *s* content, engenders a hyperfine interaction and a doublet splitting when the sp^2 carbon atom involved is ^{13}C. The splitting constant ($a = +3.83$ mT) is indeed positive and modest in magnitude.[11] A key observation is that the hypothetical pyramidal (i.e., sp^3-hybridized) structure of the methyl radical has its unpaired electron in an sp^3 orbital and is therefore expected to have a very large, positive hyperfine splitting from a ^{13}C nucleus. The sharp increase in hyperfine splitting that accompanies this rehybridization represents a sensitive tool for studying the hybridization of the reaction site. The progressive substitution of highly electronegative fluorine atoms for hydrogen, for example, tends to cause rehybridization, and the trifluoromethyl radical is considered to be sp^3 hybridized (Figure 3.21).[12] This could arise from the circumstance that fluorine bonds more strongly to less electronegative atoms, and sp^3 carbon is less electronegative than sp^2 carbon. However, the *tert*-butyl radical is also mildly pyramidalized,[13] as is the hydroxymethyl radical [$a(^{13}C) = +4.59$].[14]

Radical:	$CH_3^{\bullet}$	$FCH_2^{\bullet}$	$F_2CH^{\bullet}$	$CF_3^{\bullet}$	$(CH_3)_3C\bullet$
a(^{13}C):	+3.83	+5.48	+14.88	+27.16	+4.52

Figure 3.21 ^{13}C splittings and the hybridization of the radical center.

Both of these radicals are of the nucleophilic type, suggesting that increased electron density at the radical site could engender a carbanionlike preference for sp^3 hybridization.

Changes in the hybridization state at the radical site are actually more conveniently detected by the somewhat less dramatic, but still quite reliable, dependence of a_α (the α *proton* hyperfine splitting) on the hyridization state. Although the cyclohexyl, cyclopentyl, and cyclobutyl radicals have a_α values that are very similar to that of the isopropyl radical, the cyclopropyl radical[15] has a very sharply diminished splitting, indicating a pyramidal radical site (Figure 3.22). This suggestion and the dependence of a_α on the hybridization state are confirmed decisively by theoretical studies. The preference of the cyclopropyl radical for pyramidality is presumably based upon the diminished angle strain of an sp^3 over that of an sp^2 carbon in the cyclopropane ring.

^{13}C hyperfine splittings of the vinyl and formyl radicals, and the nonequivalence of the two β proton hyperfine splittings in the former (Figure 3.23), reflect an sp^2 hybridization state for these σ-type radicals.[16]

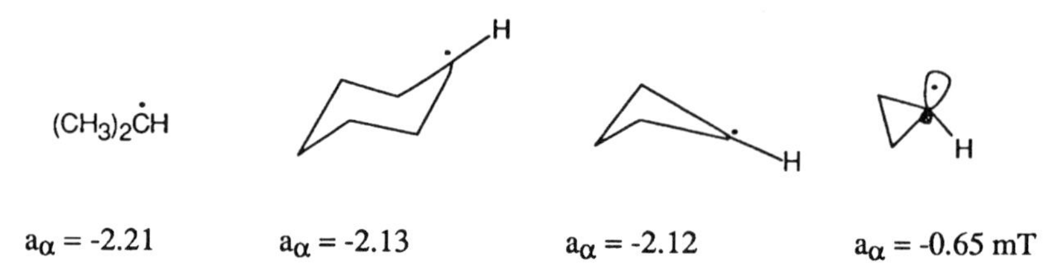

Figure 3.22 The pyramidal (sp^3) hybridization of the cyclopropyl radical.

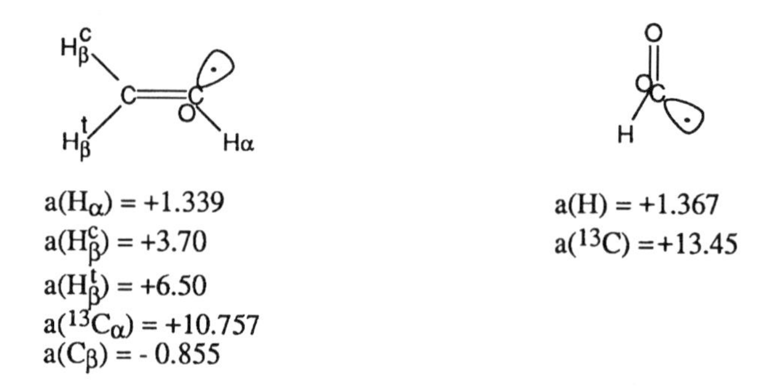

Figure 3.23 Hyperfine splittings of the vinyl and formyl radicals.

3.2 Electron–Nuclear Double Resonance (ENDOR) Spectroscopy

The ESR spectrum of a radical as simple as the ethyl radical already consists of twelve lines, assuming that all the hyperfine lines are resolved. The fully resolved spectrum of the trityl radical consists of no fewer than 196 lines. This familiar radical has just three nonequivalent sets of protons (*o*, *m*, and *p*), but there are six *ortho* and *meta* protons and three *para* protons. The *ortho* protons therefore split the signal into a septet, the *meta* protons split each member of this septet into a further septet (49 lines at this point), and finally the *para* protons split each of these 49 lines into a quartet (giving a total of 196 lines). Unlike an NMR spectrum, an ESR spectrum often has little use unless the hyperfine couplings can be resolved and unambiguously interpreted. Deuterium labeling is laborious and potentially problematic, since the deuteron has spin $S = 1$, and either contributes further lines to the spectrum or, if these splittings are too small to resolve, gives rise to inhomogeneous line broadening. A powerful technique for obtaining the desired hyperfine splitting constants is *electron–nuclear double resonance* (ENDOR).[17] The ENDOR spectrum of the trityl radical consists of just six lines, two for each of the *o*, *m*, and *p* protons. In general, any chemically equivalent set of protons gives rise to just two ENDOR lines, irrespective of the number of equivalent protons in that group. Figure 3.24 illustrates the ENDOR spectrum of the tris(4-methylphenyl)methyl radical.[18] Starting at the free proton frequency and reading to the right,

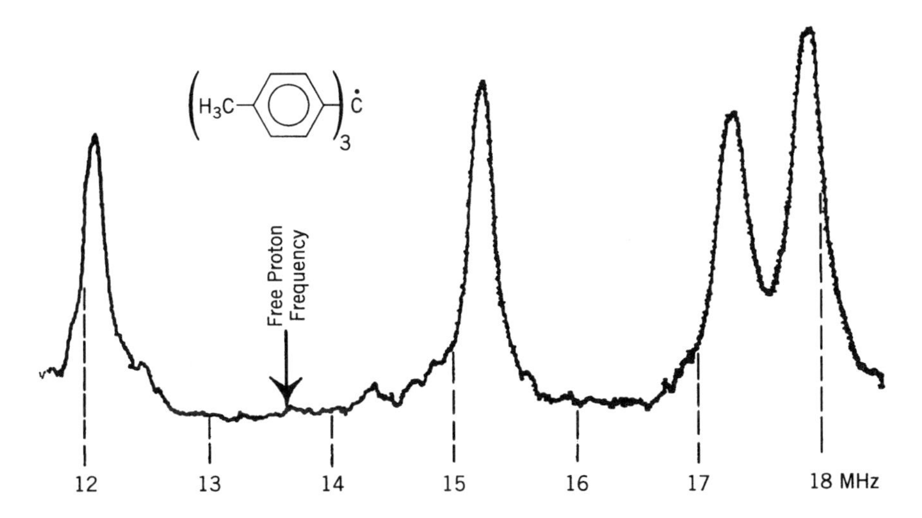

Figure 3.24 The ENDOR spectrum of the tris(4–methylphenyl)methyl radical. Proceeding to the right from the free proton frequency, the absorptions correspond to the *meta*, *ortho*, and methyl protons, respectively. The line shown to the left of the free proton frequency is the other line of the doublet corresponding to the *meta* protons. The corresponding lines for the *ortho* and methyl protons are not shown. This spectrum is from our own research files.

three ENDOR absorptions are observed that correspond to the *meta*, *ortho*, and *para* protons, in that order. Proceeding to the left from the free proton frequency, three corresponding lines would be encountered, but only one (that of the *m* proton) is shown in the illustration. In many cases, ENDOR spectra are swept only to the right of the free proton frequency, since the spectra are symmetric about this frequency. The hyperfine splittings are obtained as the frequency difference between corresponding peaks on the left and right of the free proton frequency, or twice the difference between the free proton frequency and the frequency of a single peak on the right side of the spectrum. In the present case these splittings are $a_o = 0.260\,\text{mT}$, $a_m = 0.114\,\text{mT}$, and $a(CH_3) = 0.304\,\text{mT}$.

The conditions for an ENDOR experiment are similar to those for an ESR experiment except that an additional radio-frequency field is used to promote transitions between the nuclear spin states. While constantly maintaining the resonance condition for a specific line of the ESR spectrum, the nuclear radio frequency is swept. When the resonance condition for a specific proton type is met, an ENDOR absorption occurs. This corresponds to a change in the intensity of the ESR signal being measured, as a result of the change in the population of the nuclear spin states. Two such absorptions are encountered for each proton type because the NMR spectrum of each proton has been split into a doublet by the hyperfine interaction with the electron spin. The separation of these two lines is equal to the hyperfine splitting constant.

3.3 Chemically Induced Dynamic Nuclear Polarization (CIDNP)

Chemically induced dynamic nuclear polarization (CIDNP) is a unique NMR method for detecting *radical pairs* as intermediates.[19] To illustrate the basic principles underlying the CIDNP experiment,[20] we consider the homolysis of a singlet precursor molecule (SP) to give a germinate (caged) radical pair ($\overline{R^\bullet R^\bullet}$). The radical pair is assumed to be initially formed in a singlet state, in conformity with spin conservation. Since the energies of the singlet and triplet states are now very similar (because the two radicals interact only weakly), magnetic interactions are capable of inducing the crossover of the radical pair to a triplet state (T_0; Figure 3.25). A magnetic interaction that can accelerate this conversion is the hyperfine interaction with a proton near either of the radical sites. In general, one of the nuclear spin states of the proton will accelerate the conversion to T_0 more than the

$$\text{SP} \xrightarrow[h\nu]{\Delta \text{ or}} \overline{R_1^\bullet \; R_2^\bullet} \rightleftharpoons \overline{R_1^\bullet \; R_2^\bullet}$$

Singlet (S) Triplet (T_0)

Figure 3.25 The CIDNP experiment.

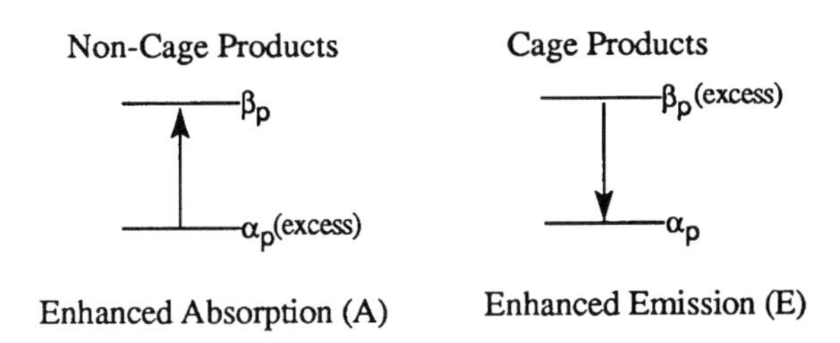

Figure 3.26 The case where spin sorting produces an excess of α states in the noncage and of β states in the cage products.

other. If we assume, arbitrarily, that the α_p state is the one that most accelerates the $S \rightarrow T$ conversion, it follows that T_0 will be formed with an excess of α_p states and the remaining S will have an excess of β_p states. This process is called *spin sorting* and results in spin state excesses much larger than in the typical thermal Boltzmann distribution of α_p and β_p states. Since coupling requires paired spins, T_0 cannot undergo cage recombination and is relatively more likely to undergo diffusive separation to produce free radicals and eventually noncage products, while S is relatively more likely to undergo cage recombination. If the cage and noncage products are different, they will carry a large excess of β_p and α_p states, respectively. If the NMR spectrum of these products is examined prior to thermal relaxation of the nuclear spin states (the time window varies, but often can be 5–30 minutes), the noncage product(s) will exhibit a greatly enhanced absorption peak for the relevant proton, while the cage product exhibits a novel enhanced emission, that is, a negative absorption (Figure 3.26).

3.3.1 CIDNP Exemplified

We consider the thermal decomposition of acetyl trichloroacetyl peroxide in the presence of the radical scavenger I_2.[21] The initially formed mixed acyloxy radical pair rapidly decarboxylates within the cage (as the CIDNP experiment will demonstrate) to give a singlet methyl–trichloromethyl radical pair. The latter radical pair then undergoes intersystem crossing to T_0 at a rate comparable to its cage recombination and diffusive separation. The scavenger is provided to prevent the subsequent coupling of free radicals after diffusion from the cage, so that cage and noncage products are sharply distinguished. Cage recombination produces 1,1,1-trichloroethane, while the uncaged methyl radical produces methyl iodide (Figure 3.27). The NMR spectrum of the products reveals that the methyl protons of 1,1,1-trichloroethane (the cage product) appear in the emission (E) mode, while the protons of methyl iodide have a strongly enhanced absorption (A) peak. It is important to note, however, that which of the two types of product (cage or noncage) appears in the absorption mode and which in the emission mode is variable and depends upon several factors. First, whether the precursor molecule is in a singlet state (normally the case for thermal homolysis) or a triplet state (as in some

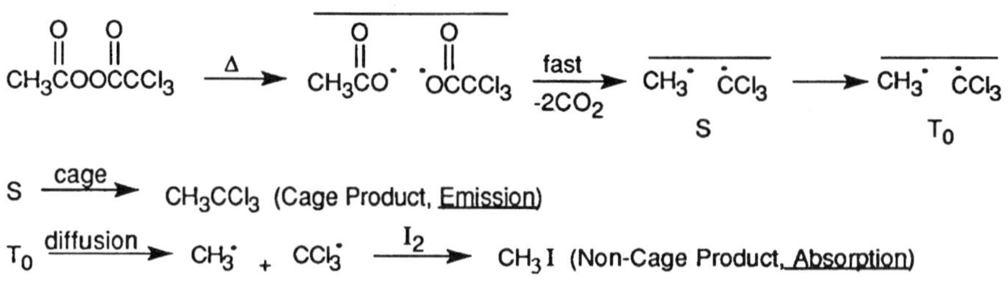

$$CH_3COOCCCl_3 \xrightarrow{\Delta} \overline{CH_3CO^{\bullet}\ {}^{\bullet}OCCCl_3} \xrightarrow[-2CO_2]{fast} \overline{CH_3^{\bullet}\ \dot{C}Cl_3}\ (S) \longrightarrow \overline{CH_3^{\bullet}\ \dot{C}Cl_3}\ (T_0)$$

$$S \xrightarrow{cage} CH_3CCl_3 \text{ (Cage Product, Emission)}$$

$$T_0 \xrightarrow{diffusion} CH_3^{\bullet} + CCl_3^{\bullet} \xrightarrow{I_2} CH_3I \text{ (Non-Cage Product, Absorption)}$$

Figure 3.27 CIDNP in the decomposition of acetyl trichloroacetyl peroxide.

photochemical reactions) determines whether the initially generated radical pair is in the S or T_0 state. If T_0 is generated initially, a rapid $T_0 \rightarrow S$ conversion will favor *increased* amounts of cage product. The caged product will therefore have an excess of the proton spin state that most accelerates the $T_0 \rightarrow S$ conversion, in contrast to the example discussed previously. Second, the sign of the hyperfine splitting constant determines whether the electron spin density at the nucleus is of the same or the opposite sign to that of the radical as a whole, and this is one factor that determines whether the α_p or β_p proton spin state most accelerates the intersystem crossing. The other factor, which is presented without discussion, is the relative magnitudes of the g value of the two radicals in the pair. The Kaptein equation is convenient for predicting the mode (E or A) of the relevant NMR peak (Figure 3.28).[22]

In the case of the thermal homolysis of acetyl trichloroacetyl peroxide, the prediction for the cage product is as follows:

$$\Gamma_n \text{ (cage)} = \mu\varepsilon\ \Delta g\ a = (-)\ (+)\ (-)\ (-) = (-), \text{ i.e., } E$$

This recognizes that the proton hyperfine splitting in the methyl radical is negative (since it is an α–type splitting) and that the g value of this radical (2.0025) is less than that for the trichloromethyl radical (2.0091). For the noncage product:

$$\Gamma_n \text{ (noncage)} = (-)\ (-)\ (-)\ (-) = (+), \text{ i.e., A}$$

It is implicit in the Kaptein equation that a caged pair of identical radicals ($\Delta g = 0$) cannot give rise to net enhancement (A or E) of the NMR signal of a proton; that

$$\Gamma_n = \mu\varepsilon a\Delta g; \quad \Gamma_n = +\text{(absorption)}, \quad \Gamma_n = -\text{(emission)}$$

μ = –for singlet precursors, + for triplet precursors

ε = + for cage product, – for noncage product

a = sign of hyperfine constant of proton in the radical intermediate

Δg = + if proton containing fragment has the higher g value, otherwise –.

Figure 3.28 Kaptein equation for a CIDNP net effect.

is, $\Delta g \neq 0$ is required for a *net effect.* However, an even more subtle CIDNP effect can be observed, even when the caged radical pair consists of identical radicals, providing that *two* nonequivalent proton hyperfine interactions are present in the radical pair. The thermal decomposition of lauroyl peroxide, which gives caged undecanyl radicals, illustrates this special type of CIDNP effect, which is designated the *multiplet effect* (Figure 3.29).[23] The noncage product, 1-iodoundecane, normally exhibits a triplet absorption at δ3.2 for the methylene protons α to the iodine atom. When generated as a product of the decomposition of lauroyl peroxide, 1-iodoundecane exhibits a very novel triplet in which the leftmost signal is positively enhanced (A) and the rightmost signal is negatively enhanced (E). The central line, which normally is the most intense line of the 1:2:1 triplet, is relatively weak (i.e., unenhanced). In contrast, one of the cage products (1-undecene, a disproportionation product) reveals a multiplet effect in the vinyl protons (δ5.0, δ5.9) in which the downfield peaks in each multiplet appear in enhanced emission and the upfield peaks appear in enhanced absorption (called the E/A mode). A second Kaptein equation predicts whether the A/E or E/A mode should be observed (Figure 3.30).[22] In the case of 1-iodoundecane (a noncage product), $\mu = (-)$, $\varepsilon = (-)$, $a_\alpha = (-)$, $a_\beta = (+)$, $J = +$ (vicinal proton/proton splitting constants are positive), and $\sigma = +$, so that the Kaptein product is negative and the A/E mode is predicted and observed.

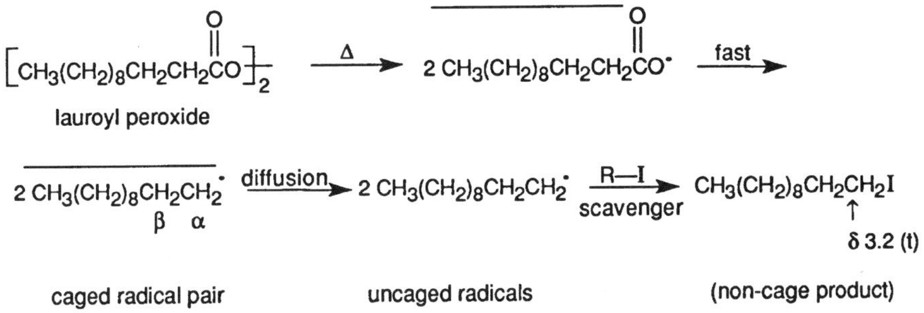

$$[CH_3(CH_2)_8CH_2CH_2C(=O)O]_2 \xrightarrow{\Delta} \overline{2\, CH_3(CH_2)_8CH_2CH_2C(=O)O^\bullet} \xrightarrow{\text{fast}}$$

lauroyl peroxide

$$\overline{2\, CH_3(CH_2)_8\underset{\beta}{CH_2}\underset{\alpha}{CH_2^\bullet}} \xrightarrow{\text{diffusion}} 2\, CH_3(CH_2)_8CH_2CH_2^\bullet \xrightarrow[\text{scavenger}]{R\text{—}I} CH_3(CH_2)_8CH_2\underset{\uparrow \atop \delta\, 3.2\, (t)}{CH_2}I$$

caged radical pair — uncaged radicals — (non-cage product)

Figure 3.29 Multiplet effect in the thermal decomposition of lauroyl peroxide.
caged radical pair uncaged radicals (noncage product)

$\Gamma_m = \mu\varepsilon a_i a_j J_{ij}\sigma_{ij}$; Γ_m = (+) corresponds to the E/A mode; (–) to the A/E mode

a_i, a_j = signs of the two hyperfine splittings (a_α, a_β in this example)

J_{ij} = sign of the NMR coupling constant of the two protons

σ_{ij} = + for the case where both protons are on the same radical fragment,
– for the case where one proton is on each radical fragment

Figure 3.30 Kaptein equation for the multiplet effect.

3.4 Chemically Induced Dynamic Electron Polarization (CIDEP)

In the previous discussion of CIDNP, it was noted that spin sorting in radical pairs results in different and abnormal nuclear spin distributions in the caged and noncaged (i.e., free) radicals. In a CIDNP experiment, the abnormal spin distributions are detected only after subsequent reactions of the radicals occur to yield stable products. In a CIDEP experiment, the free radicals themselves are examined via ESR spectroscopy. Since the electron spin populations are not abnormal, enhanced ESR absorption or emission is not observed, but multiplet effects are observed. The sense (A/E or E/A) of the multiplet effect can be used to deduce the electron spin state (singlet or triplet) of the original caged radical pair. In the example shown in Figure 3.31, the direct photolysis of AIBN was shown to yield singlet rather than triplet 2-cyano-2-propyl radical pairs initially.[24]

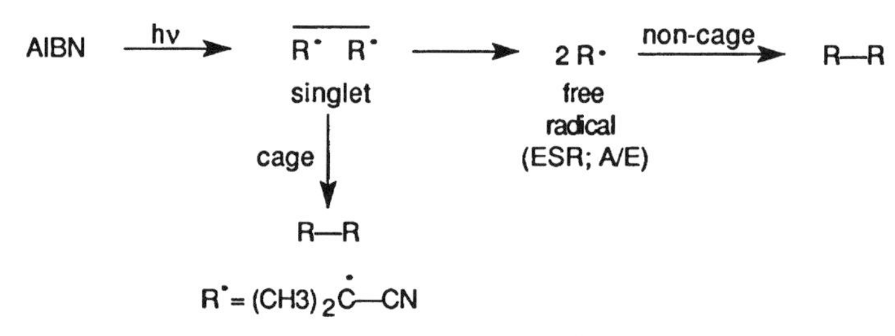

Figure 3.31 Chemically induced dynamic electron polarization.

The cyanopropyl radical ESR spectrum appears in the A/E mode. It is noteworthy that a CIDNP effect on the NMR of the dimer should not be observable, since the dimer results from both caged and noncaged radicals. A CIDEP experiment, in contrast, does not require distinct cage and noncage products, since only the uncaged radical can be observed. On the other hand, a successful CIDEP experiment normally depends upon the measurement of an ESR spectrum of a radical in fluid solution. This, in turn, would normally require a radical of at least moderate persistence.

References

General References

Bersohn, M.; Baird, J.C. *An Introduction to Electron Paramagnetic Resonance*, W. A. Benjamin: New York, 1966.

Carrington, A.; McLachlan, A. D., *Introduction to Magnetic Resonance*, Harper & Row, New York, 1967.

D. Fischer, H. in *Free Radicals*, Vol. II, Kochi, J. K., Ed., John Wiley & Sons, New York, 1973, p. 435.

Leffler, J. E. *An Introduction to Free Radicals*, John Wiley & Sons, New York, 1993, p. 11.

Lowry, T. H.; Richardson, K. S. *Mechanism and Theory in Organic Chemistry*, Third Edition, Harper & Row, New York, 1987, Appendix 1, p. 812.

Ward, H. R. in *Free Radicals*, Vol. I, Kochi, J. K., Ed., John Wiley & Sons, New York, 1973, p. 239.

Specific References

1. Fessenden, R. W.; Schuler, R. H. *J. Chem. Phys.* **1963**, *39*, 2147.
2. McConnell, H. M. *J. Chem. Phys.* **1956**, *24*, 764.
3. Kochi, J. K.; Krusic, P. J. *J. Am. Chem. Soc.* **1968**, *90*, 7157.
4. Korth, H.-G.; Trill, H.; Sustmann, R. *J. Am. Chem. Soc.* **1981**, *103*, 4483.
5. Kochi, J. K.; Krusic, P. J. *Spec. Publ. Chem. Soc. London* **1970**, *24*.
6. Heller, C.; McConnell, H. M. *J. Chem. Phys.* **1960**, *32*, 1535.
7. Skell, P. S; Shea, K. J. in *Free Radicals*, Vol. II, Kochi, J. K., Ed., John Wiley & Sons, 1973, p. 840.
8. Whiffen, D. H. *Mol. Phys.* **1963**, *6*, 223.
9. Krusic, P. J.; Rettig, T. A.; Schleyer, P. v. R. *J. Am. Chem. Soc.* **1972**, *94*, 995.
10. Forrester, A. R.; Hay, J. M.; Thomsen, R. H. *Organic Chemistry of Stable Free Radicals*, Academic Press, New York, 1968; Hoffman, B. M.; Eames, T. B. *J. Am. Chem. Soc.* **1969**, *91*, 2159; Bichutinskii A. A.; Prokof'ev, A. I.; Shalbalkin, V. A. *Russ. J. Phys. Chem.* **1964**, *38*, 534; Weiner, S.; Hammond, G. S. *J. Am. Chem. Soc.* **1968**, *90*, 1659.
11. Fisher, H., p. 443; Fessenden, R. W.; Schuler, R. H. *J. Chem. Phys.* **1963**, *39*, 2147; Fessenden, R. W. *J. Phys. Chem.* **1967**, *71*, 74.
12. Fessenden, R. W.; Schuler, R. H. *J. Chem. Phys.* **1965**, 2704.
13. Fessenden, R. W.; Schuler, R. H. *J. Chem. Phys.* **1963,** *39*, 2147; Paul, H.; Fischer, H. *Chem. Commun.* **1971**, 1938.
14. Fisher, H. p. 479.
15. Fessenden, R. W.; Schuler, R. H. *J. Chem. Phys.* **1963,** *39*, 2147.
16. Vinyl: *ibid.*; formyl: Cochran, E. L.; Adrian, T. J.; Bowers, V. A. *J. Chem. Phys.* **1966**, *44*, 4626.
17. Fehr, G. *Phys. Rev.* **1956**, *103*, 834; Hyde, J. S. *J. Chem. Phys.* **1965,** *43*, 1806.
18. Unpublished spectrum, but see: Bauld, N. L.; McDermed, J. C.; Rim, Y. S.; Hudson, C. E.; Hyde, J. S. *J. Am. Chem. Soc.* **1969**, 6666.
19. Bargon, J.; Fischer, H.; Johnsen, U. *Z. Naturforsch.* **1967**, *22*, 1551; Ward, H. R. *J. Am. Chem. Soc.* **1967**, *89*, 5517.
20. Closs, G. *J. Am. Chem. Soc.* **1969**, *91*, 4552; Kaptein, R.; Oosterhoff, J. L. *Chem. Phys. Lett.* **1969**, *4*, 195 and 214. The theoretical description of CIDNP presented by these authors is known as the CKO theory.
21. Ward, H. W. *Acc. Chem. Res.* **1972**, *5*, 18.
22. Kaptein, R. *J. Chem. Soc. D* **1971**, 732.
23. Ward H. R., p. 251.
24. Takemura, T.; Ohara, K.; Murai, H.; Kuwata, K. *Chem. Lett.* **1990**, 1635.

Exercises

3.1 The π MOs of the pentadienyl radical in the HMO approximation are listed below:

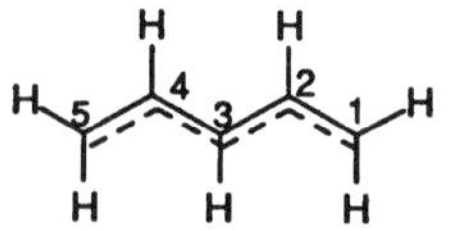

The 2,4-Pentadienyl Radical

$E_5 = \alpha - 1.732\beta$ $\quad \psi_5 = 0.289\phi_1 - 0.500\phi_2 + 0.577\phi_3 - 0.500\phi_4 + 0.289\phi_5$

$E_4 = \alpha - 1.00\beta$ $\quad \psi_4 = 0.500\phi_1 - 0.500\phi_2 + 0.500\phi_4 - 0.500\phi_5$

$E_3 = \alpha$ $\quad \psi_3 = 0.577\phi_1 - 0.577\phi_3 + 0.577\phi_5$

$E_2 = \alpha + 1.00\beta$ $\quad \psi_2 = 0.500\phi_1 + 0.500\phi_2 - 0.500\phi_4 - 0.500\phi_5$

$E_1 = \alpha + 1.732\beta$ $\quad \psi_1 = 0.289\phi_1 + 0.500\phi_2 + 0.577\phi_3 + 0.500\phi_4 + 0.289\phi_5$

E

(a) Select the SOMO of this radical and calculate the expected proton ESR hyperfine splittings (including the sign) using the appropriate McConnell equation. (b) Explain why the splittings arising from the hydrogens at C_1, C_3, and C_5 are negative. (c) Why are those splittings smaller than those of the methyl radical? (d) To the extent that the HMO SOMO is accurate, does the calculated ρ_i, for example, at C_1, C_3, or C_5, represent the total spin density at these carbon atoms? Explain your reasoning. If the answer given is "no," is the total spin density greater or less than ρ_i? Explain. (e) Is the ESR spectrum likely to be unsplit ($a = 0$) by the hydrogens at C_2,C_4? If not, explain why not and give the sign of the expected splitting and a justification for the sign.

3.2 The SOMO of the benzyl radical is given below in the HMO approximation.

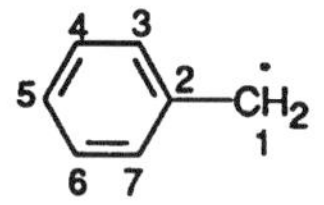

The Benzyl Radical

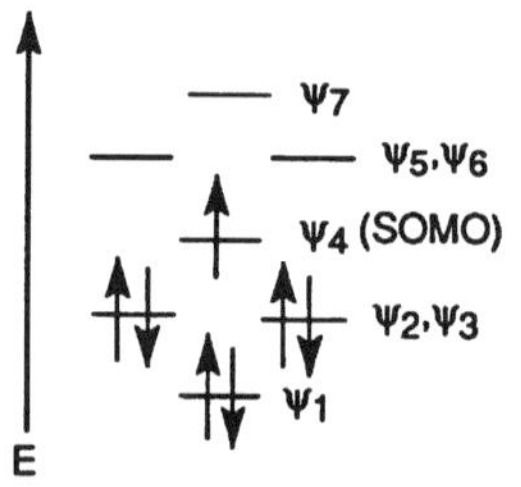

ψ_7

ψ_5, ψ_6

ψ_4 (SOMO)

ψ_2, ψ_3

ψ_1

E

$\psi_4 = 0.756\phi_1 - 0.378\phi_3 + 0.378\phi_5 - 0.378\phi_7$

Use a McConnell equation to predict the proton hyperfine splittings at the benzylic (B), *ortho* (o), *meta* (m), and *para* (p) positions. The actual splittings are $a_B = -16.35$, $a_o = -5.14$, $a_m = +1.75$, and $a_p = -6.14$. (Carrington, A.; Smith, I. C. P. *Mol. Phys.* **1965**, *9*, 137.)

3.3 The vinyl radical, confined in a solid matrix at low temperature, has the following hyperfine splitting constants:

$$\underset{\beta}{CH_2} = \underset{\alpha}{CH\bullet}$$

$$\left.\begin{aligned} a_\alpha &= 1.34\ \text{mT (1H)} \\ a_\beta &= 3.42\ \text{mT (1H)} \\ a_{\beta'} &= 6.85\ \text{mT (1H)} \end{aligned}\right\} 2^n = 2^3 = 8 \text{ lines}$$

(a) Draw the structure of the vinyl radical. Is the radical center sp^2 or sp hybridized? Explain your reasoning. (b) When the radical, adsorbed on a silica gel surface to stabilize it, is allowed to warm up to between −120 to −70°C, a six-line spectrum is obtained. Assuming that the radical species is still the vinyl radical (the spectral change is reversible), explain the change to a six-line spectrum. Using the hyperfine splittings from the solid matrix spectrum, make a prediction of the expected splitting constants in the six-line spectrum. (Fessenden, R. W.; Schuler, R. H. *J. Chem. Phys.* **1963**, *39*, 2147.)

3.4 The benzoyl radical could conceivably be either a π- or a σ-type radical. In fact, the only observable splitting in the ESR spectrum of the benzoyl radical arises from the *meta* protons [$a_m = 0.116$ mT (2H)]. Further, the ^{13}C hyperfine splitting in the carbonyl carbon-labeled benzoyl radical is $a_{13C} = 12.82$ mT.

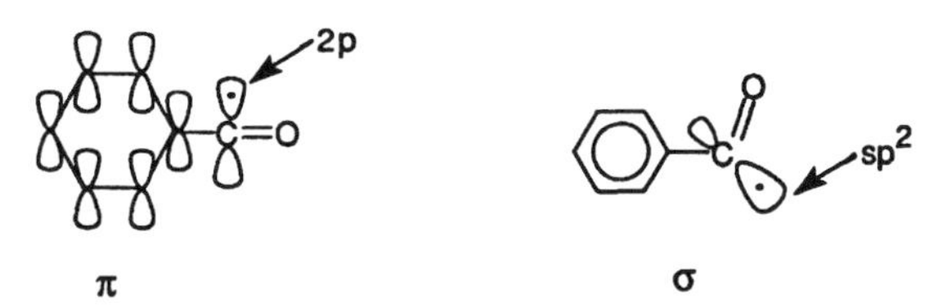

(a) Using the benzyl radical as a model for the π benzoyl radical, what qualitative pattern of proton hyperfine splittings would be expected for this radical? (b) Is the magnitude of the ^{13}C hyperfine splitting consistent with a π radical? Explain your reasoning. (c) Since the *ortho* protons are closer to the radical site (γ) than are the *meta* protons (δ), rationalize the larger a_m than a_o and a_p in the σ radical. Explain also why the two *meta* protons give rise to equivalent hyperfine splittings in this spectrum, which was observed at −86°C. (Krusic, P. J.; Rettig, T. A. *J. Am. Chem. Soc.* **1970**, *92*, 722.)

3.5 In general, β protons *anti* (*trans*) to a σ radical site give rise to larger hyperfine splittings than those *syn* (*cis*) to a σ radical site, in a manner analogous to the Karplus relationship for vicinal proton–proton couplings in NMR spectroscopy. As an example, the *anti* β proton in the vinyl radical has $a_\beta = 6.8$ mT, while the *syn* β proton has a_β = 3.4 mT. (a) Based upon a SOMO/LUMO interaction, that is, the overlap of the singly occupied SOMO with the vacant antibonding MO of the C_β–H bond, rationalize the greater *anti* than *syn* splitting. [Hint: Recall that negative overlap in the ABMO forces electron density out of the central region between carbon and hydrogen and into the region behind the C–H bond (the back lobe is enhanced in relative size in the ABMO). Your argument should then focus on the relative efficiency of overlap between the SOMO and the LUMO in the two situations.] (b) The observation of impressively large long-range (γ) hyperfine splittings in conformationally rigid systems wherein the SOMO and the relevant C_γ–H bond are in a "W-Plan" geometry can be explained in a similar manner. Rationalize, based upon a direct, through-space, interaction between the back lobe of the C_γ–H LUMO and the SOMO orbital, the preference for the W-Plan geometric arrangement.

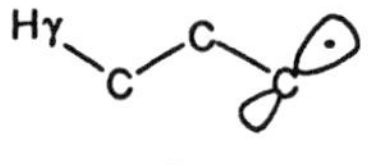

Hγ C C C

W-Plan Geometry

3.6 Iminoxy radicals, obtained by oxidation of oximes, are persistent or even isolably stable free radicals. For example, the di-*tert*-butyliminoxy radical (**A**) has been isolated. The iminoxy radicals derived from the *syn* and *anti* oximes of benzaldehyde are particularly interesting.

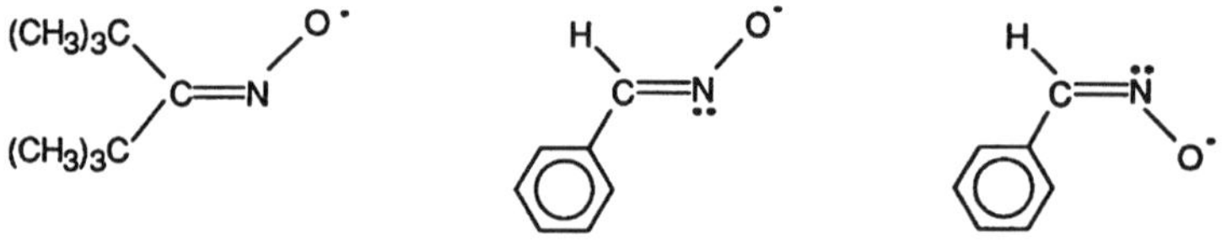

$(CH_3)_3C$ C=N O· $(CH_3)_3C$

A

H C=N O·

B

H C=N O·

C

	B	**C**
a_N =	2.92 mT	3.16 mT
a_H =	+2.69 mT (1H)	+0.62 mT (1H)
a_H =		0.14 mT (2H)

(a) Consider first the oxime **B**. There is a large nitrogen splitting (1 : 1 : 1 triplet) and a nearly equally large proton splitting from the benzylic proton. In a π iminoxy radical, the oxygen SOMO would overlap with the C–N π bond, giving an allylic-type system. In a σ-type radical, the oxygen-centered SOMO would overlap with a filled nitrogen sp^2 AO. Which of these structures, the π or σ type, is consistent with the observation of a large nitrogen splitting and a large benzylic proton splitting? Explain your reasoning. What kind of radical stabilization effect is operative here? (b) In the iminoxy radical **C**, there is again a large ^{14}N splitting, but the splitting arising from the benzylic proton is much smaller. Explain why this splitting is much smaller than that observed in **B**. (c) An additional splitting is observed in **C**, which arises from the *ortho* protons. Propose a possible basis for this splitting that is consistent with the observation that equal splittings are observed for both *ortho* protons. (Thomas, J. R. *J. Am. Chem. Soc.* **1964**, *86*, 1446. Brokenshire, J. L.; Mendenhall, G. D.; Ingold, K. U. *J. Am. Chem. Soc.* **1971**, *93*, 5278.)

3.7 The nitrogen hyperfine splitting of TEMPO is rather sensitive to solvent polarity and particularly to the hydrogen bond donating ability of the solvent. Consequently, a TEMPO structural moiety can be attached to a particular site of a complex molecule (e.g., an enzyme) and used to probe the polarity of its environment (a technique called spin labeling). The ^{14}N hyperfine splitting is found to vary from approximately 14.6 to 15.9. Predict the direction of the effect of increased H bonding on the magnitude of the TEMPO nitrogen splitting and explain in detail, using canonical resonance structures for TEMPO. (Lim, Y. Y.; Drago, R. S. *J. Am. Chem. Soc.* **1971**, *93*, 891.)

3.8 The 1-bicyclo[2.2.2]octyl radical has the following hyperfine splittings:

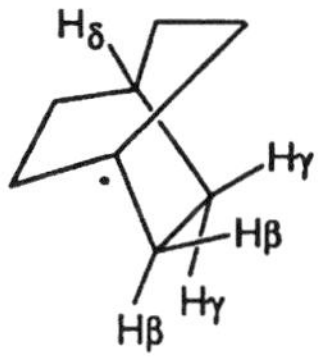

Hδ, Hγ, Hβ, Hβ, Hγ

$a_\beta = 0.66$ mT (6H)

$a_\gamma = 0.089$ mT (6H)

$a_\delta = 0.269$ mT (1H)

(a) Use the McConnell equation for β hyperfine splittings to calculate the approximate magnitude of the expected a_β for the appropriate dihedral angle (θ) in this system. (b) Explain why a_δ is much greater than a_γ. What frontier orbital interaction is presumably involved? (Krusic, P. J.; Reltig, T. A.; Schleyer, P. v. R. *J. Am. Chem. Soc.* **1973**, *94*, 995.)

3.9 (a) When N-*tert*-butyl-α-phenylnitrone (**A**) is used as a spin trap in the decomposition of AIBN in xylene at 110°C, a stable nitroxyl radical can be isolated by column chromotagraphy. Predict the structure of this stable radical.

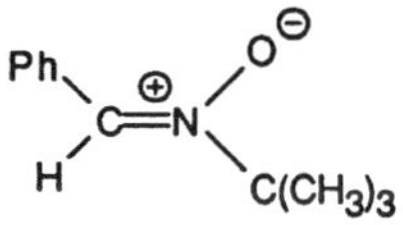

Ph, H, C=N, $O^{\ominus}$, $\oplus$, $C(CH_3)_3$

A

(b) Besides a nitrogen splitting of 1.46 mT, a proton splitting of $a_H = 0.307$ mT (1H) is observed in the ESR spectrum. Indicate on your structure which proton presumably gives rise to this splitting and generically what type of splitting this represents (α, β, γ, δ, etc. with respect to a main center of spin density). (c) Explain, in principle, how spin trapping is capable of distinguishing the nature of the radical that is trapped by ESR spectroscopy alone. (d) When

5,5-dimethylpyrroline-N-oxide (DMPO) is used as the spin trap under the same reaction conditions, the ^{14}N hyperfine splitting is identical to that observed in the previous case (1.46 mT), but the proton splitting is substantially increased (2.04 mT). Explain.

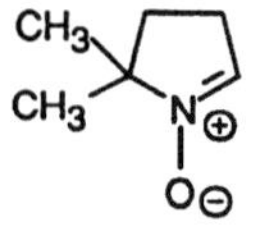

DMPO

(Iwamura, M.; Inamoto, N. *Bull. Chem. Soc. Japan* **1967**, *40*, 703.)

3.10 When methyl diazoacetate is photolyzed in carbon tetrachloride, the insertion product indicated is obtained, presumably via a carbene mechanism. The methine proton of the product is observed in the emission mode. Use the Kaptein equation for a net effect to deduce whether the insertion involved a singlet or triplet carbene. Assume that cage products arise from singlet radical pairs and that $CCl_3^\bullet$ has the larger g value in this radical pair:

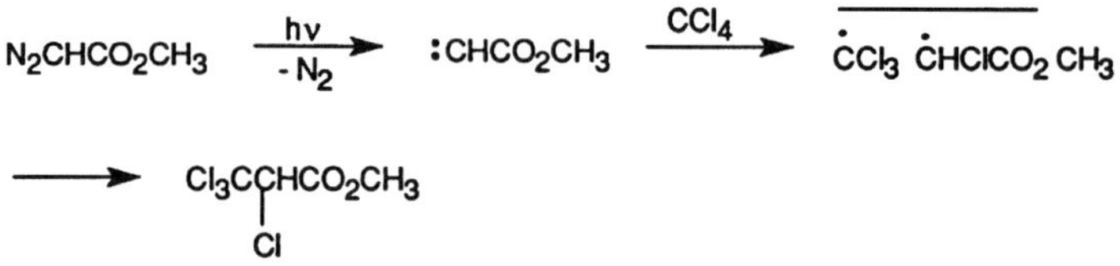

$$N_2CHCO_2CH_3 \xrightarrow[-N_2]{h\nu} :CHCO_2CH_3 \xrightarrow{CCl_4} \overline{\dot{C}Cl_3 \ \dot{C}HClCO_2CH_3} \longrightarrow Cl_3CCHCO_2CH_3 \ (\text{with Cl on CH})$$

(Cocivera, M.; Roth, H. D. *J. Am. Chem. Soc.* **1970**, *92*, 2573.)

3.11 When the sulfur ylide depicted below is heated, it undergoes a rearrangement analogous to the Stevens rearrangement. The rearrangement product reveals net CIDNP effects for both the benzylic proton and the proton α to the methylthio group.

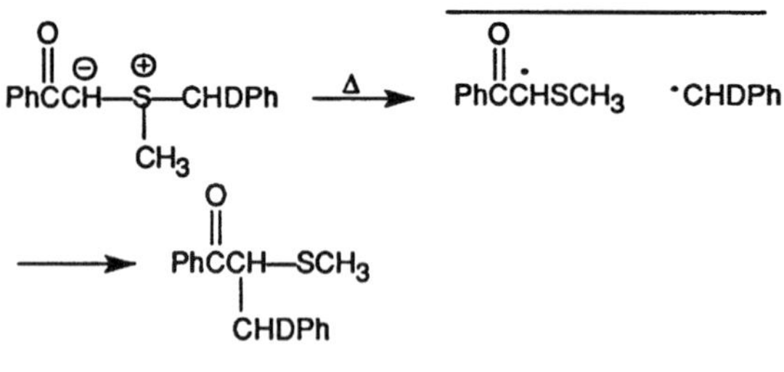

$$PhC(=O)\overset{\ominus}{C}H-\overset{\oplus}{S}(CH_3)-CHDPh \xrightarrow{\Delta} \overline{PhC(=O)\dot{C}HSCH_3 \quad {}^\bullet CHDPh} \longrightarrow PhC(=O)CH(CHDPh)-SCH_3$$

(a) Predict the sense of the CIDNP effects for both types of proton assuming that the benzyl radical has the higher g value in the radical pair. (b) The product shows 36% retention of stereochemistry when optically active ylide is used. Is this consistent with a radical pair mechanism? Explain your reasoning. (Baldwin, J. E.; Erickson, W. F.; Hackler, R. E.; Scott, R. M. *Chem. Commun.* **1970**, 576.)

3.12 (a) When diacyl peroxides are decomposed in the presence of radical scavengers (as in the example of lauroyl peroxide given in the text) by *direct irradiation*, CIDNP multiplet effects are observed that have the same multiplet sense as in the thermal decompositions. What mechanistic conclusion does this observation permit?

$$RCH_2\overset{O}{\overset{\|}{C}}OO\overset{O}{\overset{\|}{C}}CH_2R \xrightarrow[(CH_3)_2CHI]{h\nu} \text{undecene} + \text{1-iodoundecane}$$

(b) Use the appropriate Kaptein equation to predict the multiplet sense of the undecyl iodide and undecene formed in the decomposition of lauroyl peroxide by *triplet sensitization* using benzophenone triplets, in the presence of isopropyl iodide.

$$RCH_2\overset{O}{\overset{\|}{C}}OO\overset{O}{\overset{\|}{C}}CH_2R \xrightarrow[(CH_3)_2CHI]{Ph_2C{=}O,\ h\nu} \text{undecene} + \text{1-iodoundecane}$$

(Kaptein, R.; den Hollander, J. A.; Antheunis, D.; Oosterhoff, L. J. *Chem. Commun.* **1970**, 1687.)

CHAPTER

4

Anion Radicals

Addition of a single electron to a neutral molecule generates a unique chemical species that simultaneously has a unit of negative charge and an unpaired electron (Figure 4.1).[1] The electronic structure of anion radicals (also called radical anions) is readily envisioned in terms of the entry of an electron into the lowest-energy unoccupied molecular orbital (LUMO) of a neutral precursor (Figure 4.2). The

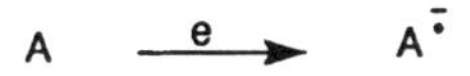

Figure 4.1 Anion radicals.

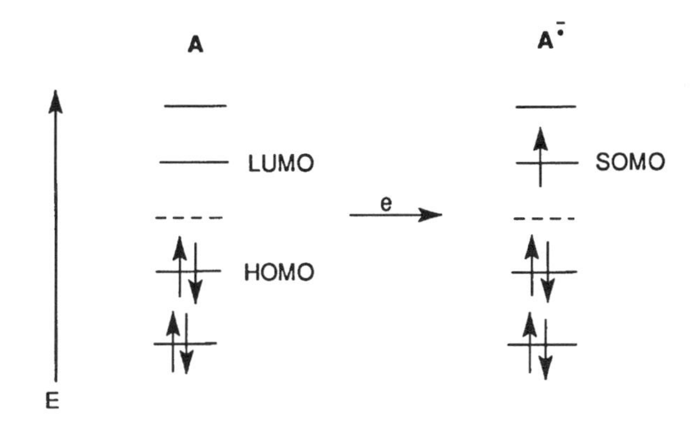

Figure 4.2 MO picture of anion radical formation.

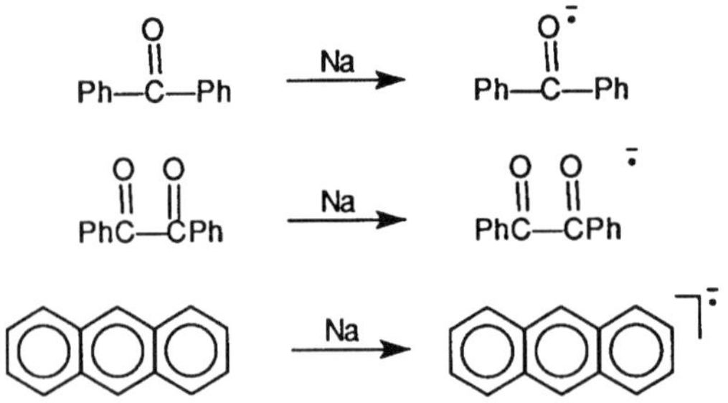

Figure 4.3 Early examples of stable anion radicals.

LUMO of the neutral then becomes the SOMO of the anion radical. The latter MO is uniquely important since it determines both the charge (Q_i) *and spin* (ρ_i) *distribution in the anion radical (via its coefficients), as well as the ease with which it is formed from the neutral (via its energy).*

4.1 Formation of Anion Radicals

The most common method for generating anion radicals is the treatment of neutral molecules with alkali metals, usually in an ethereal solvent such as tetrahydrofuran (THF). Historically, the stable anion radicals of benzophenone (benzophenone ketyl), benzil, and anthracene were among the first to be prepared (Figure 4.3).[2] Other methods of preparation include electrochemical, photochemical, chemical, and radiolytic reduction of neutrals. The reduction of organic molecules to anion radicals by photochemically induced electron transfer has become especially important and will be considered in detail in Chapter 6. Several of the other methods mentioned will also be specifically exemplified in various contexts in this chapter.

4.2 Simple π-Type Anion Radicals

The antibonding σ MOs (σ*) of ethene are substantially higher in energy than the antibonding π MO (π*). The ground-state ethene anion radical is therefore a π-type anion radical (Figure 4.4). Although net π bonding is attenuated by addition of an electron to the π* orbital, a significant residue of π bonding remains as a consequence of the double occupancy of the bonding π MO. The π bond in the ethene anion radical is therefore a three-electron bond formally analogous to those in TEMPO and other stable radicals. Two common ways of representing the structure of this anion radical are given below (**A** and **B**). The structural representation **B** uses a dotted line to indicate that the π bond is partial. A resonance theoretical description is also used sometimes (**C** ↔ **D**), but this approach has some

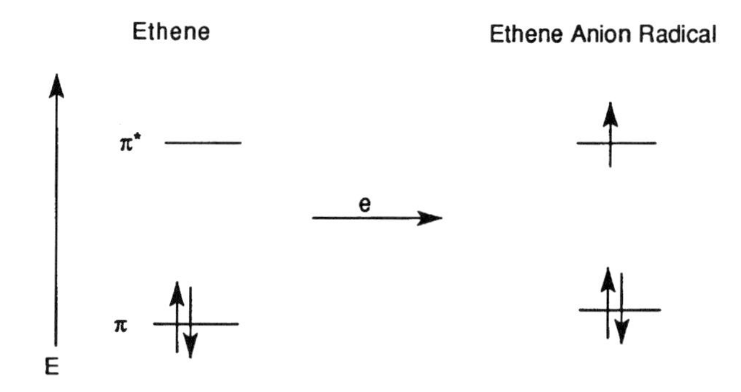

Figure 4.4 The ethene anion radical.

disadvantages in the case of ion radicals. Most important, there is no standard canonical structure that represents the partial π bond, and the resonance description [**C** $\leftrightarrow$ **D**] is therefore seriously incomplete. Structure **B**, strictly speaking, is not a

$$\left[\overset{\ominus}{C}H_2-\overset{\odot}{C}H_2 \longleftrightarrow \overset{\odot}{C}H_2-\overset{\ominus}{C}H_2\right]$$

C **D**

valid resonance structure. Second, structures **C** and **D** tend to leave the impression that charge and odd electron density are uncoupled, whereas they are normally strongly coupled. This can lead to erroneous conclusions in systems in which structures **C** and **D** are not equienergetic. For example, in the case of a styrene anion radical, it might be assumed that the odd electron density exists primarily on the β carbon and the negative charge density mainly on the α carbon (structure **E**) rather than the converse (structure **F**). In fact, both the odd electron density and charge density are greatest at the β carbon, as predicted from the squares of the SOMO coefficients at C_α (−0.39) and C_β (+0.595).

$$\left[\mathrm{Ph}\overset{-}{\underset{\alpha}{C}}H-\overset{\bullet}{\underset{\beta}{C}}H_2 \longleftrightarrow \mathrm{Ph}\overset{\bullet}{C}H-\overset{-}{C}H_2\right] \qquad \mathrm{PhCH}\overset{\bullet\,-}{=\!=}CH_2$$

E **F** **Preferred**

4.3 The Butadiene Anion Radical

The π^* MO of ethene, though lower in energy than any of the σ^* MOs of this molecule, is still of high enough energy to be relatively inaccessible to typical reducing agents. However, the LUMOs of conjugated π systems such as dienes are considerably lower in energy, and anion radical formation in these systems is correspondingly more facile. The 1,3-butadiene anion radical has been generated by electrochemical reduction in liquid ammonia at –78°C (Figure 4.5).[3] The spin density (ρ_i) is predicted, from the square of the SOMO coefficients, to be 2.6 times as great at the terminal carbons (C_1, C_4) as at the internal carbons (C_2, C_3). The ratio of the experimental hyperfine splittings of the protons at C_1, C_4 to those of the protons at C_2, C_3 is 2.7. The magnitudes of these hyperfine splittings are in reasonable agreement with those calculated from the McConnell equation for α hyperfine splittings (hfs) with $Q = -2.3$ mT (Figure 4.6). It follows that the negative charge densities are similarly greater at C_1, C_4 than at C_2, C_3. Consequently, protonation of the butadiene anion radical (involved in the Birch reduction of dienes, e.g., *vide infra*) occurs regiospecifically at C_1 and C_4. Although the addition of an electron to the antibonding LUMO of butadiene decreases the overall (net) π bonding, it strengthens the conjugation between the two ethenic links and makes rotation around the C_2–C_3 bond more difficult than in the neutral diene. This follows naturally from the circumstance that the LUMO coefficients in 1,3-butadiene (the SOMO coefficients in the anion radical) have the same sign at C_2 and C_3

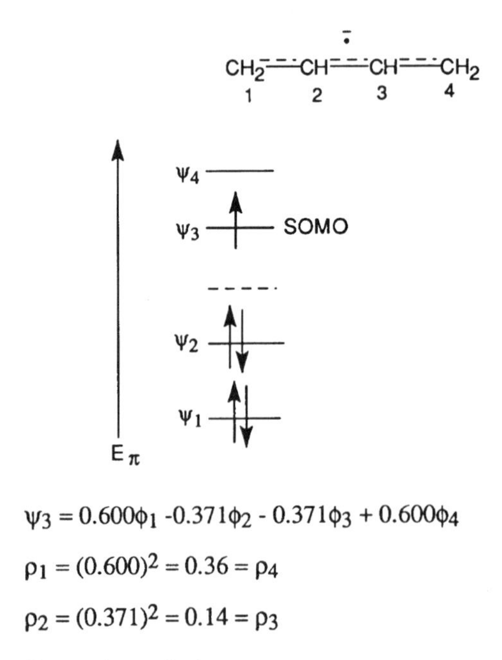

$$\psi_3 = 0.600\phi_1 - 0.371\phi_2 - 0.371\phi_3 + 0.600\phi_4$$

$$\rho_1 = (0.600)^2 = 0.36 = \rho_4$$

$$\rho_2 = (0.371)^2 = 0.14 = \rho_3$$

Figure 4.5 1,3-Butadiene anion radical.

Position	Observed Hfs (mT)	Predicted Hfs* (Q = -2.3mT)
C_1, C_4	- 0.76	- 0.83
C_2, C_3	- 0.28	- 0.32

* $a_i = Q\rho_i$ (McConnell equation)

Figure 4.6 Observed and calculated hyperfine splittings in the 1,3-butadiene anion radical.

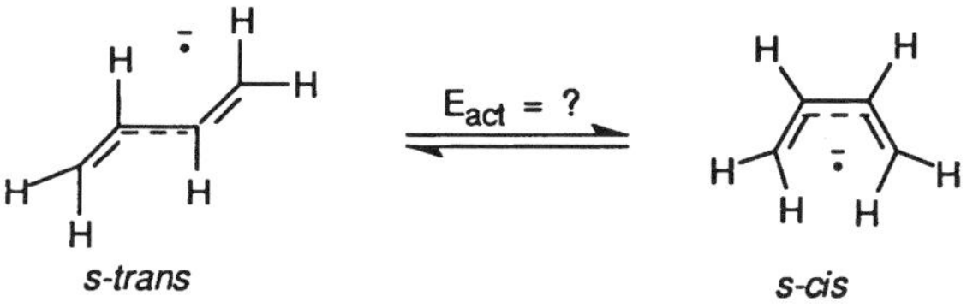

Figure 4.7 The *s-trans* and *s-cis*-1,3-butadiene anion radical.

(both −0.371) and the resulting positive overlap contributes to an increased C_2–C_3 π-bond order. On the other hand, the C_1–C_2 and C_3–C_4 bonds are both weakened in the anion radical ($a_1 = +0.600$, $a_2 = -0.371$). The exact magnitude of the barrier to the interconversion of the *s-cis* and *s-trans* anion radicals is still uncertain (Figure 4.7).

4.4 The Tetracyanoethylene Anion Radical: A Stable Anion Radical

The anion radical of the highly electron deficient alkene tetracyanoethylene (TCNE) is isolably stable, and the spin distribution in $TCNE^{\overline{\bullet}}$ has been examined by means of polarized single-crystal neutron diffraction studies on the tetrabutylammonium salt.[4] These studies provide a uniquely direct view of the SOMO of this π-type anion radical. The alkene carbons are found to contain 66% of the total spin (ρ = 0.33 at each alkene carbon), while the nitrogens each have ρ = 0.13 (a total of 52% of the total spin). The nitrile carbons have negative spin density (ρ = −0.05). These spin densities were found to be in rather good agreement with those obtained from a density-functional theoretical calculation (ρ = 0.29, 0.15, and −0.04). Perhaps most interesting is the observation (Figure 4.8) that, in the SOMO, the $2p_z$ AOs on the alkene carbons are not centered directly on the carbon atoms but are bent away from the alkene C–C bond, as would be expected for an MO that is antibonding between the two alkene carbon atoms.

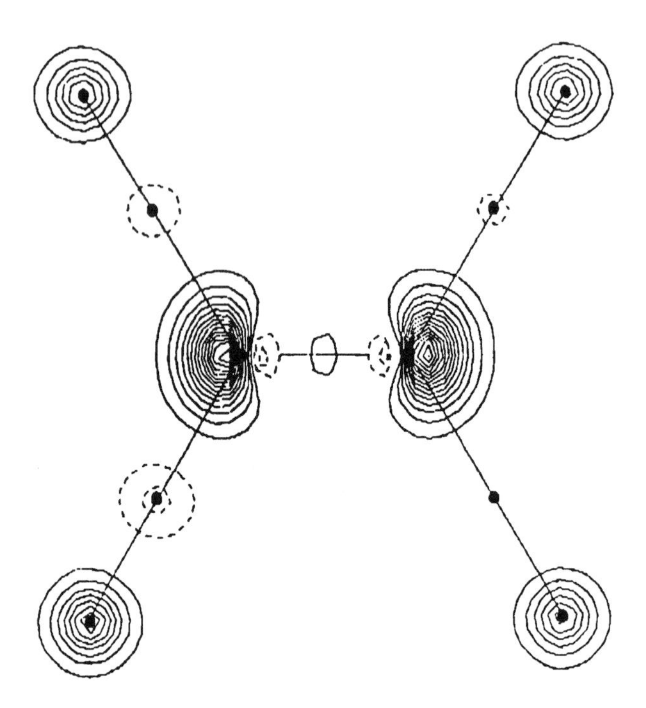

Figure 4.8 Spin density of the tetracyanoethylene anion radical projected onto the $TCNE^{-\bullet}$ molecular plane. Reprinted with permission from *J. Am. Chem. Soc.* **1994**, *116*, 7243–7249; Copyright 1994, American Chemical Society. Figure generously supplied by Professor Joel Miller.

4.5 Anion Radicals of Aromatic Systems

The naphthalene ($E_{SOMO} = \alpha - 0.618\beta$) and biphenyl ($E_{SOMO} = \alpha - 0.705\beta$) anion radicals are especially well known to organic chemists for their synthetic utility as one-electron reducing agents.[5] Solutions of both anion radicals are conveniently prepared by the reduction of the corresponding aromatic with an alkali metal in tetrahydrofuran solution (Figure 4.9). These solutions are relatively stable (persistent) in the absence of oxygen and moisture. The ESR spectrum of the naphthalene anion radical consists of a quintet of quintets, the α protons exhibiting the larger hyperfine splitting as expected theoretically on the basis of the coefficients of the α (0.425) and β (0.263) carbons, respectively, in the naphthalene LUMO (the SOMO of the anion radical).[6] The McConnell equation ($a_\alpha = Q_\alpha \rho_i$) with $Q_\alpha = -2.3$ mT, predicts $a_\alpha = 0.42$ mT (observed 0.50) and $a_\beta = 0.16$ mT (observed 0.18). The biphenyl anion radical, which has a higher-energy SOMO than the naphthalene anion radical is an even stronger one-electron reducing agent than the latter, as shown by the reduction potentials of biphenyl ($E = -2.70$ V) and naphthalene ($E = -2.50$ V, vs. SCE).[7]

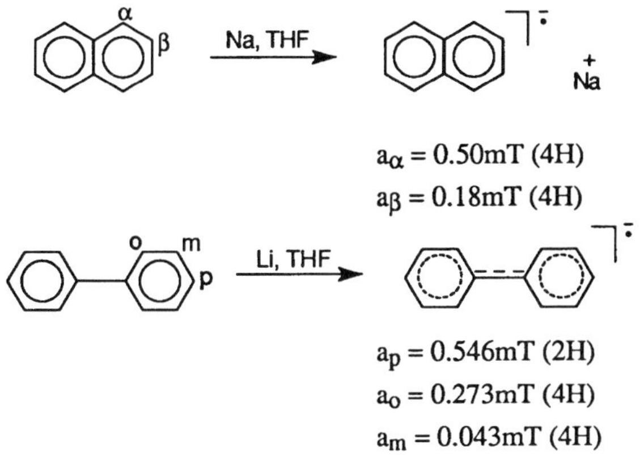

α
β
Na, THF
+
Na
aα = 0.50mT (4H)
aβ = 0.18mT (4H)
o
m
p
Li, THF
ap = 0.546mT (2H)
ao = 0.273mT (4H)
am = 0.043mT (4H)

Figure 4.9 The naphthalene and biphenyl anion radicals.

Benzene, of course, is still more difficult to reduce, but the blue benzene anion radical is readily generated in concentrations sufficient for ESR spectroscopy by reaction of potassium (or better still, sodium–potassium alloy) in dry THF or dimethoxyethane (DME) at –78°C or below (Figure 4.10).[8] An ESR septet (a = –0.375 mT) is observed, indicating that the six protons are equivalent on the ESR time scale. Since benzene has doubly degenerate LUMOs, the anion radical is subject to Jahn–Teller distortion, and two distinct benzene anion radicals are possible (Figure 4.11). In $\mathbf{A}^{\bar{\bullet}}$ for example, the SOMO is antisymmetric (A) with respect to a plane of symmetry (P/S) perpendicular to the ring and passing through two para carbons (C_1, C_4), which are therefore in a nodal plane (NP) of this SOMO.

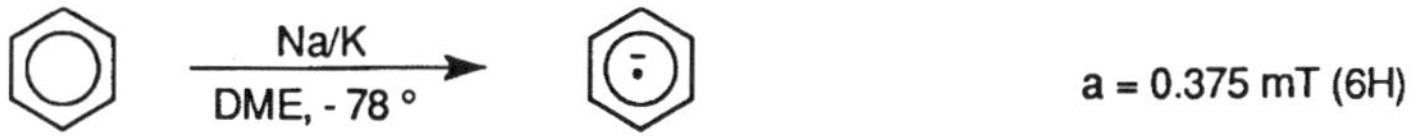

Na/K
DME, - 78 °
a = 0.375 mT (6H)

Figure 4.10 The benzene anion radical.

NP
0.0
- 0.50
+ 0.50
NP
NP
+ 0.50
- 0.50
0.0
P/S
A
+ 0.58
NP
NP
- 0.29
- 0.29
- 0.29
- 0.29
NP
NP
+ 0.58
S

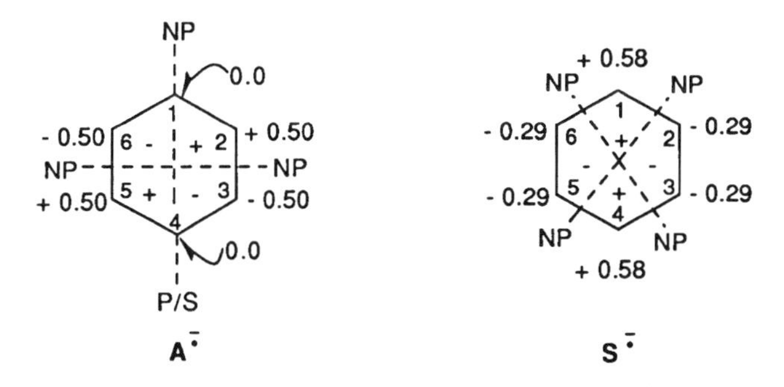

Figure 4.11 Two benzene LUMOs.

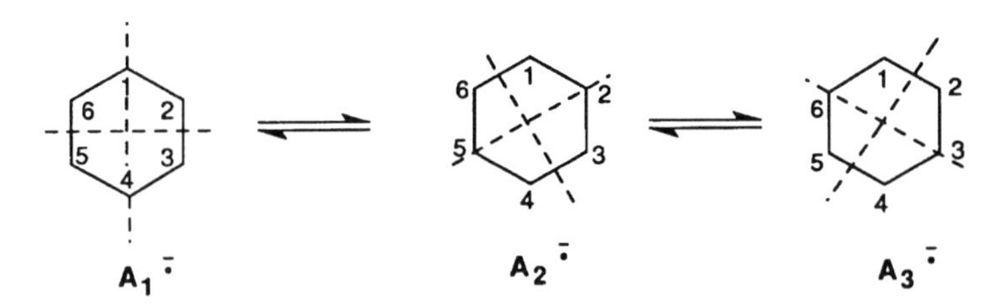

Figure 4.12 Symmetrization of spin and charge by pseudorotation.

There is also a second nodal plane perpendicular to this one and bisecting the C_2–C_3 and C_5–C_6 bonds. The numbers shown on the diagram are the coefficients in this SOMO. Spin (and negative charge) densities are therefore $\rho_1 = \rho_4 = 0$, $\rho_2 = \rho_3 = \rho_5 = \rho_6 = 0.25$ in this state. However, very slight alterations of the bond lengths (and/or other geometric parameters) can result in the rapid interconversion of this substate with two equienergetic substates $A_2^{\overline{\bullet}}$ and $A_3^{\overline{\bullet}}$ (Figure 4.12). This "pseudorotation" results in an equalization of charge and spin at all six positions, which is presumably rapid on the ESR time scale even at low temperatures. In the $S^{\overline{\bullet}}$ state (Figure 4.11), the odd electron occupies an MO that is symmetric (S) with respect to the formerly defined plane of symmetry. The two nodal planes do not pass through any of the carbon atoms, so that finite spin and negative charge density exists at all six carbons. Specifically, $\rho_1 = \rho_4 = \frac{1}{3}$ and $\rho_2 = \rho_3 = \rho_5 = \rho_6 = \frac{1}{12}$. Again, equilibration of three energy-equivalent substates by psuedorotation would be expected to engender rapid charge and spin equalization. On the basis of the experimental ESR spectrum, it is not clear whether the states of the $\mathbf{A}^{\overline{\bullet}}$ or the $\mathbf{S}^{\overline{\bullet}}$ manifold represent the ground state in solution.

The description of the benzene LUMOs given above is still approximately valid for only slightly perturbed arenes such as toluene. The electron donating character of the methyl group suggests that the preferred environment of this substituent in the toluene anion radical should be the site of lowest negative charge density, that is, C_1 (or C_4) of the $A_1^{\overline{\bullet}}$ substate (Figure 4.13). The ESR spectrum of the anion radical does indeed reflect large and approximately equal *o* and *m* proton splittings and very small *p* and methyl splittings.[9] In the context of the Birch reduction (*vide*

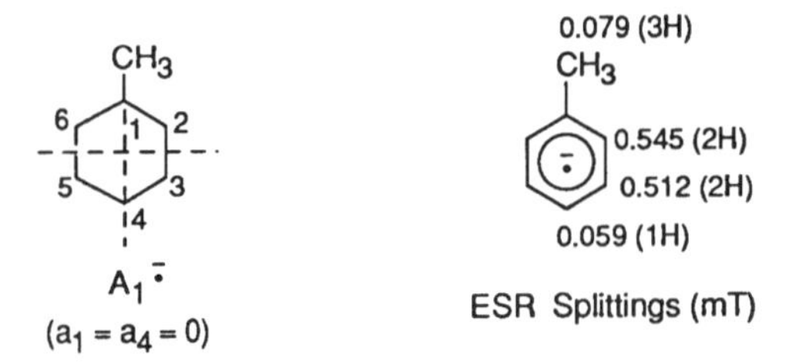

Figure 4.13 The toluene anion radical.

infra), this anion radical would be expected to protonate primarily at the *m* and *o* positions.

4.6 Anion Radicals of Nonbenzenoid Cyclic Conjugated Systems: The Cyclooctatetraene Anion Radical

The hypothetical planar form of cyclooctatetraene (COT) has a nonbonding LUMO ($E = \alpha$) and should be especially easily reduced to the corresponding anion radical ($COT^{\bar{\bullet}}$)and the $4n + 2$ (aromatic) π electron dianion (COT^{-2}). Neutral COT, of course, is nonplanar and has a tublike shape that minimizes the angle and torsional strain of the planar form. The four ethenic moieties are very nearly orthogonal to each other, so that tub $COT^{\bar{\bullet}}$ is expected to be a high-energy species with four ethene like SOMOs. The reduction of COT is nevertheless easily effected, for example, by alkali metals, and as expected, the resulting persistent anion radical is planar (Figure 4.14).[10] The difference in SOMO energies of the two forms ($\Delta E_{SOMO} = \beta$) is evidently sufficient to overcome the torsional and angle strain present in the planar structure. However, as a result of this expenditure of energy in the first reduction step, the second reduction step is, relatively, unusually favorable.

This is indicated by the extensive disproportionation of $COT^{\bar{\bullet}}$ to COT^{-2} (Figure 4.15).[11] This tendency of $COT^{\bar{\bullet}}$ to disproportionate is considered to be a direct result

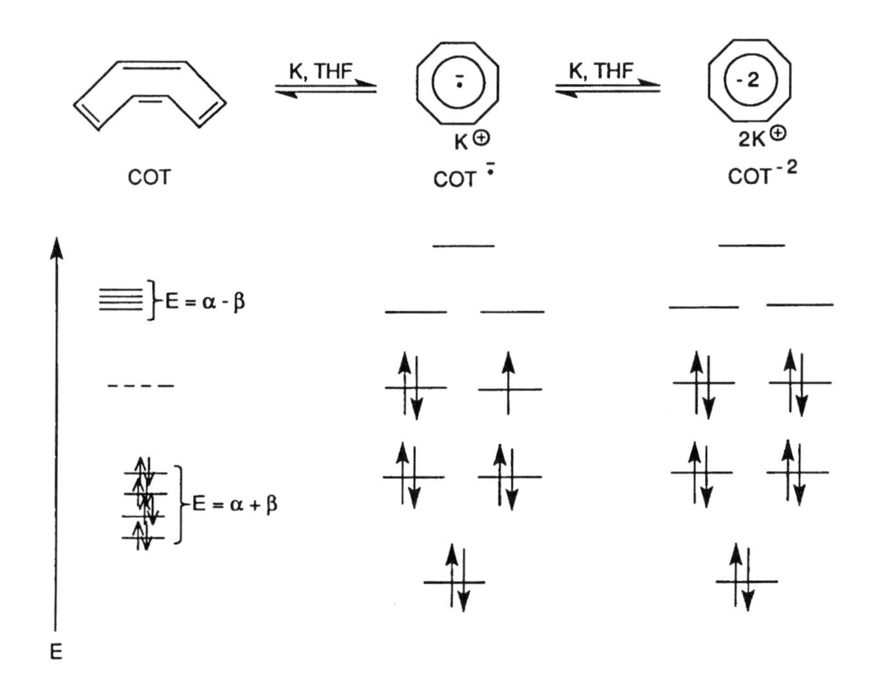

Figure 4.14 The cyclooctatetraene anion radical.

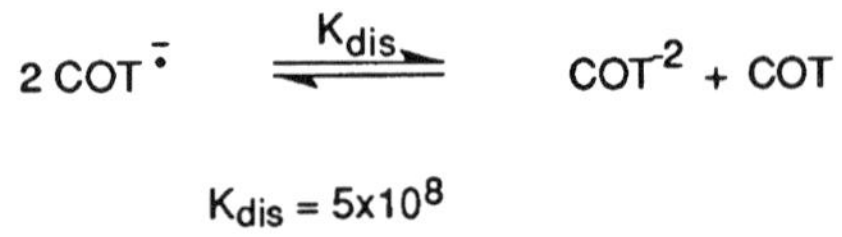

$$2\,COT^{\bar{\bullet}} \underset{}{\overset{K_{dis}}{\rightleftharpoons}} COT^{-2} + COT$$

$$K_{dis} = 5\times10^{8}$$

Figure 4.15 Disproportionation of $COT^{\bar{\bullet}}$.

of the planarity of $COT^{\bar{\bullet}}$. Thus, in the disproportionation equilibrium, only one product molecule (COT^{-2}) is strained as a result of planarity, but two reactant molecules ($2COT^{\bar{\bullet}}$) are strained. Quite possibly the aromaticity of COT^{-2} could also contribute to this tendency. The planarity of $COT^{\bar{\bullet}}$ is also indicated by the much higher rate of electron transfer between $COT^{\bar{\bullet}}$ and COT^{-2} (both are planar and have similar geometries) than between $COT^{\bar{\bullet}}$ and neutral COT (very different geometries). Perhaps the most decisive proof of planarity of $COT^{\bar{\bullet}}$ is its ESR spectrum [a = –0.32 mT (8H)], which is in rather good agreement with that predicted by the McConnell equation (–0.29 mT). The latter equation is only valid for planar conjugated systems of sp^2-hybridized carbon atoms. In particular, tub $COT^{\bar{\bullet}}$ is expected to have a very much smaller ESR splitting constant.

The behavior of the COT anion radical exemplifies, albeit to a rather extreme extent, an important general behavior of anion radicals, viz., the tendency to disproportionate to dianions.[12] Moreover, the profound structural change that accompanies the addition of an electron to COT is an appropriate reminder that anion radical structures are not necessarily closely analogous to those of the corresponding neutrals. Anion radical formation is, indeed, necessarily accompanied by at least modest, but sometimes even quite drastic, structural changes, including possible σ bond cleavage, as will be seen in subsequent sections of this chapter.

4.7 Multianion Radicals

Anion radical formation occurs when the LUMO of a neutral molecule accepts an additional electron. Conjugated organic anions also have vacant ABMOs. Acceptance of an electron by the LUMO of an anion would yield a *dianion radical*. The first recorded instance of dianion radical formation was the formation of the pentacyanoallyl dianion radical by anodic reduction of the corresponding anion (Figure 4.16).[13] The first hydrocarbon dianion radical observed was the tropenide dianion radical.[14] The tropenide anion (a 4N electron cyclic system) is relatively

Figure 4.16 The first dianion radical.

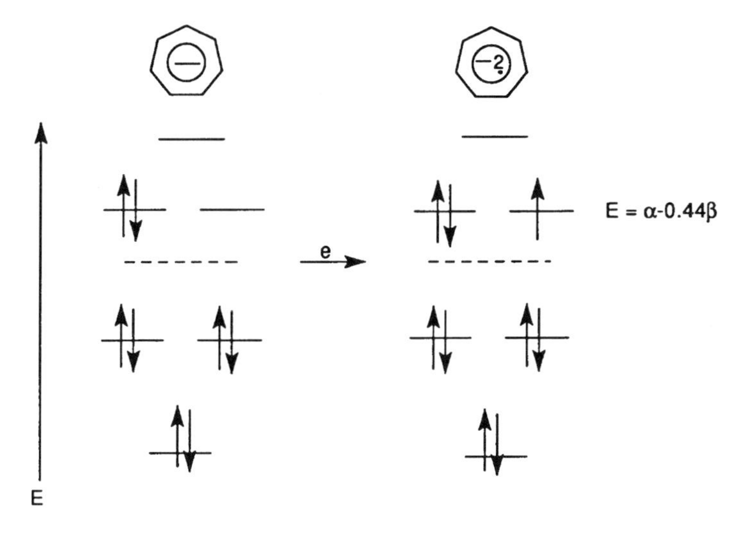

Figure 4.17 Electronic configuration of the tropenide anion and dianion radical.

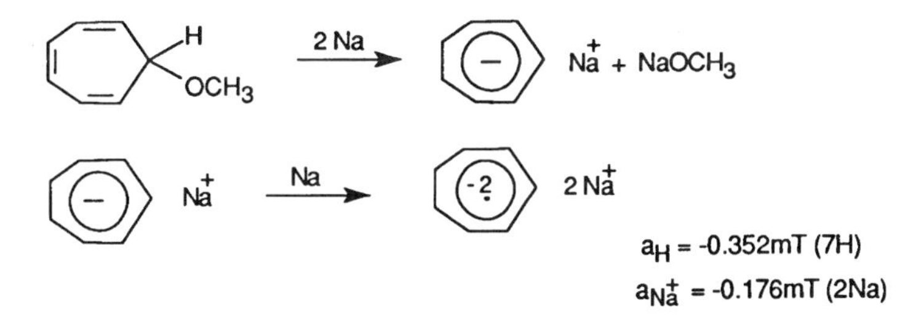

Figure 4.18 The formation of the tropenide dianion radical.

stable and has an unusually low-lying LUMO for an anion ($E_{\mathrm{LUMO}} = \alpha - 0.44\beta$; Figure 4.17). Reduction of tropenyl methyl ether by potassium or sodium in THF proceeds first to the diamagnetic anion and then subsequently to the persistent blue dianion radical (Figure 4.18). Interestingly, the disodium salt reveals rather strong hyperfine splittings from two sodium ions, an observation that clearly demonstrates the dianionic nature of the radical.

Subsequently, a number of other hydrocarbon dianion radicals, including the fluorenide dianion radical, have been prepared (Figure 4.19).[15] Similarly the reduction of the enolate of dibenzoylmethane yields a relatively stable dianion radical.[16] A number of *trianion radicals* have also been prepared by reduction of dianions. An especially interesting example of a stable trianion radical is that of heptafulvalene (Figure 4.20).[17] This species has the odd electron localized on a single tropenyl ring, in contrast to the corresponding anion radical, which is fully delocalized over both rings. Apparently, the strong coulombic attractions of the

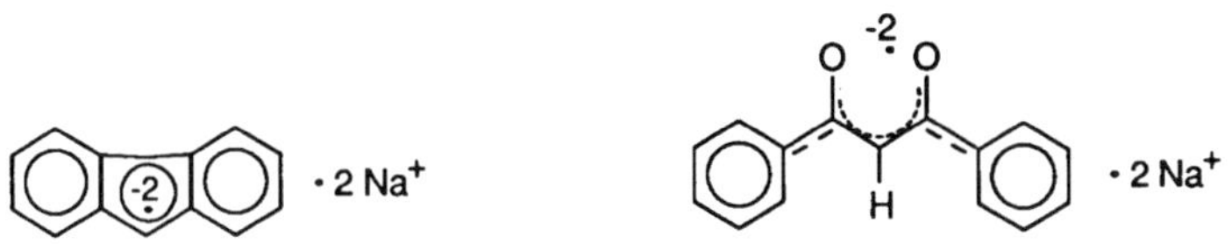

Figure 4.19 Some additional dianion radicals.

Figure 4.20 The heptafulvalene anion radical and trianion radical.

dianionic ring to the two metal counterions prevents rapid equilibration of the electron between the two rings.

4.8 Birch Reduction: Protonation of Anion Radicals

The deep blue solutions of alkali metals in liquid ammonia contain solvated electrons and, as such, are very effective reducing media for a variety of conjugated systems. The Birch reduction[18] of naphthalene to 1,4-dihydronaphthalene is an excellent example (Figure 4.21). Electron transfer to naphthalene yields the corresponding anion radical, which is protonated exclusively at the α position, as expected on the basis of previously noted charge distribution in the anion radical. The same mode of protonation is also favored by product character development in the transition state, since the resulting radical is more stable than the isomeric one generated by β protonation. A second electron transfer, to the radical intermediate,

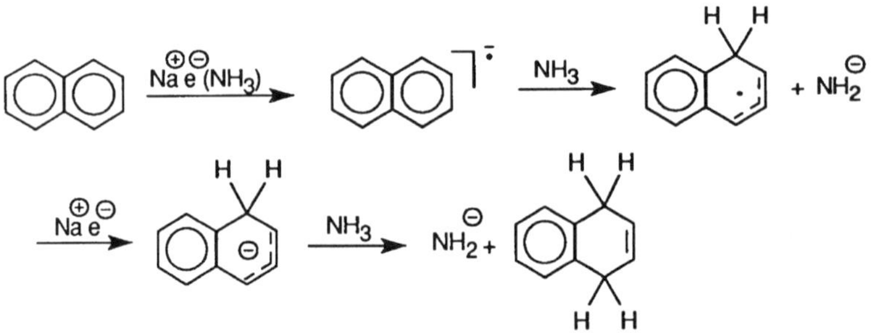

Figure 4.21 Birch reduction of naphthalene.

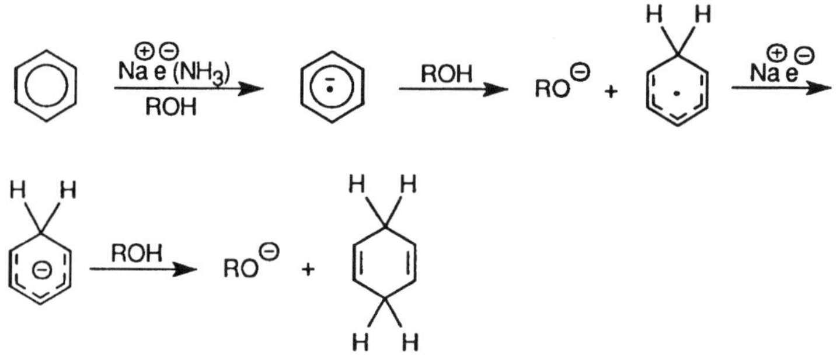

Figure 4.22 Birch reduction of benzene.

yields the corresponding carbanion, which is also preferentially protonated at the α position, giving 1,4-dihydronaphthalene instead of 1,2-dihydronaphthalene. Since product development effects would tend to favor 1,2-dihydronaphthalene (which has a styrenelike conjugated system), it is apparent that charge control is again dominant in the regiospecific protonation of the carbanion.

Birch reduction of benzene is more difficult, but can be smoothly accomplished when an appropriate proton donor such as *tert*-butyl alcohol or ammonium chloride is included with the substrate when it is added to the reducing solution (Figure 4.22).[19] Electron transfer to a benzene LUMO is apparently sufficiently endergonic as to require a stronger acid than ammonia, in order to protonate the benzene anion radical at a rate that is at least comparable to the very rapid back electron transfer to the medium. The intermediate radical is once again reduced to a carbanion, which, as a pentadienyl anion-like conjugated system, is preferentially protonated at the central carbon, giving 1,4-dihydrobenzene.

The Birch reduction of arenes having EDG-type substituents, including toluene and anisole, regiospecifically gives 3,6-dihydrobenzene derivatives as predicted for *o* or *m* protonation of an $A^{\bar{\bullet}}$ type of anion radical (Figure 4.23).[18,19] Initial protonation at either the *o* or *m* position, incidentally, eventually yields the same

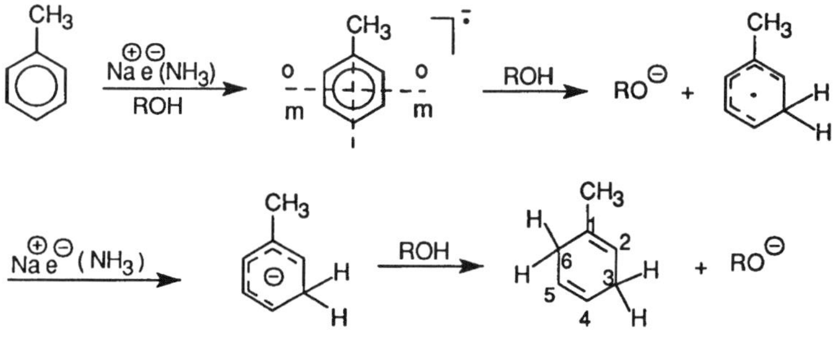

Figure 4.23 Birch reduction of toluene.

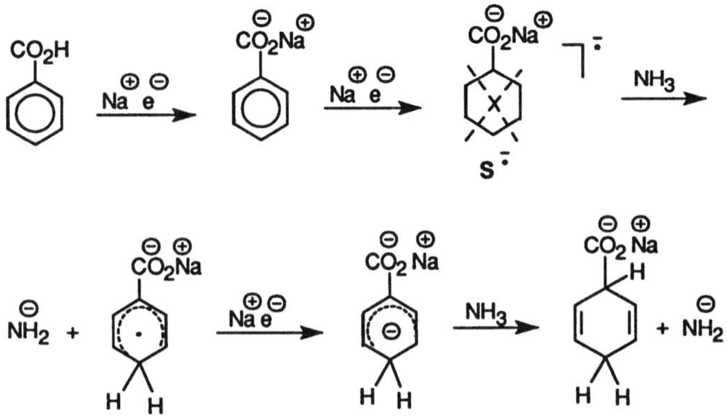

Figure 4.24 Birch reduction of benzoic acid.

product, although *m* protonation might be expected to be somewhat favored by steric effects.

The Birch reduction of benzoic acid, in contrast, yields the 1,4-dihydroarene and is considered to proceed via the $S^{\bullet-}$-type anion radical, which prefers protonation at the *p* or *ipso* positions (Figure 4.24). Once again, the same product is accessible irrespective of whether the initial protonation is *ipso* or *para*, but steric effects and product development would suggest preferential *para* protonation. The reaction has the additional complication of proceeding via the benzoate anion, which is formed in a rapid metal–acid reaction. The carboxylate anion, somewhat surprisingly, is nevertheless a mild EWG ($\sigma_p = 0.06$).

Conjugated dienes are easily reduced to internal alkenes (Figure 4.25). In the case of 1,3-butadiene, electron transfer apparently produces both the *s-cis* and *s-trans* anion radicals, since *cis* and *trans*-2-butene are produced in comparable amounts.[20] Protonation of both isomeric diene anion radicals occurs regiospecifically at a terminal carbon, as expected from charge effects. Protonation of the intermediate methylallyl carbanion also occurs regiospecifically at the terminus.

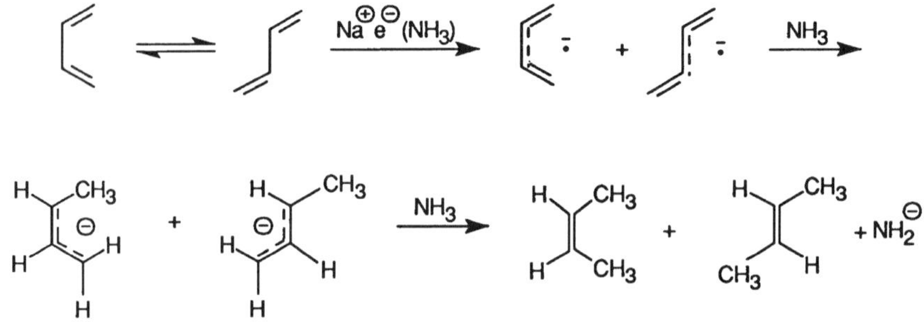

Figure 4.25 Birch reduction of 1,3-butadiene.

This may reflect a greater negative charge density at the primary allylic than at the secondary allylic position, but steric effects and product development effects would also appear to favor protonation at the primary carbon.

4.9 The Pinacol Coupling Reaction

The inherent basicity of anion radicals is nicely underscored by the Birch reaction. In effectively nonacidic media, however, a different reaction tendency, *viz.* coupling, is often featured. A familiar and useful example is the pinacol coupling reaction of ketones (Figure 4.26).[21] The reduction of benzophenone to its persistent blue anion radical (a ketyl) by metals has been mentioned previously. The ketyls of simpler ketones are, naturally, much less stable kinetically, and would be expected to undergo rapid and essentially irreversible radical-type coupling. It is instructive to consider the energetics of electron transfer to a ketone and the spin and charge

$$CH_3-C(=O)-CH_3 \xrightarrow[Et_2O]{Mg} [CH_3-C(=O^{\cdot -})-CH_3] \quad Mg^{+\cdot}$$

Acetone Ketyl

$$2\ CH_3-C(=O^{\cdot -})-CH_3 \longrightarrow (CH_3)_2C(O^{\ominus})-C(O^{\ominus})(CH_3)_2 \xrightarrow{H_3O^+} (CH_3)_2C(OH)-C(OH)(CH_3)_2$$

Pinacol

Figure 4.26 Pinacol coupling of acetone.

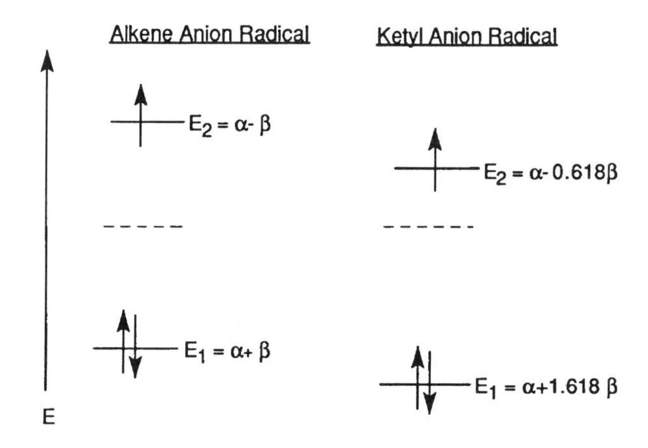

Figure 4.27 Ketyls versus alkene anion radicals.

distribution in ketone anion radicals (ketyls). The LUMO energy of a carbonyl π bond is lowered considerably from that of an alkene π bond as a result of the increased electronegativity of oxygen relative to carbon. In a typical HMO approach, the coulomb integral of oxygen (α_O) is taken as that of carbon (α_C) plus an increment of the reasonance integral (β), that is, $\alpha_O = \alpha_C + \beta$. At this level of approximation, the carbonyl LUMO energy is $E_{LUMO} = \alpha - 0.618\beta$, compared to the LUMO energy of ethene ($E_{LUMO} = \alpha - \beta$; Figure 4.27). That electron transfer to the carbonyl group should be more favorable than to an alkene is therefore eminently reasonable. The π moiety of a ketyl, like that of an alkene anion radical, is still viewed essentially as a three electron π bond. Although the bonding carbonyl MO (ψ_{BMO}) is heavily weighted towards oxygen, the antibonding carbonyl MO (ψ_{ABMO}) is equally heavily weighted toward carbon (Figure 4.28). Consequently the spin density, which is primarily determined by ψ_{ABMO}, which contains the odd electron, is greater on carbon than on oxygen ($\rho_C/\rho_O = 2.6$). The coupling of two ketone anion radicals at carbon is therefore eminently plausible. Of course, product development effects and the weakness of the O–O bond would also render O–O coupling unfavorable, although C–O coupling would not be affected quite as adversely.

It has been emphasized that spin density is controlled, exclusively at the HMO level and predominantly even in more sophisticated calculations, by the electron distribution of the SOMO. It has also been stressed that spin and charge are tightly coupled in simple hydrocarbon anion radicals, so that the negative charge

$$\psi_{BMO} = 0.525\phi_C + 0.851\phi_O$$

$$\psi_{ABMO} = 0.851\phi_C - 0.525\phi_O$$

Figure 4.28 The carbonyl BMO and ABMO.

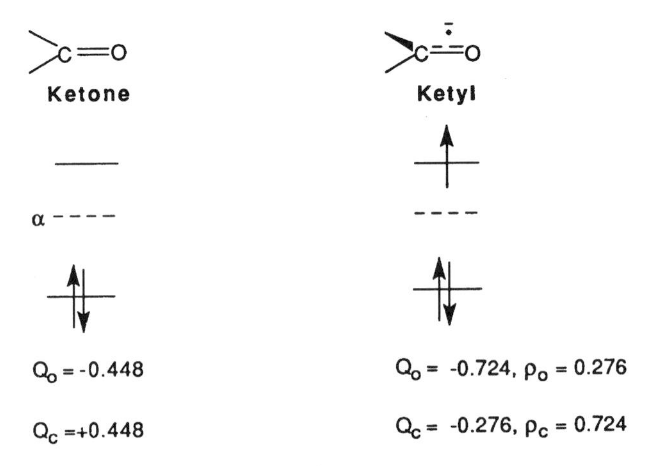

Figure 4.29 Spin and charge density in ketyls.

distribution is also controlled by the LUMO. This is, however, not generally the case with such heteroatom-containing anion radicals as ketyls. Thus the circumstance that the ketyl LUMO is heavily concentrated on carbon does not result in a larger negative charge density on carbon (Q_C) than on oxygen (Q_O). This follows from the fact that the heavy concentration of the ketone bonding MO on oxygen leaves the carbonyl carbon with a large partial positive charge in the neutral ketone (Figure 4.29). Consequently, the ketyl has more spin on carbon and more (negative) charge on oxygen.

4.10 The Acyloin Condensation

The reductive coupling of esters to α-hydroxyketones (acyloins) by means of an alkali metal is a traditional carbon–carbon bond forming reaction that is closely related to the pinacol coupling reaction (Figure 4.30).[22] In this case, ester anion radicals couple to give an intermediate dianion, which eliminates two alkoxide anions, affording the corresponding α-diketone. Electron transfer to the α-diketone yields first the α-diketone anion radical (called a semidione) and then the corresponding dianion. Aqueous workup then gives the acyloin.

4.11 Semidiones and Semiquinone

The anion radical of an α-diketone is often referred to as a semidione by analogy to semiquinone, the anion radical of quinone (Figure 4.31).[23] As in the case of conjugated dienes, addition of an electron to an α-diketone increases the bond order between the two carbonyl carbons, so that discrete *cis* and *trans* semidiones are

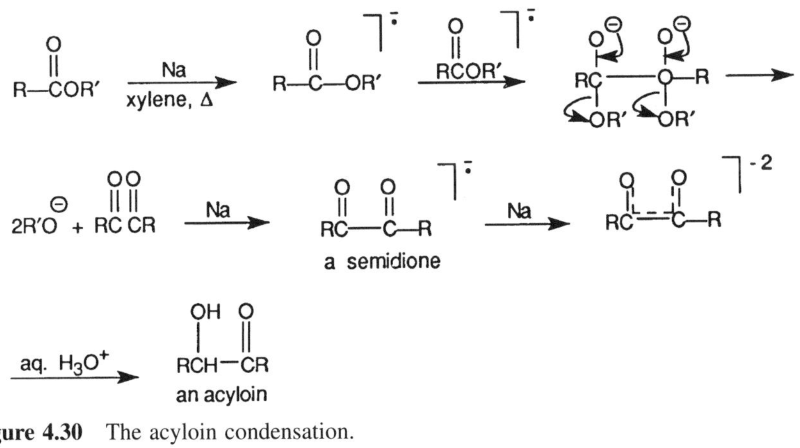

Figure 4.30 The acyloin condensation.

often discernible in ESR spectral studies. Russell's classic studies of semidiones have established that these persistent anion radicals are conveniently generated by the treatment of acyloins with *tert*-butoxide in dimethyl sulfoxide (Figure 4.32).[23] The reaction is presumed to involve the double deprotonation of the acyloin by the strongly basic medium to give the α-diketone dianion, which then *comproportionates* with the neutral diketone to yield the semidione (Figure 4.33).[24] With potassium *tert*-butoxide as the base and acetoin as the reactant, both *trans*- and *cis*-semidiones are generated, with the former being predominant ($\Delta\,\Delta H° = 2$ kcal/mol).[25] However, lithium *tert*-butoxide gives only the *cis*-semidione, owing to stronger chelation of this isomer to the smaller lithium ion. In THF or DME solutions, even potassium counterions strongly favor the *cis*-semidione as a result of ion pairing.

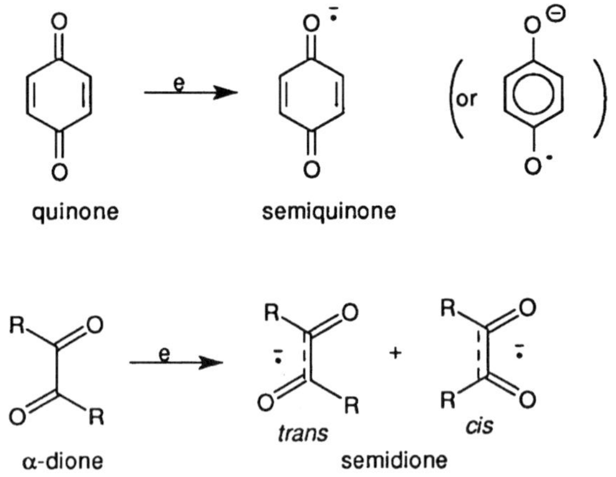

Figure 4.31 Semidiones and semiquinone.

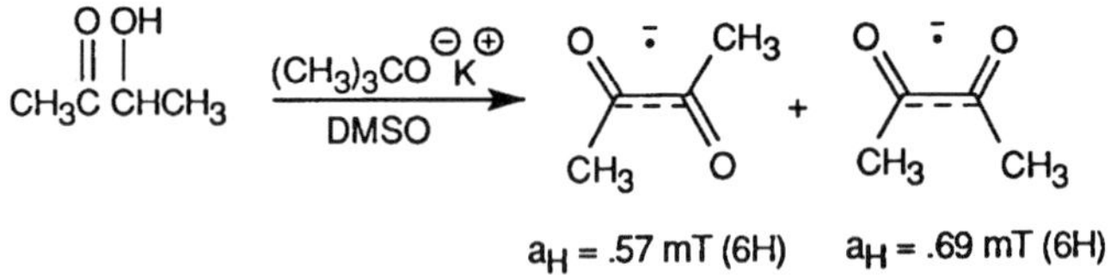

Figure 4.32 Generation of semidiones.

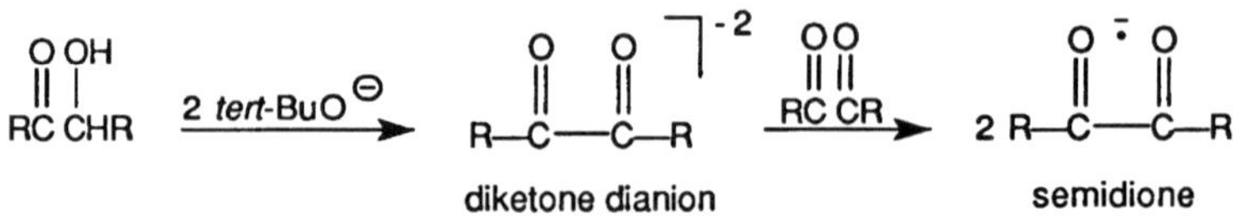

Figure 4.33 Mechanism of semidione formation.

Acyloins are especially desirable precursors for semidiones, but simple ketones can also be oxidized to semidiones in basic solution. Symmetrical ketones are preferred because they avoid the formation of semidione regioisomers. Cyclohexanone, for example, is converted to cyclohexane-1,2-semidione. At 25°C, this semidione shows a quintet splitting deriving from the four α hydrogens (a_H = 0.97 mT).[26] Upon cooling to −96°C in DME, discrete splittings from quasiaxial [a_H = 1.42 mT (2H)] and quasiequatorial [a_H = 0.69 mT (2H)] hydrogens are observed (Figure 4.34). The large difference between the hyperfine splittings of axial and equatorial β-hydrogens is a natural consequence of the a_β dependence upon $<\cos^2 \theta>$.

The ESR spectrum of bicyclo [2.2.1] heptane-2,3-semidione is especially interesting (Figure 4.35).[27] The long-range splitting from the H_7 (*anti*) proton is quite large (0.65 mT), especially in comparison to the small β splitting (a_{H1} = a_{H4} = 0.25 mT), and the H_7 (*syn*) proton (0.04 mT). A substantial long-range interaction with H_5(exo) and H_6(exo) is also apparent, and this is also highly stereospecific (no *endo* splitting is observed). That the homohyperconjugative interaction between the semidione moiety and the *anti*-C_7 hydrogen gives rise to a splitting as large as the β (hyperconjugative) splittings in the butane-2,3-semidiones is no less than startling. A relatively large long-range splitting would have been anticipated on the basis of the rigid and favorable geometry of the C_7–H (*anti*) bond for homohyperconjugative interaction with the semidione moiety and the circumstance that the spin density at both semidione carbons (C_2, C_3) can be transmitted to C_7–H (*anti*) by the simultaneous interaction of the $2p_z$ AOs at both C_2,C_3 with the backlobe of the C_7–H (*anti*) σ* ABMO. Further analysis reveals that the magnitude of the splitting is doubled by the presence of a quantum-mechanical effect (the Whiffen effect; see Chapter 3) that critically depends upon the fact that the semidione SOMO is symmetric (S).[28]

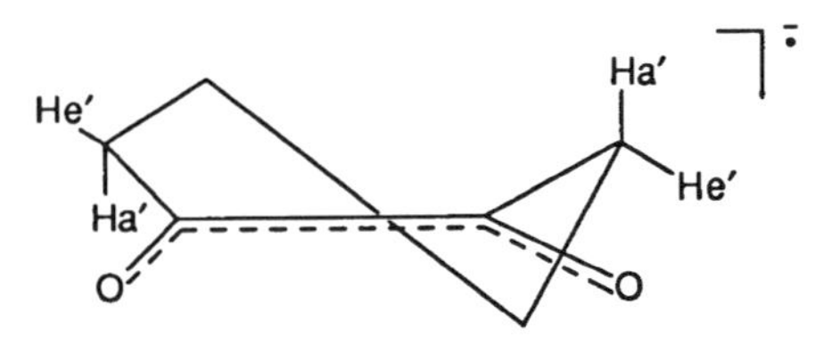

Figure 4.34 The cyclohexane-1,2-semidione.

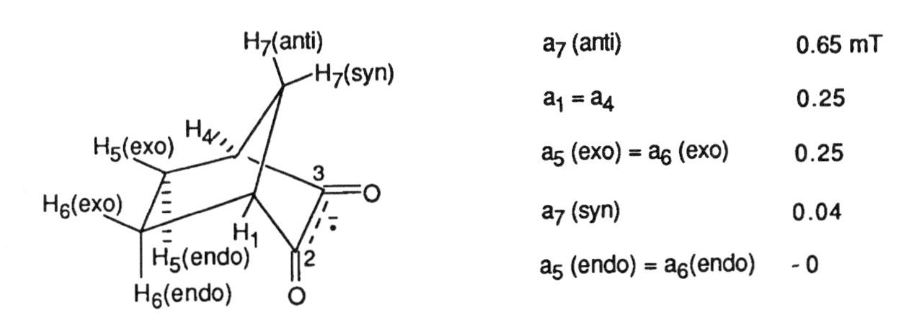

Figure 4.35 Bicyclo[2.2.1]heptane-2,3-semidione.

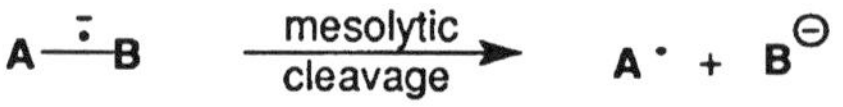

$$A \overset{\overline{\cdot}}{—} B \xrightarrow[\text{cleavage}]{\text{mesolytic}} A^{\cdot} + B^{\ominus}$$

Figure 4.36 Anion radical fragmentation.

$$R—X + \cdot\dot{Mg} \xrightarrow{Et_2O} [R\text{-}\overset{\overline{\cdot}}{-}\text{-}X]^{\ddagger}\ Mg^{\dot{+}} \longrightarrow \dot{R} + \overset{\ominus}{X}Mg^{\dot{+}} \longrightarrow R—\overset{\oplus\ \ominus}{MgX}$$

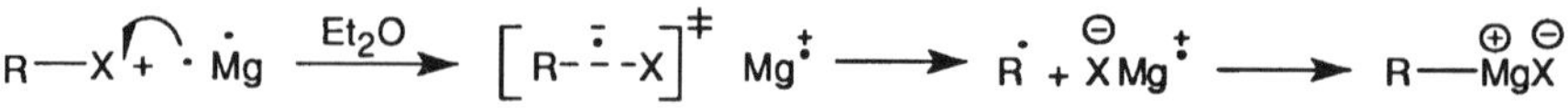

Figure 4.37 The making of a grignard reagent.

4.12 Fragmentation Reactions and Their Reversal

The fragmentation of an anion radical into an anion and a radical (called a mesolytic cleavage; Figure 4.36) is a plausible consequence of the bond weakening caused by the presence of an antibonding electron. Especially facile fragmentation occurs where a relatively stable anion or radical or both are formed, and where the SOMO is highly antibonding. The reaction of organic halides with magnesium is a particularly cogent example (Figure 4.37).[29] The magnesium surface contributes an electron to the antibonding (σ^*) orbital of the carbon–halogen bond.

Because of the highly antibonding character of the σ^* orbital and the possibility of forming a stable halide ion, the incipient anion radical probably fragments without activation; that is, electron transfer and dissociation are concerted.

4.13 The $S_{RN}1$ Reaction

An important addition to the mechanistic repertoire of organic chemistry, which involves the anion radical as a key intermediate, was discovered independently by Kornblum and by Russell in 1966.[30] The reaction of Figure 4.38 illustrates this novel mechanistic type, which has subsequently been designated the $S_{RN}1$ reaction (substitution, reductively initiated, nucleophilic, unimolecular).[31] Reaction is initiated by electron transfer (ET), presumably from the 2-nitro-2-propyl carbanion, to the LUMO of 4-nitrocumyl chloride. The powerfully electron-attracting nitro group strongly stabilizes this LUMO and facilitates the necessary ET. Subsequent loss of halide ion from the anion radical yields the 4-nitrocumyl radical. The designation $S_{RN}1$ refers to the unimolecular cleavage of the carbon–halogen bond in this anion radical. The next step of the reaction sequence (Step 3) is especially interesting in that it is the reverse of an anion radical fragmentation; that is, an anion

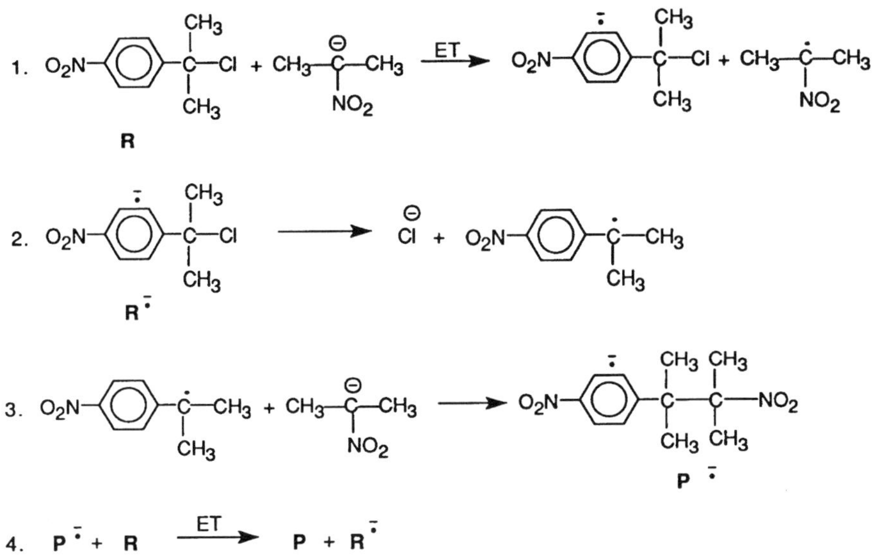

1. $O_2N{-}C_6H_4{-}C(CH_3)_2{-}Cl$ (**R**) $+ \; CH_3{-}\overset{\ominus}{C}(NO_2){-}CH_3 \xrightarrow{ET} [O_2N{-}C_6H_4{-}C(CH_3)_2{-}Cl]^{\bar{\cdot}} + CH_3{-}\dot{C}(NO_2){-}CH_3$

2. $[O_2N{-}C_6H_4{-}C(CH_3)_2{-}Cl]^{\bar{\cdot}}$ ($R^{\bar{\cdot}}$) $\longrightarrow Cl^{\ominus} + O_2N{-}C_6H_4{-}\dot{C}(CH_3){-}CH_3$

3. $O_2N{-}C_6H_4{-}\dot{C}(CH_3){-}CH_3 + CH_3{-}\overset{\ominus}{C}(NO_2){-}CH_3 \longrightarrow [O_2N{-}C_6H_4{-}C(CH_3)_2{-}C(CH_3)_2{-}NO_2]^{\bar{\cdot}}$ ($P^{\bar{\cdot}}$)

4. $P^{\bar{\cdot}} + R \xrightarrow{ET} P + R^{\bar{\cdot}}$

Figure 4.38 The $S_{RN}1$ reaction.

and a radical combine to give an anion radical, that of the product. The latter is neutralized in Step 4 by ET to a reactant 4-nitrocumyl chloride molecule. Steps 2, 3 and 4 of the reaction sequence of Figure 4.38 therefore constitute an anion radical chain reaction. The reaction is, apropriately, inhibited by dioxygen, cupric ion, and various radical scavengers. The $S_{RN}1$ reaction is not highly sensitive to leaving group ability and has been accomplished even with such unconventional leaving groups as NO_2^- and $PhSO_2^-$.

The $S_{RN}1$ reaction has also been observed with aryl halides, particularly in their reactions with potassium in liquid ammonia containing amide ion (Figure 4.39).[32] In this case solvated electrons are transferred to the aromatic LUMO, setting up the fragmentation. Reaction of the aryl radical with amide anion yields a substituted aniline anion radical. The latter is neutralized by ET to a reactant molecule, setting up the three-step propagation cycle of the chain process. Radical scavengers such as 2-methyl-2-nitrosopropane inhibit the reaction and instead allow substitution to occur via a benzyne mechanism (Figure 4.40). Once again, the use of leaving groups other than halogen, such as PhS^- and Me_3N, is feasible.

4.14 Pericyclic Reactions

All the anion radical reactions discussed involve the conversion of anion radicals to other species such as anions or radicals or both. A unique aspect of anion radical

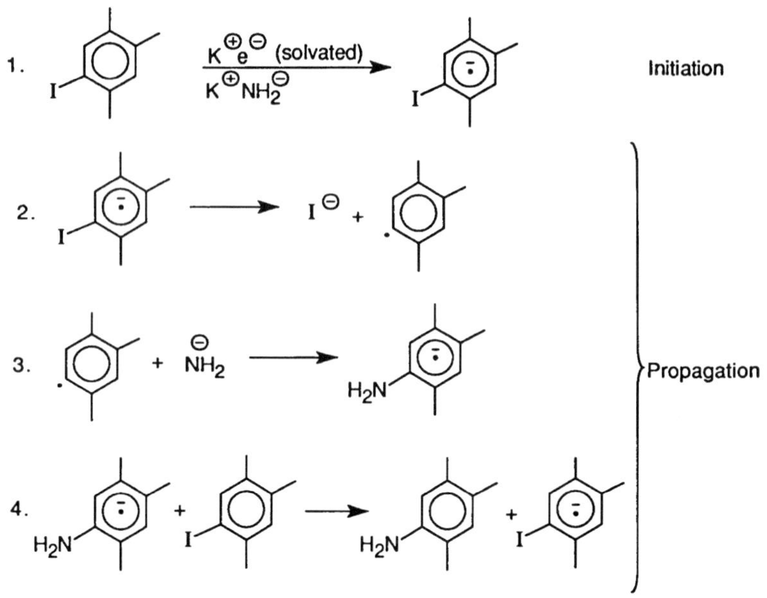

Figure 4.39 The $S_{RN}1$ reaction in the aromatic series.

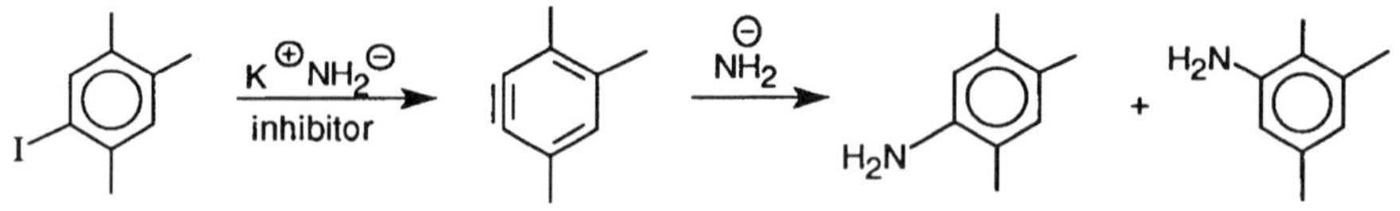

Figure 4.40 The benzyne substitution mechanism.

pericyclic chemistry is that it occurs entirely on the anion radical energy surface, converting a reactant anion radical to a product anion radical. If the SOMO of the product anion radical is lower in energy than the SOMO of the reactant anion radical, the anion radical pericyclic reaction has greater thermodynamic driving force than the corresponding reaction of the neutral reactant. The anion radical reaction may then be correspondingly accelerated to the extent the transition state has developed product character. Additional, purely kinetic, impetus for reaction may be available if, for some reason, the transition state has an unusually low SOMO energy for reasons other than product character development. The pioneering example of this kind was the retroelectrocyclic cleavage of the *cis*- and *trans*-3,4-diphenylbenzocyclobutene anion radicals (Figure 4.41).[33] The cleavages are conrotatorily stereospecific, as predicted by orbital correlation diagrams. The cleavage is extemely fast even at –78°C and produces the corresponding dianions by further reduction.

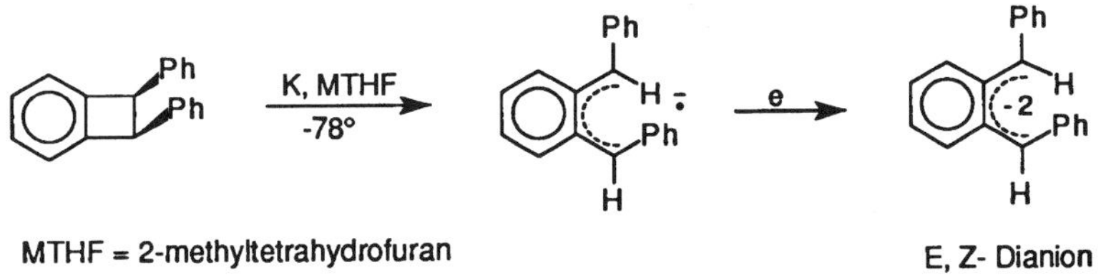

MTHF = 2-methyltetrahydrofuran

E, Z- Dianion

Figure 4.41 A retroelectrocyclic anion radical reaction.

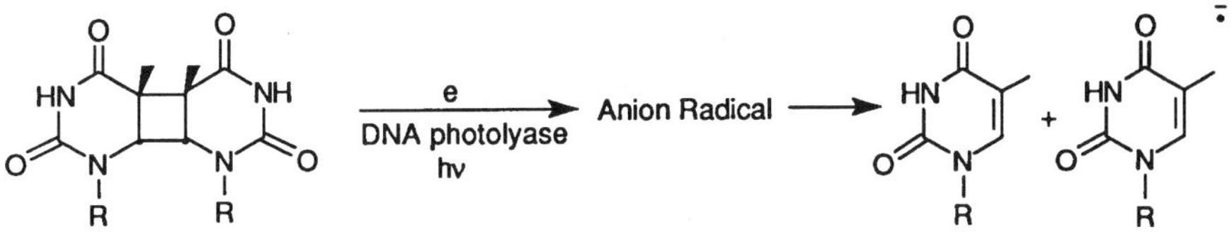

Figure 4.42 Anion radical retrocyclobutanation.

The most significant examples of anion radical pericyclic chemistry may be the retrocyclobutanation of the pyrimidine cyclodimers formed when DNA absorbs uv light. Repair of these photolesions is accomplished by DNA photolyase in the presence of *visible* light and appears to involve anion radical retrocyclobutanation (Figure 4.42).[34]

An example of an anion radical Diels-Alder addition reaction has also been proposed.[34]

References

General Reference

Kaiser, E. T.; Kevan, L. *Radical Ions*, John Wiley & Sons, New York, 1968.

Specific References

1. Weitz, E. Z. *Electrochem.* **1928**, *34*, 538. Weitz coined the term *anion radicals* and was the first to understand the true nature of ion radicals.
2. Schlenk, W.; Weickel, T. *Chem. Ber.* **1911**, *44*, 1182.
3. Levy, D. H.; Myers, R. J. *J. Chem. Phys.* **1964**, *41*, 1062.
4. Zhedulev, A.; Grand, A.; Ressouche, E.; Schweizer, J.; Morin, B. G.; Epstein, A. J.; Dixon, D. A.; Miller, J. S. *J. Am. Chem. Soc.* **1994**, *116*, 7243.
5. Garst, J. F. *Acc. Chem. Res.* **1971**, *4*, 400; Bank, S.; Juckett, D. A. *J. Am. Chem. Soc.* **1976**, *98*, 7742; House, H. O. *Modern Synthetic Reactions*, W. A. Benjamin, London, 1972.
6. Atherton, N. M.; Weissman, S. I. *J. Am. Chem. Soc.* **1961**, *83*, 1330.

7. Streitwieser, A., Jr. *Molecular Orbital Theory for Organic Chemists*, John Wiley & Sons, New York, 1961, p. 178.

8. Tuttle, T. R., Jr.; Weissman, S. I. *J. Am. Chem. Soc.* **1958**, *80*, 5342; Pople, J. A.; Beveridge, D. L.; Dobosh, P. A. *J. Am. Chem. Soc.* **1968**, *90*, 4201.

9. Bolton, J. R.; Carrington, A.; Forman, A.; Orgel, L. E. *Mol. Phys.* **1962**, *5*, 43.

10. Katz, T. J.; Strauss, H. L. *J. Chem. Phys.* **1960**, *32*, 1873.

11. Strauss, H. L.; Katz, T. J.; Fraenkel, G. K. *J. Am. Chem. Soc.* **1963**, *85*, 2360.

12. Garst, J. F. in *Free Radicals*; Vol. I, Kochi, J. K., Ed., John Wiley & Sons, New York, 1973, p. 518–519.

13. Rieger, P. H.; Bernal, I.; Reinmuth, W. H.; Fraenkel, G. K. *J. Am. Chem. Soc.* **1963**, *85*, 683.

14. Bauld, N. L.; Brown, M. S. *J. Am. Chem.. Soc.* **1965**, *87*, 5417.

15. Bauld, N. L.; Zoeller, J. H., Jr. *Tetrahedron Lett.* **1967**, *10*, 885.

16. Bauld, N. L. *J. Am. Chem. Soc.* **1964**, *86*, 2305.

17. Bauld, N. L.; Chang, C.-S.; Eilert, J. N. *Tetrahedron Lett.* **1973**, 153.

18. Birch, A. J. *Quart. Revs.* **1950**, *4*, 69.

19. Krapcho, A. P.; Bothner-By, A. A. *J. Am. Chem. Soc.* **1959**, *81*, 3658.

20. Bauld, N. L. *J. Am. Chem. Soc.* **1962**, *84*, 4347.

21. Walling, C. *Free Radicals in Solution*, John Wiley & Sons, New York, 1957, p. 584; March, J. *Advanced Organic Chemistry*, Fourth Edition, John Wiley & Sons, New York, 1992, p. 1225.

22. March, J. *Advanced Organic Chemistry*, Fourth Edition, John Wiley & Sons, New York, 1992, p. 1228.

23. Russell, G. A. in *Radical Ions*, Kaiser, E. T.; Kevan, L. Eds., John Wiley & Sons, New York, 1968, p. 87.

24. *Ibid.*, p. 95.

25. *Ibid.*, p. 104.

26. *Ibid.*, p. 109.

27. *Ibid.*, p. 140.

28. For analogous Whiffen effects in anion and cation radicals, see: Bauld, N. L.; Farr, F. R.; Hudson, C. E. *J. Am. Chem. Soc.* **1974**, *96*, 5634; Bauld, N. L.; Cessac, J. *Tetrahedron Lett.* **1975**, 3677; Cessac, J.; Bauld, N. L. *J. Am. Chem. Soc.* **1976**, *98*, 2712.

29. March, J. *Advanced Organic Chemistry*, Fourth Edition; John Wiley & Sons, New York, 1992, p. 624.

30. Kornblum, N.; Michel, R. E.; Kerber, R. C. *J. Am. Chem. Soc.* **1966**, *88*, 5662; Russell, G. A.; Danen, W. C. *J. Am. Chem. Soc.* **1966**, *88*, 5663.

31. Kornblum, N. *Angew. Chem. Int. Ed. Engl.* **1975**, *14*, 734; Bunnett, J. F. *J. Am. Chem. Soc.* **1981**, *103*, 7140.

32. Bunnett, J. F. *Acc. Chem. Res.* **1978**, *11*, 413; Galli, C.; Bunnett, J. F. *J. Am. Chem. Soc.* **1981**, *103*, 7140.

33. Bauld, N. L.; Chang, C.-S.; Farr, F. R. *J. Am. Chem. Soc.* **1972**, *94*, 7164; Bauld, N. L.; Hudson, C. E. *Tetrahedron Lett.* **1974**, 3147; Bauld, N. L. Cessac, J. *J. Am. Chem. Soc.* **1975**, *97*, 2284; Bauld, N. L. ; Cessac, J.; Chang. C.-S.; Farr, F. R.; Holloway, R. L. *J. Am. Chem. Soc.* **1976**, *98*, 4561.

34. Witmer, M.; Altmann, E.; Young, H.; Sancar, A.; Begley, T. P. *J. Am. Chem. Soc.* **1989**, *111*, 9264.

35. Borhani, D.W.; Greene, F. D. *J. Org. Chem.* **1986**, *51*, 1563; Bauld, N. L. in *Advances in Electron Transfer Chemistry*, Vol. 2; Mariano, P. S., Ed., JAI Press, Greenwich, 1992, p. 61.

Exercises

4.1 Benzophenone ketyl having a potassium counterion shows a ^{13}C hyperfine splitting of 0.93 mT from the carbonyl carbon. When the counterion is a divalent metal such as Mg^{+2}, the ^{13}C hyperfine splitting increases sharply to 1.58 mT.(a) Using resonance theory and canonical structures for the ketyl, rationalize the shift in spin density at the carbonyl carbon. (b) Rationalize this change using MO theory, assuming that the association of Mg^{+2} with the oxygen of the carbonyl group effectively converts it to a much more electronegative atom. Consider the qualitative effect of increased oxygen electronegativity on the coefficients of the carbon and oxygen atoms of the carbonyl group in the HOMO and in the SOMO. (c) Given this effect, would you expect the *cis*- or *trans*-biacetyl semidione to have the larger methyl hyperfine splitting? (Hirota, N. in *Radical Ions*, Kaiser, E. T., Keval, L., Eds., John Wiley & Sons, New York, 1968, p. 43.)

4.2 Methyl cyclopropyl semidione has a methyl hyperfine splitting of 0.588 mT, but the cyclopropyl methine splitting, which is also a β hyperfine splitting, is only 0.057 mT.

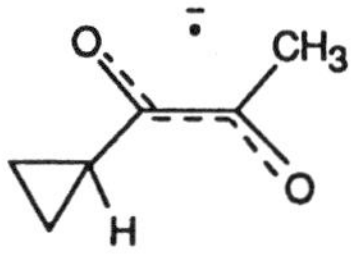

O CH3 O H

Explain the small methine splitting based upon a specific conformational preference, and suggest why this conformation might be preferred. (Russell, G. A. in *Radical Ions*, Kaiser, E. T., Kevan, L., Eds., John Wiley & Sons, New York, 1968, p. 108.)

4.3 The reduction of nonterminal alkynes to *trans*-alkenes under Birch conditions is a synthetically useful adjunct to the catalytic hydrogenation of alkynes, which yields *cis*-alkenes. Write a detailed mechanism for the reaction of 4-octyne with sodium in liquid ammonia and propose an explanation for the preference for the *trans*-alkene.

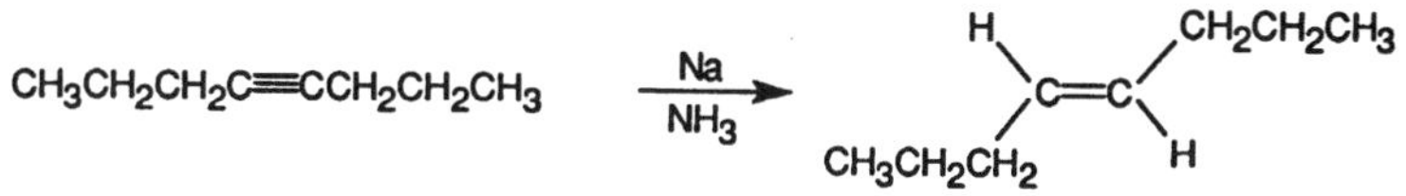

$$CH_3CH_2CH_2C{\equiv}CCH_2CH_2CH_3 \xrightarrow[NH_3]{Na} (H)(CH_3CH_2CH_2)C{=}C(H)(CH_2CH_2CH_3)$$

(Campbell, K. N.; Eby, L. T. *J. Am. Chem. Soc.* **1941**, *63*, 216 and 2683.)

4.4 The reaction given below is catalyzed by small amounts (1%) of sodium naphthalene or sodium metal, and is considered to be an anion radical chain reaction. Propose a detailed mechanism for the reaction. What kind of anion radical pericyclic reaction might be involved?

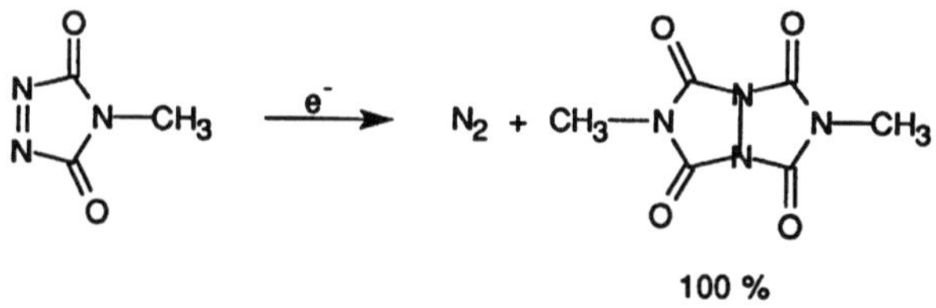

N_2 + 100 %

(Borhani, D. W.; Greene, F. D. *J. Org. Chem.* **1986**, *51*, 1563.)

4.5 As noted in the text, Birch reduction of naphthalene gives 1,4-dihydronaphthalene (**A**). However, further reduction is possible, especially when an alcohol or ammonium chloride proton source is used in conjunction with an excess of alkali metal. Propose a specific structure for $\mathbf{A}^{\bar{\bullet}}$, based upon a symmetric or antisymmetric benzenelike LUMO, and write a plausible mechanism for the reduction of **A** to a single tetrahydronaphthalene isomer, specifying and explaining the specific structure of the tetrahydro product.

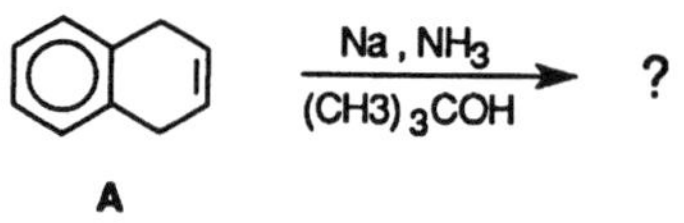

Na, NH_3 / $(CH3)_3COH$ → ?

A

4.6 (a) Birch reduction of *cis*- and *trans*-diphenylbenzocyclobutene gives a quantitative yield of *o*-dibenzylbenzene in each case. Write a plausible mechanism for these cleavage reactions.

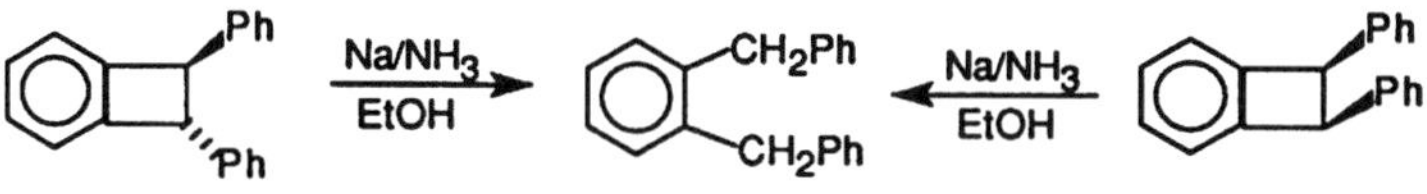

Na/NH_3, EtOH; CH_2Ph, CH_2Ph; Na/NH_3, EtOH

(b) When these benzocyclobutenes are reduced by potassium in 2-methyltetrahydrofuran (MTHF) at –78°C, no anion radicals are spectroscopically detected. Instead, bright red solutions that prove (NMR) to be isomeric dianions of diphenyl *o*-xylylene are formed. Quenching these dianions with dimethyldichlorosilane yields cyclic silanes of different *cis*/*trans* stereochemical composition depending upon the *cis* or *trans* geometry of the original benzocyclobutene. From the composition of the cyclic silanes, deduce the predominant stereochemistry of the dianion obtained from each benzocyclobutane. Why are the silanes not formed completely stereospecifically? What is the preferred stereochemistry of the electrocyclic cleavages of these benzocyclobutene anion radicals?

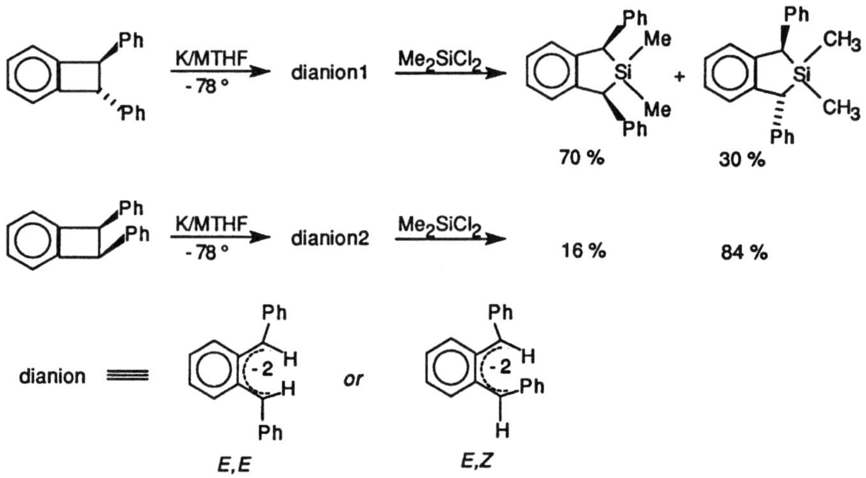

K/MTHF, −78°: diastereomer → dianion1 → Me_2SiCl_2 → 70 % + 30 %

K/MTHF, −78°: diastereomer → dianion2 → Me_2SiCl_2 → 16 % + 84 %

dianion ≡ E,E or E,Z

(Bauld, N. L.; Chang. C.-S.; Farr, F. R. *J. Am. Chem. Soc.* **1972**, *94*, 7164.)

4.7 The $S_{RN}1$ mechanism represents a plausible possible mechanism for the Sandmeyer and related reactions. Write an $S_{RN}1$ chain mechanism for the specific reaction shown below, including initiation and propagation steps.

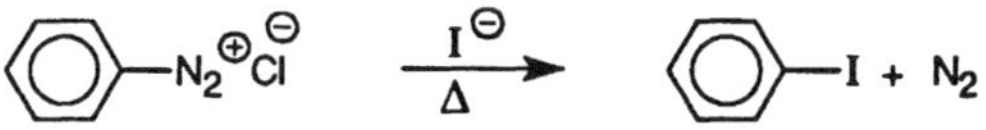

$$C_6H_5N_2^{\oplus}Cl^{\ominus} \xrightarrow[\Delta]{I^{\ominus}} C_6H_5I + N_2$$

4.8 The $S_{RN}2$ reaction is a reductively initiated (anion radical) version of the S_N2 reaction. While much less common than the S_N2 reaction, it represents a potential mechanistic alternative in certain rather specific circumstances. The $S_{RN}2$ reaction involves a nonchain, caged diradical pair mechanism:

$$N:^{\ominus} + R\text{—}L \xrightarrow{ET} \overline{N\cdot\ R\text{—}L^{\overset{-}{\cdot}}} \longrightarrow \overline{N\cdot\ R\cdot} + L^{\ominus} \xrightarrow{\text{coupling}} R\text{—}N$$

(a) The reaction of 2-bromooctane with trimethylstannyl sodium yields the expected substitution product, but the stereochemical result is 77% inversion. Provide an $S_{RN}2$ mechanism that is capable of rationalizing this stereochemical result.

$$CH_3(CH_2)_5CH(Br)CH_3 + Me_3Sn^{\ominus}Na^{\oplus} \longrightarrow CH_3(CH_2)_5CH(SnMe_3)CH_3 + Na^{\oplus}Br^{\ominus}$$

77% inversion
(23% racemization)

(c) Suggest a radical probe that should be able to detect the operation of an $S_{RN}2$ mechanism for a structurally similar secondary bromide. Indicate the structure of the product(s) that would be expected in an application of this probe. (Ashby, E. C. *Acct. Chem. Res.* **1988**, *21*, 414.)

CHAPTER 5

Cation Radicals

Ionization of an electron from a neutral molecule generates a unique chemical species known as a *cation radical* (or *radical cation*). The MO from which the electron is removed is usually the HOMO of the neutral precursor, which then becomes the SOMO of the cation radical (Figure 5.1). Essentially, the removal of an electron from the HOMO of a neutral molecule can be considered to leave a "hole" in the electron density. The "hole density" at atom i is controlled by $a_{\mathrm{SOMO},i}{}^2$ and

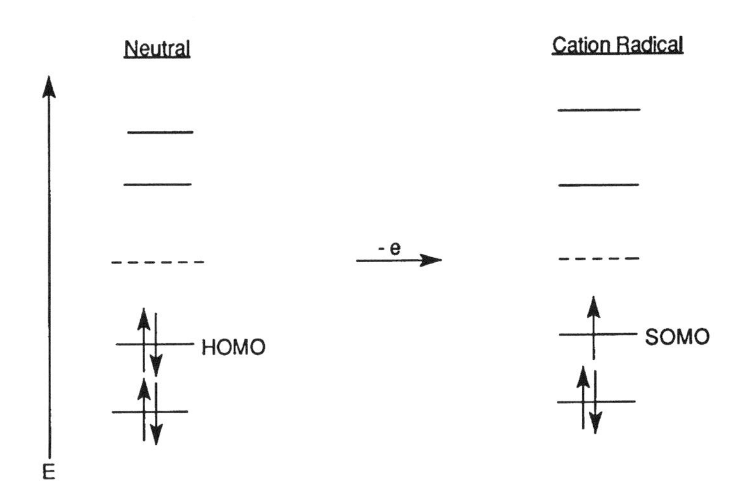

Figure 5.1 MO picture of cation radical formation.

consists, *simultaneously*, of *both* positive charge and odd electron density. However, as previously noted for anion radicals, spin and charge densities are substantially uncoupled (i.e., unequal) when the neutral precursor molecule is polar, since MOs other than the SOMO contribute substantially to the charge density in these cases.

5.1 Analogy to Anion Radicals: The Pairing Theorem

The corresponding cation radicals and anion radicals of alternant hydrocarbons are especially closely related (Figure 5.2). In such systems, the orbital energies of the HOMO and LUMO of the neutral molecules are exactly the negative of each other ($E_{HOMO} = -E_{LUMO}$) in the simplest MO approximation (HMO). Further, the absolute magnitudes of the coefficients of given atoms are identical in the HOMO and LUMO (i.e. $a_{i,HOMO} = \pm a_{i,LUMO}$). The HOMO and LUMO are considered to be paired orbitals, and their interrelationships are described as the Pairing theorem.[1] As a consequence of this theorem, the spin and charge densities in alternant hydrocarbon cation radicals are predicted to be the same as those in the

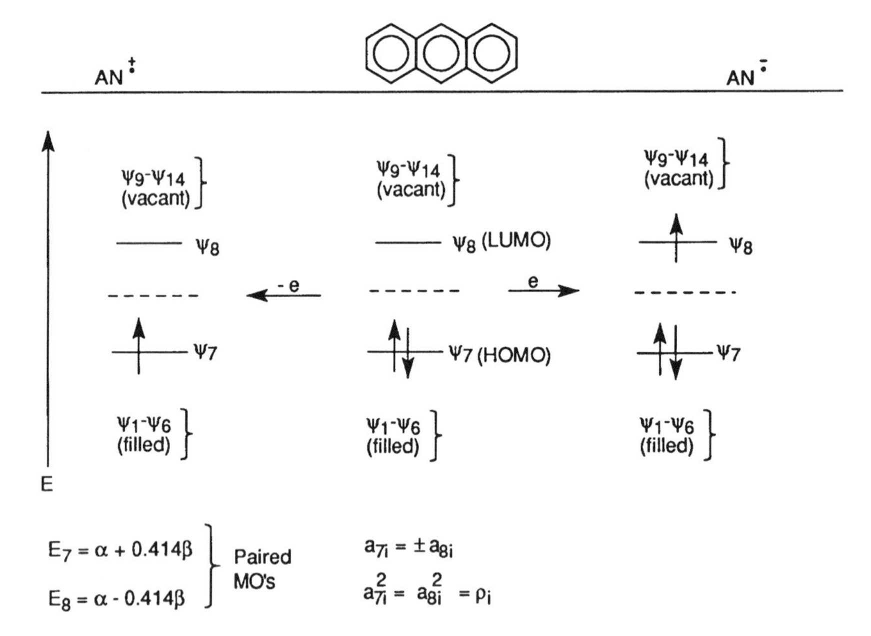

Figure 5.2 Pairing relationships between cation and anion radicals of alternant hydrocarbon conjugated systems.

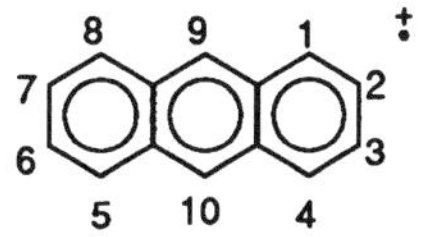

8 9 1 +·
7 2
6 3
5 10 4

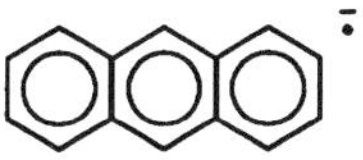

$a_1 = a_4 = a_5 = a_8 = -0.306$ mT
$a_2 = a_3 = a_6 = a_7 = -0.138$
$a_9 = a_{10} = -0.653$

−·

$a_1 = a_4 = a_5 = a_8 = -0.266$ mT
$a_2 = a_3 = a_6 = a_7 = -0.153$
$a_9 = a_{10} = -0.528$

^{13}C Hyperfine Splittings

$a_9 = 0.848$ mT
$a_1 = 0.450$ mT

$a_9 = 0.880$ mT
$a_1 = 0.459$ mT

SOMO Coefficients (HMO), Squared

$a^2_{SOMO,9} = 0.192 = \rho_9$
$a^2_{SOMO,1} = 0.096 = \rho_1$
$a^2_{SOMO,2} = 0.047 = \rho_2$

Figure 5.3 Proton hyperfine splittings (a_i) of the anthracene cation and anion radicals: the pairing theorem.

corresponding anion radical (i.e., $a^2_{i,\mathrm{HOMO}} = a^2_{i,\mathrm{LUMO}} = \rho_i$). However, the value of the proportionality constant (Q_i) in the McConnell equation is somewhat larger for cation than anion radicals, so that the cation radicals tend to have somewhat larger α proton hyperfine splittings (Figure 5.3).[2] Nevertheless, the hyperfine splittings at particular positions in the cation and anion closely parallel the squares of the SOMO coefficients. The ^{13}C hyperfine splittings make the same point, but even more impressively.[3] The uv-visible absorption spectra of paired cation and anion radicals, incidentally, are also strikingly similar.[4]

5.2 Historical

The first stable cation radical salt was isolated well before the discovery of the first stable radical or carbocation. Wurster isolated his now well-known salts (**1a,b**; Figure 5.4) in 1879,[5] fully 21 years before Gomberg discovered the trityl radical[6] and 23 years before von Baeyer isolated the trityl carbocation.[7] Wieland prepared triarylaminium tribromides (**2a**) in 1907,[8] but it was not until 1926 that Weitz's research on the corresponding aminium perchlorate (**2b**) clarified the actual nature of these substances as monomeric species in which the cationic component has both an unpaired electron and a single unit of positive charge.[9] Weitz coined both of the now common terms *cation radical* and *aminium* ion. Although the perchlorate salts

Wurster's Salts

Triarylaminium Salts

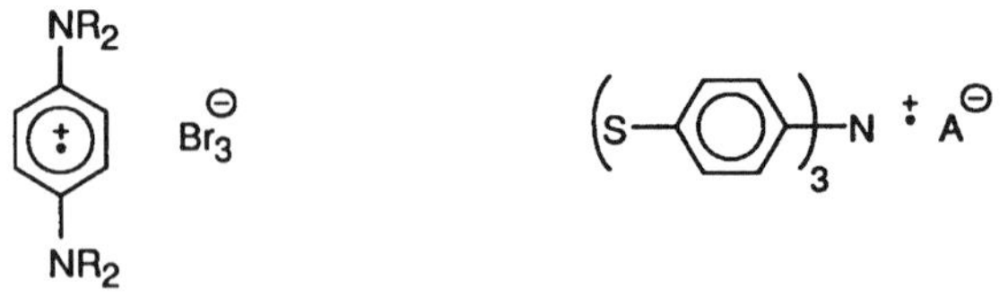

1a: R = H (Red)
1b: R = Me (Blue)

2a: S = CH_3, $A^{\ominus}$ = $Br_3^{\ominus}$ (Wieland)
2b: S = CH_3, $A^{\ominus}$ = $ClO_4^{\ominus}$ (Weitz)
2c: S = Br, $A^{\ominus}$ = $SbCl_6^{\ominus}$ (Walter)

Figure 5.4 Historically important cation radicals.

of *para*-substituted triarylaminium salts are isolably stable,[9, 10] the corresponding hexachloroantimonate salts are much more shelf stable, and the tris(4-bromophenyl) aminium hexachloroantimonate salt (**2c**) is commercially available and conveniently preparable.[11]

5.3 Scope of Cation Radical Formation

Ionization or, in solution, one-electron oxidation is one of the most basic of all chemical processes. Given a suitably energetic chemical or physical agent, virtually any neutral molecule can be ionized. A simple 1:1 correspondence of neutral molecules and their corresponding cation radicals is not quite maintained, however. Some molecules undoubedly ionize dissociatively, yielding a carbocation fragment and a neutral radical fragment. This may well be the case for even such a simple molecule as neopentane (which fragments to the *tert*-butyl cation and methyl radical; Figure 5.5). The cleavage of a bond of a cation radical to give a radical and a cationic fragment has been termed a *mesolytic* cleavage to distinguish it from the homolytic and heterolytic cleavage modes.[12] On the other hand, cation radicals can

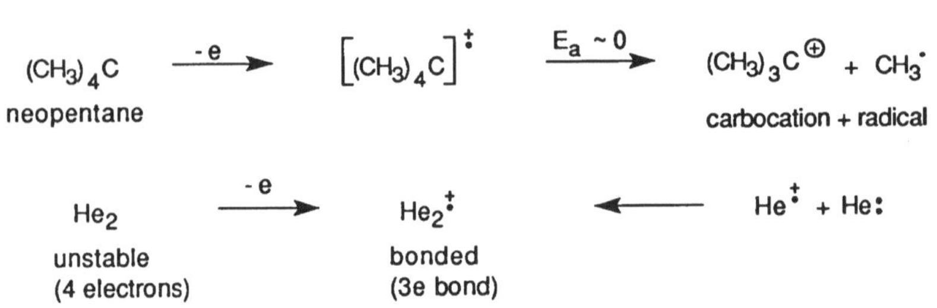

$$(CH_3)_4C \xrightarrow{-e} [(CH_3)_4C]^{+\cdot} \xrightarrow{E_a \sim 0} (CH_3)_3C^{\oplus} + CH_3^{\cdot}$$

neopentane — carbocation + radical

$$He_2 \xrightarrow{-e} He_2^{+\cdot} \longleftarrow He^{+\cdot} + He:$$

He_2: unstable (4 electrons); $He_2^{+\cdot}$: bonded (3e bond)

Figure 5.5 Dissociative and associative ionization.

potentially exist for species that would be dissociative on the neutral energy surface. The dihelium cation radical ($He^{\overset{+}{\bullet}}$), which has a three-electron bond, is presumably such an example.

5.4 Simple Inorganic Cation Radicals

The dihydrogen cation radical is the prototype example of a molecular cation radical. It also provides a strikingly simple illustration of the stabilization provided by a one-electron bond (Figure 5.6). The ionization of an electron from the BMO of molecular hydrogen naturally results in a lengthening and weakening of the bond, but the bond dissociation energy (*D*) for the mesolytic dissociation to a proton and a hydrogen atom is nevertheless impressive (D = 64.4 kcal/mol). In a sense, the dioxygen cation radical is even more impressive than $H_2^{\overset{+}{\bullet}}$, since it can be prepared as stable salts (e.g., $O_2^{\overset{+}{\bullet}}\ SbF_6^-$).[13]

Cation radical formation from *triplet ground state* O_2 involves ionization of an electron from one of the two π^* (antibonding) MOs, thus *increasing* π bonding. The hydrazine cation radical is impressively highly stabilized by a three-electron π bond.[14] The maximization of π bonding apparently results in sp^2 hybridization for both nitrogen atoms and a fully planar cation radical.[15] This inference is strongly supported by the small nitrogen hyperfine splittings in the ESR spectrum of the hydrazine cation radical [a = 1.15 mT (2N)], indicating that the SOMO has little or no *s* character. The cation radical of ammonia has a nitrogen hyperfine splitting of 1.97 mT and is also considered to have an sp^2-hybridized nitrogen atom with a $2p_z$ SOMO.[16] If the SOMO were an sp^3 orbital, a hyperfine splitting of ca. 8.0 mT would be expected.[17] Since the SOMO of the hydrazine cation radical is delocalized over two nitrogen atoms, the nitrogen hyperfine splitting would be expected to be approximately one-half that in the ammonia cation radical. The exceptional stability of many stable organic cation radicals is based directly or indirectly upon the prototype three-electron bonding in the hydrazine cation radical. In fact, the Wurster salts can be considered *benzologues* of a hydrazine cation radical.

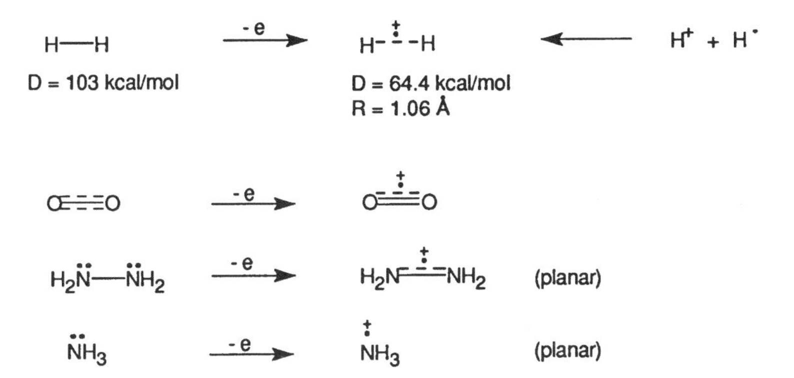

Figure 5.6 Inorganic cation radicals.

5.5 Classification of Cation Radicals

Ionization of an electron from an atom (such as He) that has an even number of electrons is capable of generating an atomic cation radical ($He^{\overset{+}{\bullet}}$). An example of an atomic cation radical that is of especial importance in synthetic organic chemistry is the magnesium cation radical (Mg^{I}), which is considered to be an intermediate in Grignard-forming reactions (Figure 5.7). Molecular cation radicals (also called molecular ions) are of primary interest in this chapter, and here two subtypes are often distinguished. Most of the more familiar cation radicals are of the π type, in which the SOMO is a π MO (*vide supra*). Cation radicals of the σ type have a SOMO that corresponds to a σ-type MO or an n (nonbonding) atomic orbital in the σ plane of a π-bond-containing system (Figure 5.8). A cation radical in which the positive charge is more or less completely uncoupled from the spin is termed *distonic*. An especially interesting example of a distonic cation radical is the "isomethanol" cation radical, which is found to be more stable than the cation radical of methanol.[18] Cation radicals in which a carbocation site is "insulated" (e.g., by one or more methylene groups) from a radical site (Figure 5.8) are also often considered to be distonic.

Finally, cation radicals are sometimes characterized as *localized* or *delocalized*, depending upon whether the SOMO is highly concentrated in a bond between two atoms (or even on a single atom) or whether it is extensively delocalized over more than two atoms.

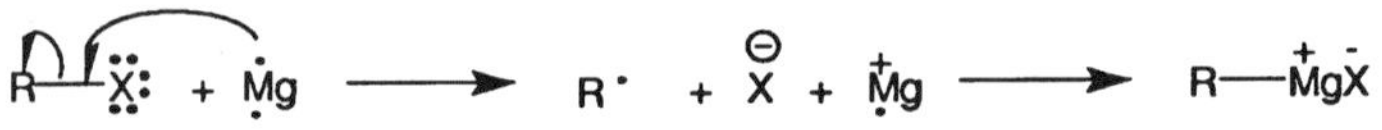

$$R\text{—}\ddot{X}\text{:} + \dot{Mg} \longrightarrow R^{\bullet} + \overset{\ominus}{X} + \overset{+}{\dot{Mg}} \longrightarrow R\text{—}\overset{+}{Mg}\overset{-}{X}$$

Figure 5.7 Mg^{I}: An atomic cation radical.

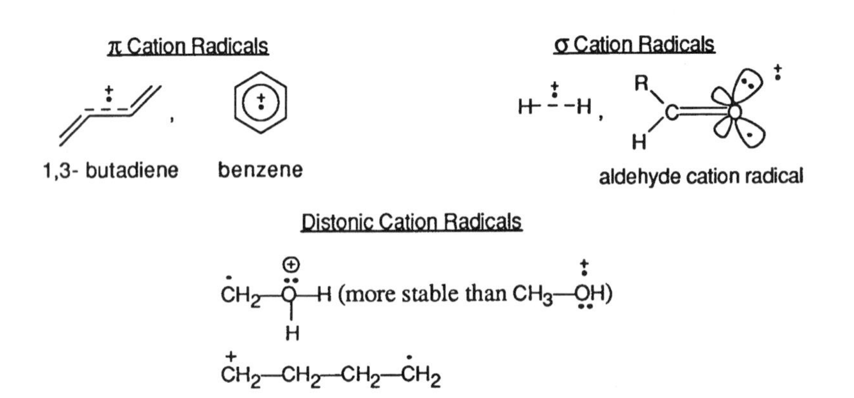

Figure 5.8 Classes of molecular cation radicals.

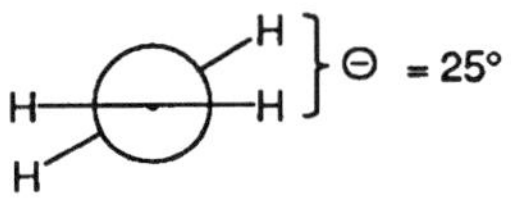

Figure 5.9 Novel structural features of cation radicals.

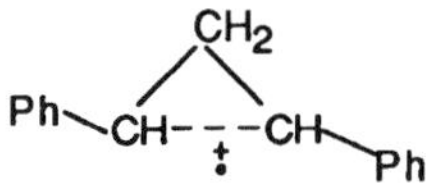

5.6 Structure

The removal of an electron from a bonding MO is inevitably structure-weakening. In the extreme case, as noted previously, there may be no energy minimum at all on the cation radical surface for a geometry having the connectivity of the reactant, so that dissociation or rearrangement occurs without activation. At the other extreme are large, highly delocalized π cation radicals (e.g., the anthracene cation radical), where only relatively small changes in bond lengths and angles accompany ionization. In intermediate cases, where the SOMO is relatively localized, the cation radical structure may differ significantly from that of the neutral precursor, even though the basic connectivity remains the same. Dramatic examples of unique structural features of cation radicals are the ethene cation radical, which has the highly twisted structure **3**,[19] and the 1,2-diphenylcyclopropane cation radical, which has the long (ca. 1.9 Å), one-electron bond structure **4** (Figure 5.9).[20] Another complication that arises when cation radical equilibrium geometries differ substantially from those of the corresponding neutrals is that the molecular orbital that is the HOMO of the neutral at its equilibrium geometry may not be the HOMO at the cation radical geometry. In such cases, the HOMO of the neutral does not correspond to the SOMO of the cation radical (see the discussion on the ethane cation radical, *vide infra*).

5.7 Reactivity

Since ionization is structure-weakening and energy-increasing, it is not unexpected that cation radicals are much more reactive and less stable than corresponding neutrals. They also appear, generally, to be more reactive than corresponding anion radicals. For example, the 1,3-butadiene anion radical has been observed (ESR) in solution at –78°C, but the corresponding cation radical apparently can only be observed in frozen matrices. Similarly, the benzene anion radical is easily observed (even visually) under a variety of conditions, but the benzene cation radical requires matrix isolation at low temperatures. The naphthalene anion radical is persistent enough to be useful as a reducing agent in organic synthesis, but, in solution, the corresponding cation radical is transient and usually is observed spectroscopically as its dimeric complex with neutral naphthalene.[21] On the other hand, the reactivity of cation radicals would appear to be more nearly comparable to that of radicals or of carbocations. Like carbocations, cation radicals react readily with nucleophiles, and, like radicals, they can undergo coupling (to give dications). However, the most interesting and useful chemistry of cation radicals may be their unique pericyclic chemistry. This latter chemistry is especially effective in that it often occurs via a cation radical chain or catalytic mechanism, and therefore does not require stoichiometric generation of cation radicals.

5.8 Chemical Methods of Preparation

The famous Wurster's salts were originally prepared by the oxidation of the appropriate 1,4-bis(dialkylamino)benzene with excess bromine. Since then, many other chemical oxidants have been used to generate cation radicals, of both the stable and reactive varieties, from neutral substrates. The triarylaminium salts are most readily prepared by the oxidation of a triarylamine with antimony pentachloride in dichloromethane, as shown by Walter. The tris(4-bromophenyl)a-minium hexachlorantimonate salt (**2c**) is by far the most readily available (synthetically or commercially), and also probably the most shelf stable, of all the stable cation radical salts. Using this salt, cation radicals of a variety of other substrates, even including many whose oxidation potentials are as much as 0.5 eV greater than that of tris(4-bromophenyl)amine (E_{ox} = 1.05 V vs. SCE),[22] can be generated in amounts sufficient to sustain cation radical chain or catalytic processes. By analogy to the triphenylmethyl radical, which is isoelectronic with the triphenylaminium cation radical, isolably stable triarylaminium salts require that all three *para* positions be substituted in order to retard radical coupling at this position. Tris(2,4-dibromophenyl)aminium hexachloroantimonate (**5**) is a higher potential (E_{ox} = 1.50 V)[22] aminium salt that is conveniently prepared in a manner similar to **2c**, but that is somewhat less shelf stable (Figure 5.10). The hexachloroantimonate counterion is particularly effective in achieving stabilization of these aminium salts in comparison, for example, with the tetrafluoroborate and perchlorate salts, which are not especially shelf stable. The perchlorate salt of thianthrene (**6**) is, however, another example of a readily prepared, isolably stable cation radical salt, and this, too, can be used to generate the cation radicals of other substrates (Figure 5.10).[23]

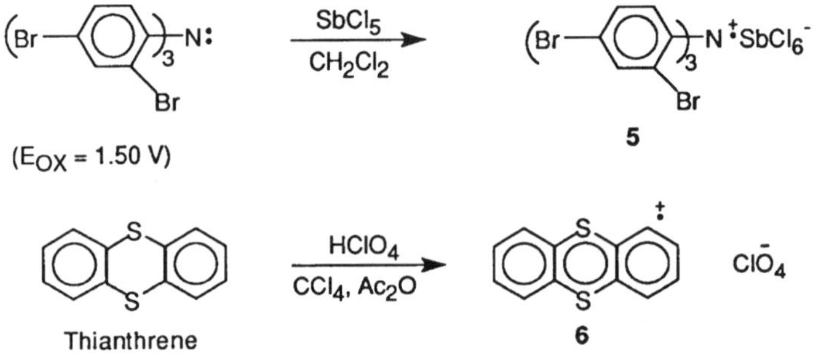

(Br–C₆H₃(Br)–)₃N: —SbCl₅, CH₂Cl₂→ (Br–C₆H₃(Br)–)₃N⁺• SbCl₆⁻ (5)

(E_{OX} = 1.50 V)

Thianthrene —$HClO_4$, CCl_4, Ac_2O→ **6** ClO_4^-

Figure 5.10 Other conveniently prepared, stable cation radical salts.

Persistently (but usually not isolably) stable cation radicals of larger aromatics such as anthracene and perylene are generated in many (Brønsted or Lewis) acidic media such as concentrated sulfuric acid, trifluoroacetic acid (especially in daylight), $AlCl_3/CH_2Cl_2$, $SbCl_3/CH_2Cl_2$, etc.[24] Reactive, transient cation radicals can often be generated by treatment of the substrate with ferric ion (or other oxidizing ions) or metal complexes such as tris(phenanthrolene)iron^{III} complexes.[25]

A variety of cation radicals have been generated on zeolite surfaces,[26] where they often are surprisingly long-lived, or on montmorillonite clays,[27] on semiconductor surfaces (TiO_2),[28] and by cation radical polymers.[29]

5.9 Physical, Photochemical, and Electrochemical Methods

The ionization of virtually any molecule can be accomplished by γ radiation of the substrate in a frozen matrix. The immobilization of the cation radical often permits the study (ESR, optical spectroscopy) of even very reactive cation radicals. Pulse radiolysis (bombardment with accelerated electrons) of a substrate in solvents such as dichloromethane yield ionized solvent molecules (solvent holes) that, in turn, ionize the substrate by "hole transfer." Cation radicals are also effectively generated by electrochemical oxidation, laser excitation, and photosensitized electron transfer, among other methods. Several of these methods to generate and study cation radical structures will be illustrated in this chapter. Photosensitized electron transfer, in particular, will be considered in detail in Chapter 6.

5.10 Cation Radicals of Small Organic Molecules

A combination of ESR and MO studies of the cation radical of the parent alkane firmly establishes the structure as **7** (Figure 5.11), which resembles a carbene fragment complexed to a hydrogen molecule. The H_b–C–H_b bond angle (116°) calculated by MO theory suggests an sp^2-hybridized carbon, with the SOMO being confined to the H_a–C–H_a fragment.[30] These latter C–H bonds are elongated, and the H_a–C–H_a angle is sharply compressed (59°), so that the two hydrogens of this fragment approach each other rather closely (1.16 Å). The SOMO (**8**) involves only

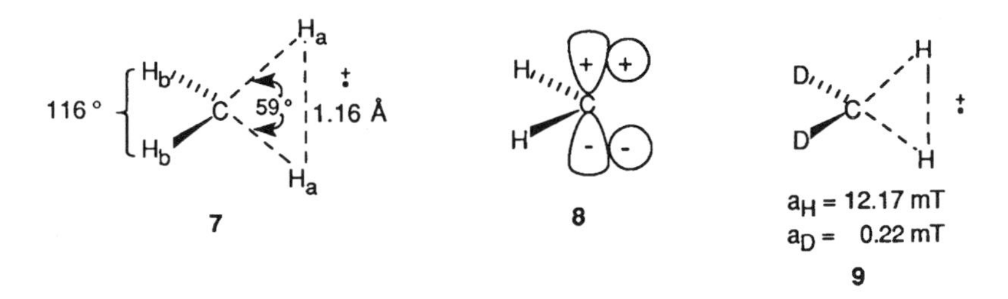

Figure 5.11 The methane cation radical.

the $2p_z$ carbon AO and the two H_a AOs. Consequently, a large (positive) triplet hyperfine splitting (13.0 mT) is predicted from the H_a's. Because the spin on carbon is in a $2p_z$ AO, the α-type H_b splittings are predicted to be small and negative (−1.7 mT). The ESR spectrum of $CH_4^{+\bullet}$, generated by γ radiolysis in a frozen matrix, actually exhibits a quintet splitting (a = 5.48 mT), from four equivalent protons. That this is the result of a dynamic equilibration process is shown by the spectrum of CD_2H_2 (**9**). In this case, the "stronger" C–D bonds preferentially assume the b positions, and dynamic equilibration is not a factor. The protons now exhibit the predicted large triplet splitting (a_H = 12.17 mT), while the deuterons give rise to a much smaller splitting (a_D = ±0.22 mT). This deuterium hyperfine splitting constant (which is presumably negative) is equivalent to a splitting a = −1.46 mT on the proton scale. The average of these two splittings, viz. [2(12.17) + 2(−1.46)]/4, is 5.5 mT, virtually exactly that observed for the dynamically equilibrating methane cation radical.

The structure of the ethane cation radical in a frozen matrix at 4.2 K somewhat resembles the transition state for an *anti* E2 elimination (**10**; Figure 5.12).[31] The SOMO is mainly concentrated in two *anti* C–H_a bonds, which are distinctly elongated, while the corresponding H_a–C–C bond angles are compressed in a manner analogous to diborane. The C–C bond is slightly shortened, corresponding to the removal of an electron from an MO that is antibonding between the carbon atoms. The spectrum is a triplet, with a very large splitting from two protons. Molecular-orbital calculations of very high quality indicate that **10** (which is designated as the ^2A_u state) and the $^2A_{1u}$ state **11** are no more than a few *tenths* of a kilocalorie different in energy.[32] The latter state would be especially interesting since the SOMO is primarily concentrated in the C–C bond, which is substantially elongated (1.90 Å), thus engendering a one electron C–C bond analogous to that in $H_2^{+\bullet}$. The extremely small energy difference between the ^2A_u and $^2A_{1u}$ states suggests that either could be the ground state in the gas phase, and a dynamic equilibrium between the two is a plausible possibility. It has been suggested that the ^2A_u state (**10**) may be favored in a matrix environment because the matrix host resists the C–C bond elongation of the $^2A_{1u}$ state, with the concomitant increase in the volume occupied by the guest.[33] Further, the charge density in the favored ^2A_u

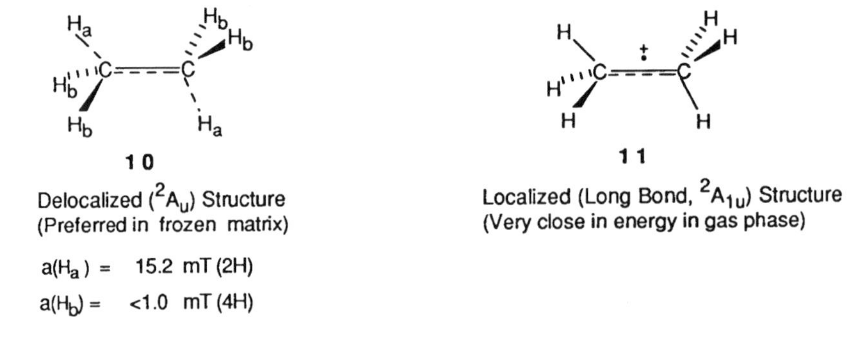

Figure 5.12 Two possible states of the ethane cation radical.

state is primarily concentrated on the periphery of the molecule, where stabilizing host–guest interactions should be strongest. It is of interest that the HOMO of neutral ethane at the optimal ethane geometry is the SOMO of **10**, rather than **11**. Nevertheless, the relaxation of the two states to their equilibrium geometries involves substantial geometric changes, and these cause a reordering of the MO energies such that the two states have different SOMOs but are still extremely close in energy. The question of *delocalized* versus *long-bond* (localized) structures arises frequently in studies of alkane cation radicals.

The ethene cation radical also has an unusual structure. Removal of an electron from the π HOMO of ethene weakens the π bond substantially and permits twisting to take place, affording a nonplanar cation radical structure.[19] This is encouraged by relief of torsional strain between vicinal C–H bonds, and is partially compensated for by the development of hyperconjugative overlap between the $2p_z$ AOs of each carbon with the C_β–H bonds of the other, which interaction is absent in the planar cation radical. The ground-state structure has a 25° twist, but nevertheless resists further torsion to the perpendicular state, which would allow the interconversion of *cis-* and *trans-*alkene isomers. That is, the one-electron π bond is easily twisted but not so easily rotated. More extensive delocalization of the SOMO, as in 1,3-butadiene and styrene, favors a planar cation radical structure. Actually, twisted structures appear not to be very general even for simple alkenes.[19]

5.11 Cation Radical Reactions: Electron (Hole) Transfer

Perhaps the most characteristic reaction of cation radicals is the acceptance of an electron by the SOMO to form a neutral molecule, that is, the reverse of the usual method of cation radical generation. When the source of the electron is another neutral molecule, the original cation radical is neutralized and a new cation radical is formed (Figure 5.13). This reaction is a key to efficient cation radical chain chemistry in that the energy released in neutralizing a product cation radical ($P^{\dot{+}}$) can be used to effect the ionization of a new reactant molecule (R). The process can be described as an electron transfer (ET) from R to $P^{\dot{+}}$ or, more specifically, for the case of a cation radical–neutral electron transfer, as a hole transfer (HT) from $P^{\dot{+}}$ to R. The feasibility of a particular hole transfer reaction can be quantitatively assessed from the reversible oxidation potentials (E_{ox}) or approximately from peak oxidation potentials (E_P) of the neutral reactant and

$$P^{\dot{+}} + R \xrightarrow[\text{HT}(P^{\dot{+}} \rightarrow R)]{\text{ET}(R \rightarrow P^{\dot{+}})} P + R^{\dot{+}}$$

$$\Delta G^0 = E_{OX}(R) - E_{OX}(P)$$

Figure 5.13 Electron (hole) transfer.

product molecules, as indicated in Figure 5.13. HT is especially efficient when the reactant is more easily oxidized than the product. It is generally assumed that hole transfers that have $-\Delta G^0 \geq 0.5$ V (12 kcal/mol) are diffusion controlled, that is, are essentially activationless.

5.12 Acidity: Thermodynamic and Kinetic

Cation radicals of molecules that have allylic or benzylic hydrogens have dramatically enhanced acidities in comparison to their neutral precursors. Typically, in fact, such cation radicals are appropriately classified as *superacids* (Figure 5.14). An example is provided by 4-benzylanisole, which has a $pK_a \approx -57$ in acetonitrile.[34] The enhanced acidities of cation radicals can be related directly to the much greater oxidation potentials of the neutral molecules (HA in Figure 5.14) than of the conjugate bases (A^-) of the neutral molecules.[35]

Aniline is a notable exception, however. In part because of the relatively low oxidation potential of this substrate, the aniline cation radical is actually a weak acid. Nevertheless, the acidity of the cation radical is approximately 24 pK_a units greater than that of neutral aniline. The preference for allylic or benzylic hydrogens over vinylic or phenyl-type hydrogens is presumably based upon the much higher oxidation potentials of conjugate bases (A^-) of the latter, which have sp^2-hybridized carbanion sites.

$$\mathrm{HA} \underset{}{\overset{K(\mathrm{HA})}{\rightleftharpoons}} \mathrm{H^+} + \mathrm{A^-}$$

$$E_{OX}(\mathrm{HA}) \updownarrow \qquad\qquad \updownarrow E_{OX}(\mathrm{A^-})$$

$$\mathrm{HA^{\cdot+}} \overset{K(\mathrm{HA^{\cdot+}})}{\rightleftharpoons} \mathrm{H^+} + \mathrm{A^{\cdot}}$$

$$pK(\mathrm{HA^{\cdot+}}) = pK_{HA} + [E_{OX}(\mathrm{A^-}) - E_{OX}(\mathrm{HA})]\ 23.06/1.37$$

HA	pK(HA)	$E_{OX}(A^-)$	$E_{OX}(HA)$	$pK(HA^{\cdot+})$ in DMSO
Ph—CH_3	43	-1.06	2.70	-20
Ph_2CH_2	32.2	-0.665	2.72	-25
Cyclopentadiene	18.0	0.153	2.23	-17
Ph—OH	18.0	0.550	2.10	-8.1
Ph—NH_2	30.6	-0.117	1.32	+6.5

Figure 5.14 The superacidity of cation radicals. Reprinted with permission from *J. Am. Chem. Soc.* **1989**, *111*, 1792–1795; Copyright 1989, American Chemical Society.

In spite of the thermodynamic superacidity of many cation radicals, deprotonation is often surprisingly slow. In the case of the 4-benzylanisole cation radical, for example, the cation radical is readily observed via cyclic voltammetry in a reversible process, in spite of its astounding thermodynamic acidity. The relatively *low kinetic acidity* of cation radicals appears to be general for cation radicals that are "carbon acids" (i.e. that dissociate a C–H proton) and has a close parallel in the relatively low kinetic acidity of neutral molecules of this type. Since alternate reaction pathways are usually available for cation radicals, and are often extremely fast, the acidity of cation radicals does not play as large a role in cation radical chemistry, as would be suggested by their powerful thermodynamic acidity. The addition of bases to the reaction medium, however, can sharply accelerate their deprotonation. The deprotonation of the hexamethylbenzene cation radical by pyridine bases (Figure 5.15) occurs smoothly, with rate constants of *ca.* $10^7\ M^{-1}\ s^{-1}$.[36] Even so, these proton transfers occur at a rate *ca.* 1000 times slower than the maximum (diffusion-controlled) rate.

The oxidation of alkyl side chains of readily ionizable arenes by manganic acetate is a classic example of cation radical deprotonation (Figure 5.16).[37] The deprotonation step is actually slow enough to be rate determining, since the rate is inversely dependent upon the Mn^{II} concentration. The resulting 4-methoxybenzyl radical is then easily oxidized by Mn^{III} to the corresponding carbocation, which reacts with acetate ion or acetic acid. When cobaltic trifluoroacetate is used as the oxidant, even toluene undergoes side-chain oxidation, and the mechanism appears to be entirely analogous to that given in Figure 5.16.[38]

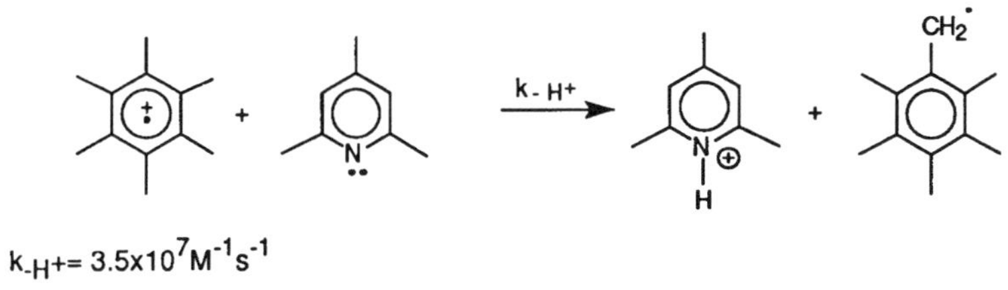

$k_{-H^+} = 3.5\times10^7 M^{-1}s^{-1}$

Figure 5.15 Deprotonation of the hexamethylbenzene cation radical.

$$CH_3O-C_6H_4-CH_3 \xrightleftharpoons{Mn^{III}} Mn^{II} + CH_3O-C_6H_4-CH_3^{\cdot+} \xrightarrow{\ominus OAc,\ rds} CH_3O-C_6H_4-\dot{C}H_2 + AcOH$$

$$CH_3O-C_6H_4-\dot{C}H_2 + Mn^{III} \longrightarrow Mn^{II} + CH_3O-C_6H_4-\overset{\oplus}{C}H_2 \xrightarrow{\ominus OAc} CH_3-C_6H_4-CH_2OAc$$

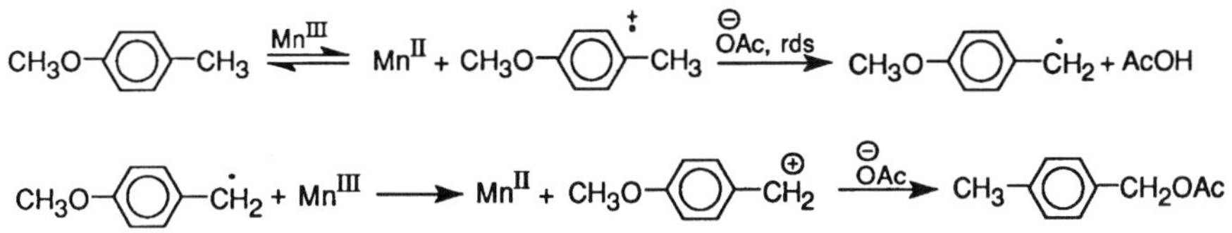

Figure 5.16 Manganic acetate oxidation of readily ionizable toluene derivatives.

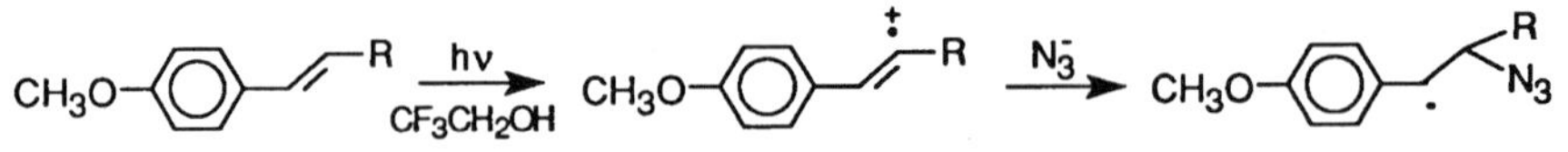

R = H, CH3

Figure 5.17 Rapid cation radical–nucleophile reactions.

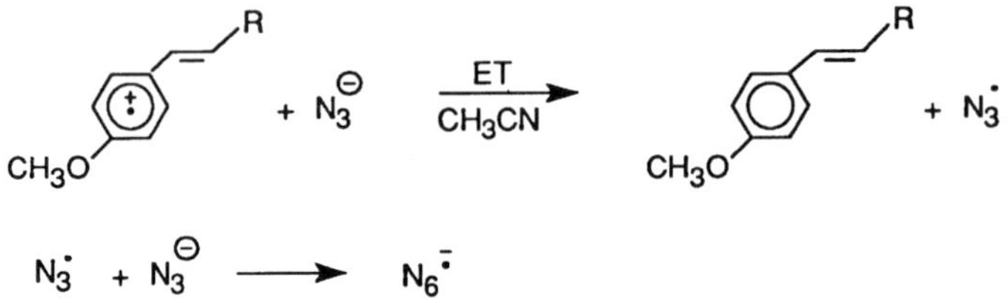

Figure 5.18 ET reactions between cation radicals and nucleophiles.

5.13 Reactions with Nucleophiles

Cation radicals are also powerful electrophiles. Their reactions with strong or weak nucleophiles are usually extremely rapid and can even reach the level of diffusion control. The reaction of the 4-vinylanisole cation radical with azide ion (Figure 5.17), for example, occurs at the diffusion-controlled rate ($7.0 \times 10^9\,M^{-1}\,s^{-1}$) in trifluoroethanol solvent.[39] The corresponding reaction of the *trans*-anethole cation radical with azide ion has a rate constant that is within a factor of 2 of the diffusion-controlled rate ($3.5 \times 10^9\,M^{-1}\,s^{-1}$). The cation radical is generated by flash laser photolysis, and reaction occurs at the β carbon of the styrene moiety, as expected from the SOMO coefficients. Even the anthracene cation radical reacts (at the 9 position) at a rate of $3.3 \times 10^9\,M^{-1}\,s^{-1}$.

A potential competitor in cation radical–nucleophile reactions is electron transfer. In fact, when the reactions of Figure 5.17 are attempted in acetonitrile solvent (Figure 5.18), electron transfer dominates and occurs at the diffusion-controlled rate ($3 \times 10^{10}\,M^{-1}\,s^{-1}$).[39] In acetonitrile solvent, the ET reaction is substantially exergonic and therefore rapid. However, in the relatively acidic solvent trifluoroethanol, the azide ion is strongly stabilized by hydrogen bonding and its oxidation potential is increased by ca. 0.5 V, making ET endergonic. Only under these conditions is covalent bond formation dominant.

A reaction that illustrates the potential synthetic value of cation radical–nucleophile reactions is the reaction of chloride ion with the phenothiazine cation radical (Figure 5.19).[40]

The reactions of cation radicals with alkene π bonds are of special importance and will be discussed in detail in a subsequent section.

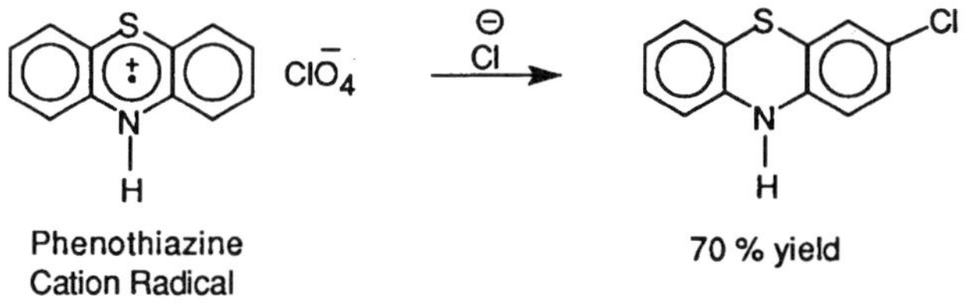

Phenothiazine Cation Radical

70 % yield

Figure 5.19 Cation radical–nucleophile reactions of synthetic utility.

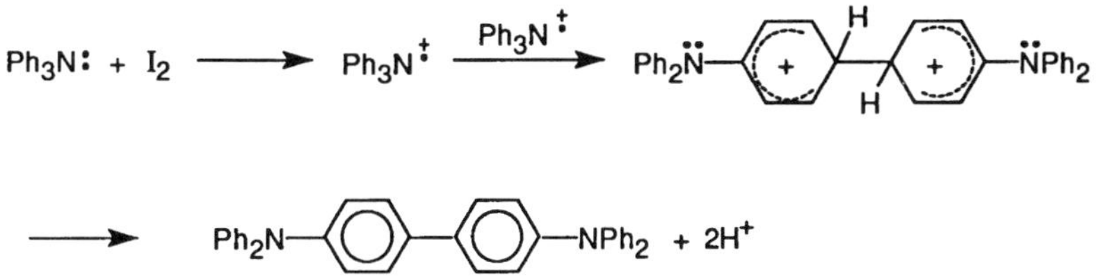

Figure 5.20 Coupling of triphenylaminium cation radicals.

5.14 Radical Coupling

The reactions of cation radicals with nucleophiles and bases are analogous to the corresponding reactions of carbocations. Other reactions of cation radicals reveal their "radical character." Radical coupling is perhaps the most common of these. An especially interesting example is the coupling of triphenylaminium cation radicals (Figure 5.20). Although triarylaminium ions are isolably stable when the three *para* positions are all blocked by a substituent, they are unstable with respect to coupling when any one of these positions is unblocked. Consequently, the triphenylaminium ions couple to give a dication, which then deprotonates, leading to tetraphenylbenzidine.[41] It is an important general attribute of cation radical coupling that Brønsted acid is generated by deprotonation of the intermediate dications. When a basic functionality is not present, the acid generated is a strong Brønsted acid, which can potentially engender acid catalyzed competing or secondary reactions. Thus not only are cation radicals Brønsted superacids themselves; they can produce strong Brønsted acids by coupling and subsequent deprotonation of the intermediate dications. It is further noted that triarylamines are effectively nonbasic molecules even in the presence of such strong acids.

5.15 Cation Radical–Radical Coupling: The ET Mechanism for Aromatic Nitration

The oxidation potential of NO_2 is considerably greater than that of naphthalene (Figure 5.21). Consequently electron transfer from naphthalene to the nitronium ion is significantly exergonic. The cation radical–radical pair that would result (NO_2 is

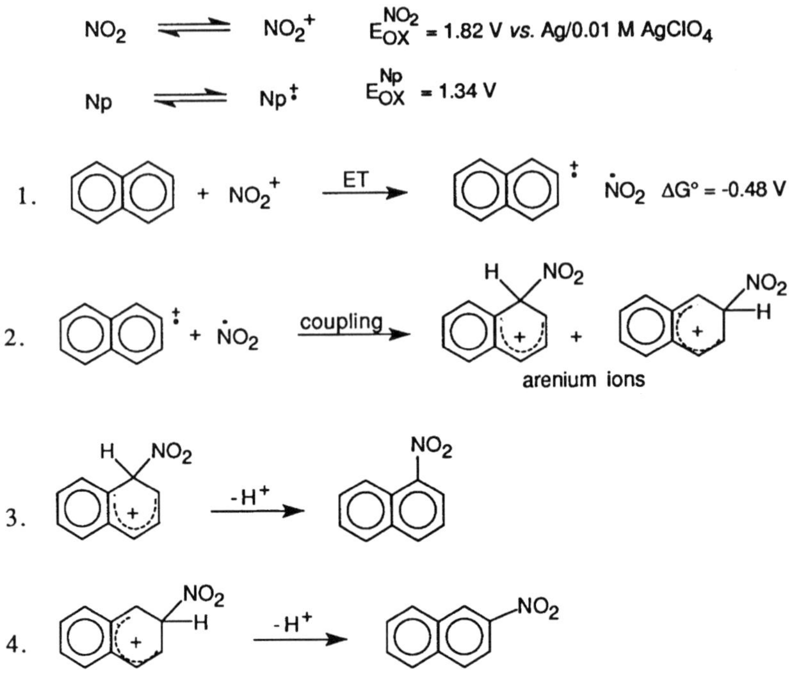

$$NO_2 \rightleftharpoons NO_2^{+} \qquad E_{OX}^{NO_2} = 1.82\ \text{V}\ vs.\ \text{Ag}/0.01\ \text{M AgClO}_4$$

$$\text{Np} \rightleftharpoons \text{Np}^{\cdot +} \qquad E_{OX}^{\text{Np}} = 1.34\ \text{V}$$

1. Naphthalene + NO_2^{+} —ET→ naphthalene cation radical + $\dot{N}O_2$ $\Delta G° = -0.48$ V

2. Naphthalene cation radical + $\dot{N}O_2$ —coupling→ arenium ions

3. 1-arenium ion —$-H^{+}$→ 1-nitronaphthalene

4. 2-arenium ion —$-H^{+}$→ 2-nitronaphthalene

Figure 5.21 The proposed ET mechanism for the nitration of naphthalene.

a radical) could then undergo radical coupling to give the familiar arenium ions, which are conventionally considered to be formed by a direct electrophilic reaction of NO_2^{+} with the aromatic. The distinction between polar and ET mechanisms of nitration and other electrophilic aromatic substitutions is a challenging one. It would, of course, appear highly unlikely that ET mechanisms can effectively compete with polar mechanisms in cases where the requisite ET is endergonic. In the case of nitration, this would definitely include benzene itself and any arene with a higher oxidation potential than benzene. Toluene is probably also included, but may possibly be a borderline case. Although the ET mechanism cannot be said to have been definitively established in any case, it is now well established that (1) aromatic cation radicals do, in fact, couple with NO_2 to yield arenium ions, and (2) the positional selectivities are impressively alike in ordinary (thermal) nitration and nitration via the cation radical route (Figure 5.22).[42] The cation radicals can be generated by irradiation of appropriate charge-transfer complexes at the charge-transfer band. Positional selectivities have been examined for both thermal and ET nitration of a wide variety of electron-rich arenes, and the results are impressively similar. Positional selectivities in the thermal nitrations using the nitropyridinium salts as mild nitrating agents also appear to be very similar to those obtained with more conventional thermal nitrating agents (HNO_3; $NO_2^{\oplus}BF_4^{\ominus}$). Since the charge-transfer excitation procedure undoubtedly leads to arene cation radicals, this

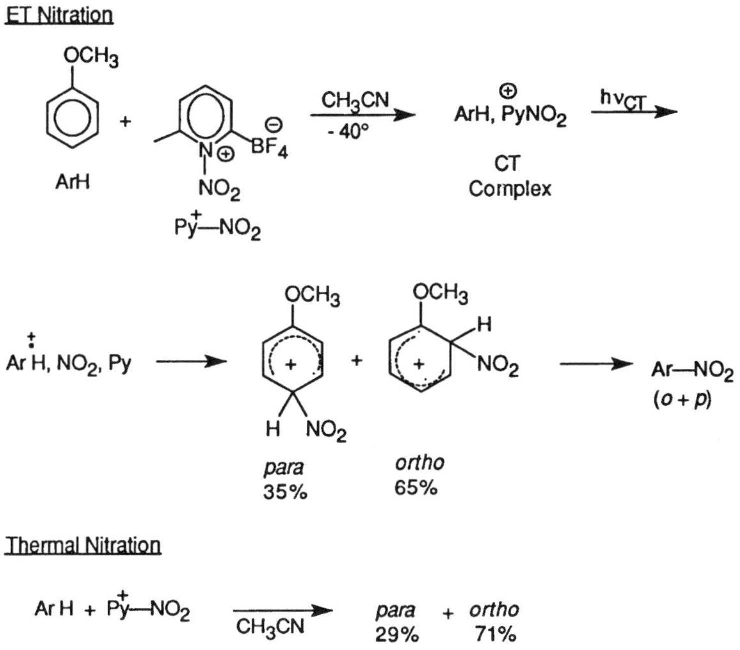

ET Nitration

OCH_3 ArH + $\overset{+}{Py}$—NO_2 $BF_4^{\ominus}$ $\xrightarrow[-40^\circ]{CH_3CN}$ $\overset{\oplus}{ArH, PyNO_2}$ CT Complex $\xrightarrow{h\nu_{CT}}$

$\overset{+\bullet}{ArH}, NO_2, Py$ ⟶ *para* 35% + *ortho* 65% ⟶ Ar—NO_2 (*o* + *p*)

Thermal Nitration

ArH + $\overset{+}{Py}$—NO_2 $\xrightarrow{CH_3CN}$ *para* 29% + *ortho* 71%

Figure 5.22 The cation radical route for aromatic nitration.

correspondence of positional selectivities would appear to represent strong support for the ET mechanism in thermal nitrations of substrates for which ET is exergonic. It should be noted, however, that in cases where the relevant ET is endergonic, the initial generation of arene cation radicals via charge-transfer excitation does not assure a cation radical–radical coupling mode for the formation of arenium ions. This is because back electron transfer from NO_2 to the arene cation radical would then be exergonic, and this could lead to a polar nitration mechanism. Consequently the correspondence of product distributions in charge transfer and thermal nitrations loses its mechanistic significance when ET from the arene to nitronium ion is endergonic.

5.16 Mesolytic Cleavages

The mesolytic cleavage of C–H bonds of cation radicals, yielding protons and organic radicals, have already been considered. The removal of an electron from a bonding molecular orbital was seen to provide a powerful thermodynamic driving force for this important variety of mesolytic cleavage. Cleavage of other bonds, including especially C–C bonds, is also greatly enhanced thermodynamically. A

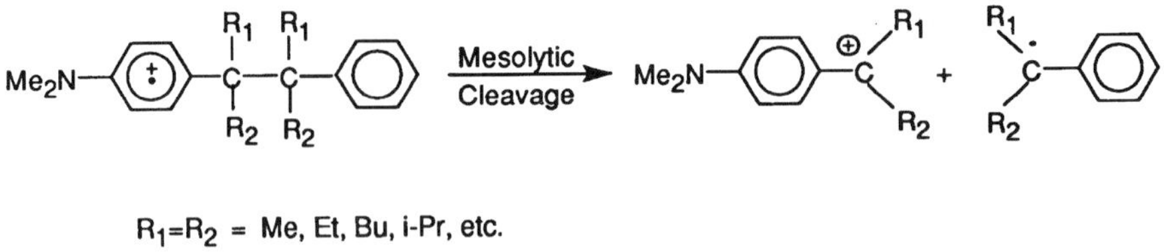

Figure 5.23 Mesolytic versus homolytic cleavages.

study of the mesolytic cleavages of the doubly benzylic C–C bonds of a series of 1,2-diarylethanes is especially instructive (Figure 5.23).[43] In this series, the thermodynamic driving force for mesolytic cleavage is fully 28 kcal/mol greater than for homolytic cleavage of the corresponding neutrals. The kinetic driving force for mesolytic cleavage is 23 kcal/mol greater than for homolytic cleavage. The data further suggest that for unstrained members of the series (e.g., $R = H$), essentially the *full amount* of the thermodynamic driving force (28 kcal/mol) is available kinetically. The cleavage reactions of these latter substrates are extremely slow and were not amenable to study, but the conclusion appears plausible given the large endergonicities of these reactions and in view of the Hammond Postulate. Of further interest is the observation that the thermodynamic driving forces for the mesolytic cleavages of these cation radicals is ca. 11–12 kcal/mol greater, on the average, than for the mesolytic cleavages of analogous anion radicals (where the anion radical moiety is of the nitroarene type).

5.17 Abstractions

Radical abstraction reactions are not often observed for delocalized cation radicals, but the abstraction of hydrogen atoms by a localized cation radical is nicely illustrated by a familiar organic reaction, the Hoffmann–Löffler–Freitag reaction (Figure 5.24).[44] Localized aminium ions are formed by thermal or photochemical homolysis of the N–Cl bond of the protonated chloramine, or by ferrous chloride reduction of this bond (Step 1). These aminium ions abstract hydrogen from the δ-carbon preferentially, yielding a distonic cation radical having a primary alkyl radical center (Step 2). The latter radicals then abstract chlorine atoms from the protonated chloramine, generating a 5-chloroalkylammonium salt and regenerating an aminium ion (Step 3). Steps 2 and 3 constitute the propagation cycle of the chain reaction. The pyrolidine product is not formed until the product is treated with base. When the alkyl group is longer than butyl, abstraction of a δ-hydrogen is still favored because the 6-membered ring transition state (including the hydrogen) provides a more nearly linear N–H–C geometry, which is favored for such hydrogen abstractions.

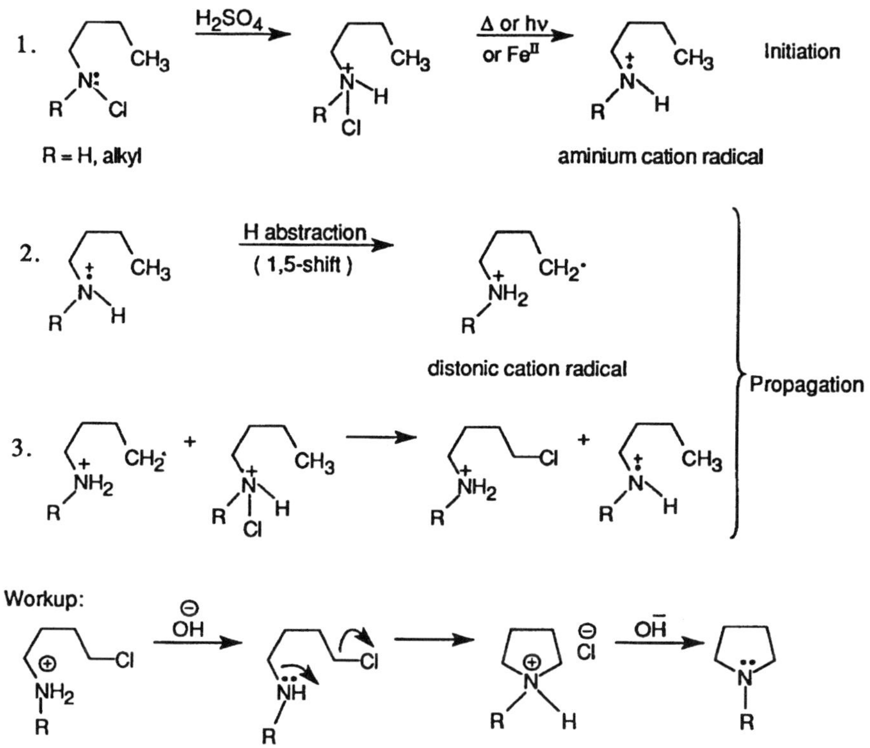

Figure 5.24 The Hoffmann–Löffler–Freitag reaction.

5.18 Rearrangements

The removal of an electron from a neutral molecule (R) elevates the molecule from a potential surface where reaction barriers ($\Delta G\ddagger$) are relatively high even for so-called symmetry-allowed reactions to a relatively flat surface where many reactions possess quite minuscule barriers and symmetry allowedness/forbiddenness is of very little import (Figure 5.25).[45] Interestingly, the profound acceleration of reaction rates is not usually derived, either directly or indirectly via product character in the transition state, from increased thermodynamic driving force, as might be thought for such highly energetic species, so long as the reactant and product are both cation radicals (i.e., the reaction occurs on the cation radical surface). Rather, the structure-weakening effect of the removal of a bonding electron allows many structural deformations to occur with diminished activation. Stated in a slightly different, but more quantitative way, the decrease in activation energy occasioned by ionization is equal to the difference in ionization potentials of the reactant and transition state (Figure 5.25). Since transition states are less strongly bound than stable molecules, they are more readily ionized, resulting in decreased activation energies on the cation radical surface.

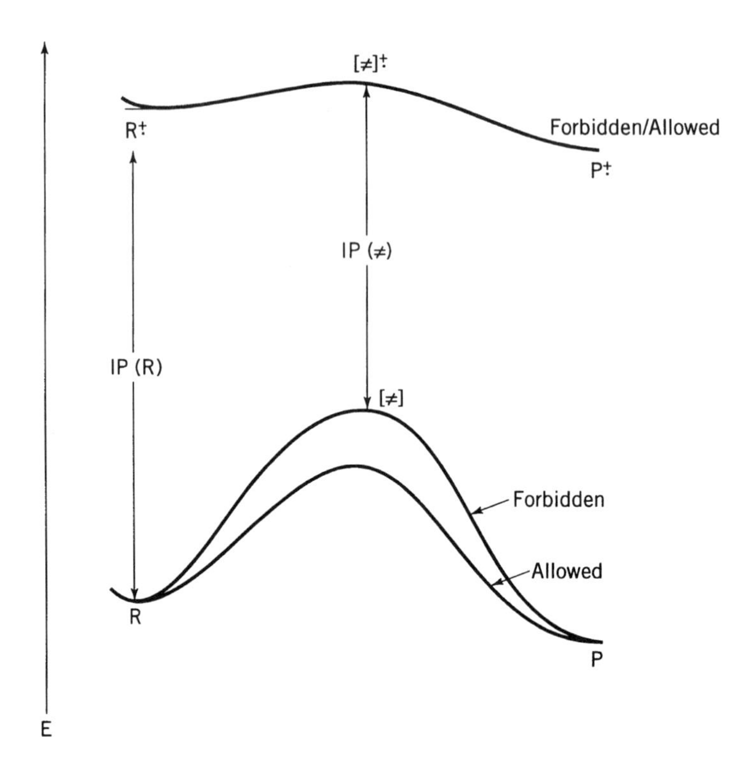

Figure 5.25 The cation radical potential surface.

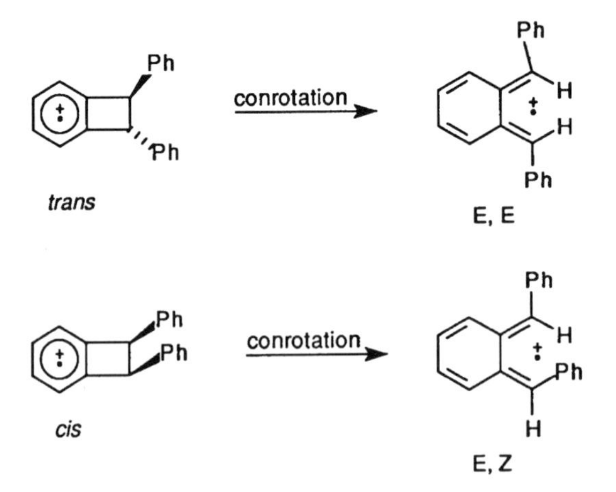

Figure 5.26 A stereospecific cation radical retroelectrocyclic reaction.

Cation radical rearrangements represent the simplest type of reaction that occurs exclusively on the cation radical potential surface. An elegant example is the retroelectrocyclic reaction of Figure 5.26.[46] The conrotatorily stereospecific retroelectrocyclic reaction of the corresponding anion radical has been mentioned previously. In these cases, at least, the stereochemical preferences of the anion and cation radicals are identical and are the same as found for the neutral molecule.

The vinylcyclopropane[47] and vinylcyclobutane[48] rearrangements are both powerfully promoted by cation radical formation. Many of these reactions are produced by catalytic amounts of aminium salts, providing a synthetically convenient and efficient means of accomplishing these transformations under extremely mild thermal conditions (Figure 5.27). In cases where relatively easily ionizable groups are present (e.g., *p*-anisyl), the milder tribromo salt (**2c**) is

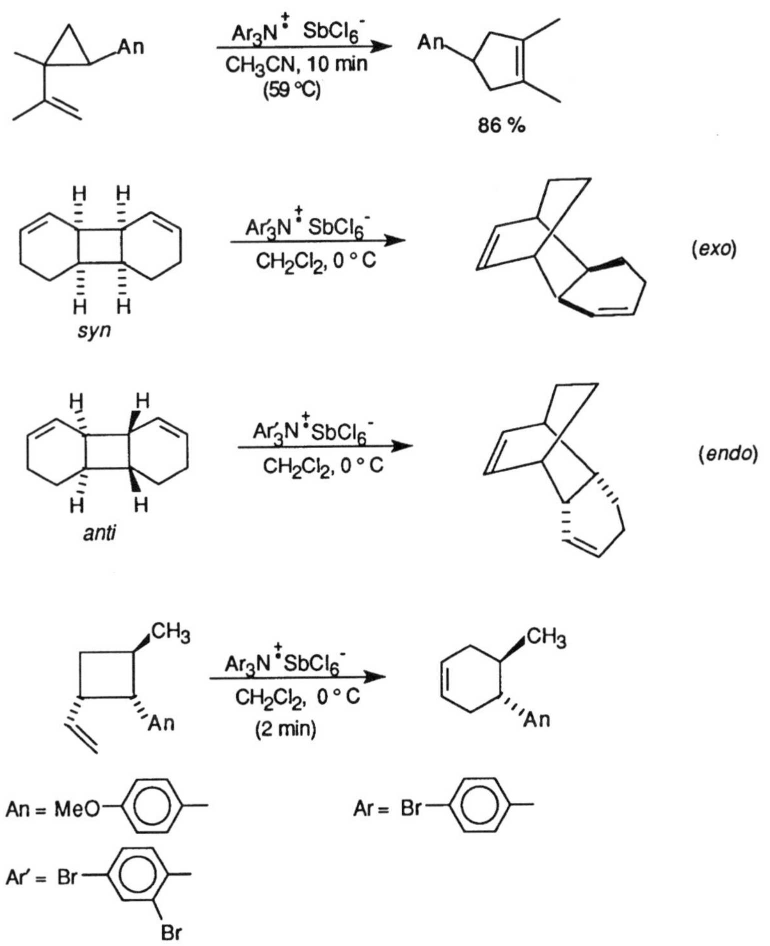

Figure 5.27 Vinylcycloalkane rearrangements.

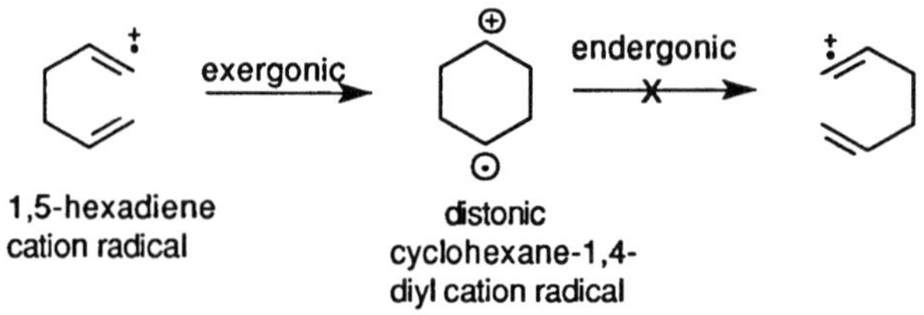

Figure 5.28 Partial cation radical "Cope" reaction.

sufficient to induce rapid reaction. In the case of simple alkene functionality, the more powerful hexabromo salt (**5**) is required. The vinylcyclopropane rearrangement is considered to occur in a stepwise fashion, via a distonic cation radical, which subsequently cyclizes to the product cation radical.[47] In contrast, at least some of the vinylcyclobutane rearrangements are stereospecific, indicating a concerted (i.e., pericyclic) mechanism.[48] The catalytic aspect of these reactions will be discussed in a subsequent section.

Another cation radical rearrangement of special interest is the Cope rearrangement. The simple 1,5-hexadiene cation radical cyclizes efficiently to the 1,4-cyclohexanediyl cation radical (Figure 5.28), a distonic cation radical, but the latter fails to complete the Cope reaction.[49] When the temperature is raised slightly, rearrangement to the cyclohexene cation radical occurs. Since the cyclized, distonic cation radical is more stable than the 1,5-hexadiene cation radical, the former cannot reopen to the rearranged 1,5-hexadiene cation radical. However, a novel means of bringing such "hemi-Cope" cation radical reactions to a successful conclusion has been devised.[50] When a 1,5-diene cation radical is generated by photosensitized electron transfer, the intermediate distonic cation radical is reduced to the diradical by the sensitizer anion radical. Ring opening of the diradical is then exergonic and rapid (Figure 5.29).

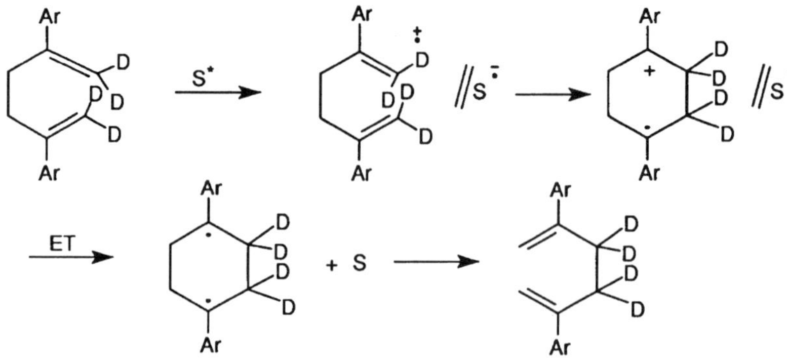

Figure 5.29 Cation radical Cope reaction under photosensitized electron transfer conditions.

A few examples of the full cation radical Cope reactions under both aminium salt and matrix isolation conditions are now available for specialized systems in which the cleavage step is exergonic.[51]

5.19 Chain versus Catalytic Mechanisms

The observation that catalytic quantities (5–10 mol %) of an aminium salt are often sufficient to effect cation radical reactions in high yield implies the operation of either a catalytic or chain mechanism (Figure 5.30).[45] In the chain mechanism (Steps 1–3), ionization of the reactant substrate (R) is effected by product cation radicals ($P^{\overset{+}{\bullet}}$) via a hole–transfer reaction. The aminium salt acts as an initiator in sustaining a small, steady state concentration of $R^{\overset{+}{\bullet}}$. In the closely related catalytic mechanism (Steps 1, 2, and 4), the product cation radicals are neutralized by hole transfer to the neutral triarylamine formed in the first step. In this mechanism, all reactant molecules are ionized by the aminium salt. When the milder aminium salt (E_{ox} = 1.05 V) is used, this ionization step is usually substantially endergonic, and therefore more selective than ionization by product cation radicals. The relative slowness of the ionization is appropriate since, if cation radicals are generated too rapidly, termination by cation radical–cation radical coupling can be problematic. Since the neutral triarylamine concentration actually builds up during the reaction as a result of termination steps that consume cation radicals, and since the triarylamine tends to inhibit the chain reaction and convert it to a catalytic process, both mechanisms may be operative in a single reaction system. Catalytic mechanisms are, however, less likely when the more potent aminium salt (E_{ox} = 1.50 V) is used, since hole transfer to this triarylamine is much less favorable.

$$1.\quad Ar_3N^{\overset{+}{\bullet}} + R \longrightarrow Ar_3N + R^{\overset{+}{\bullet}}$$

$$2.\quad R^{\overset{+}{\bullet}} \longrightarrow P^{\overset{+}{\bullet}}$$

$$3.\quad P^{\overset{+}{\bullet}} + R \longrightarrow P + R^{\overset{+}{\bullet}}$$

$$4.\quad P^{\overset{+}{\bullet}} + Ar_3N \longrightarrow P + Ar_3N^{\overset{+}{\bullet}}$$

Chain Mechanism: Step 1 (Initiation), 2 and 3 (Propagation)

Catalytic Mechanism: Steps 1, 2, and 4

Figure 5.30 Chain and catalytic mechanisms.

5.20 Cycloadditions

The cyclodimerization of N-vinylcarbazole (Figure 5.31) was the first reported instance of a cation radical pericyclic reaction in solution.[52] The ionization of the substrate could be effected either by metal ions, such as ferric ion, or by photosensitized electron transfer. The cation radical chain mechanism found for this reaction represented the first observation of this fundamental mechanistic type. The reaction is strongly inhibited by dioxygen, which reacts efficiently with the relatively long-lived NVC cation radicals (*vide infra*; reactions of cation radicals with triplet O_2). The proposed mechanism involves stepwise addition of an N-vinylcarbazole cation radical to a molecule of N-vinylcarbazole, proceeding via a distonic cation radical intermediate. Chain lengths of up to 90 were found for this reaction.

The γ radiolysis of 1,3-cyclohexadiene in solution was subsequently found to result in the formation of the Diels–Alder cyclodimer of this diene, and an analogous cation radical chain mechanism was proposed.[53] A synthetically much more convenient, and efficient, means of effecting this cyclodimerization uses tris(4-bromophenyl)aminium hexachloroantimonate (**2c**) as the catalyst for ionizing the diene substrate (Figure 5.32).[54] The reaction is extremely fast even at 0°C, requiring less than 5 minutes for completion. The cycloaddition step has been found

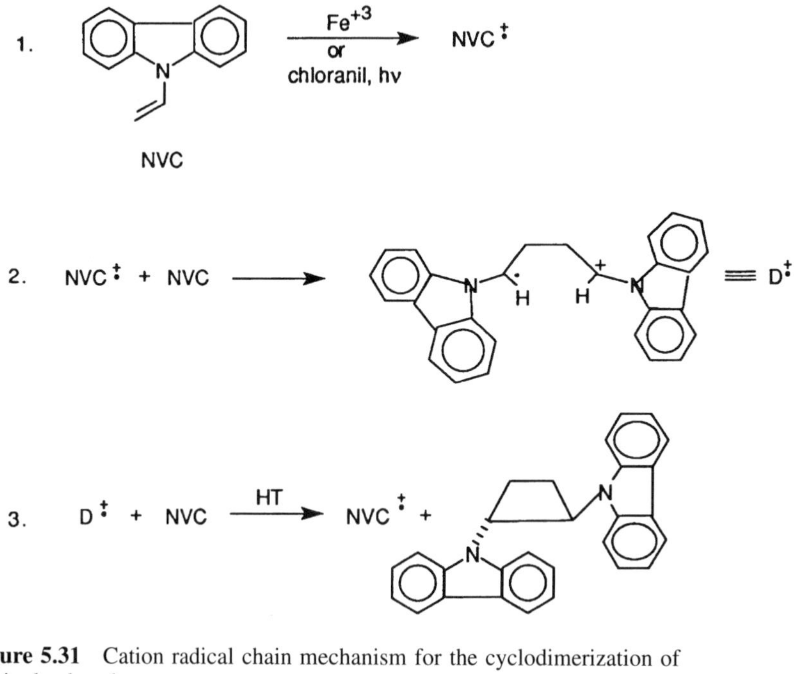

1. NVC $\xrightarrow[\text{chloranil, hv}]{Fe^{+3} \text{ or}}$ NVC$^{\cdot+}$

2. NVC$^{\cdot+}$ + NVC ⟶ [distonic cation radical] ≡ D$^{\cdot+}$

3. D$^{\cdot+}$ + NVC $\xrightarrow{HT}$ NVC$^{\cdot+}$ + [cyclobutane dimer]

Figure 5.31 Cation radical chain mechanism for the cyclodimerization of N-vinylcarbazole.

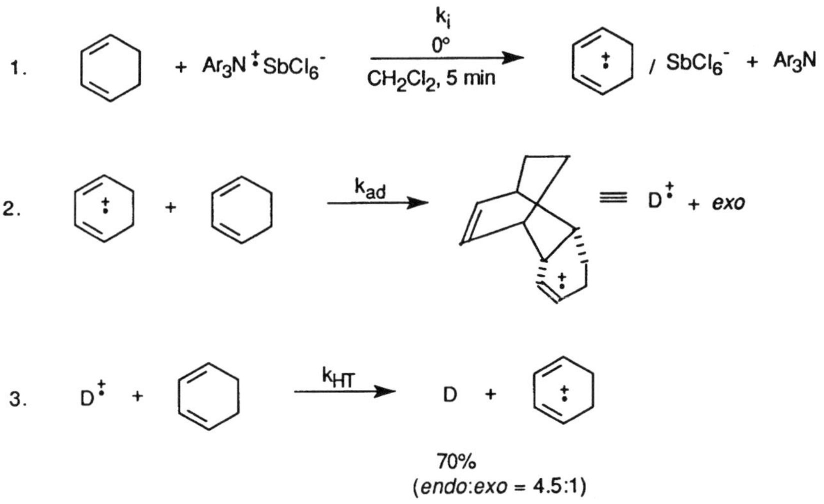

Figure 5.32 The triarylaminium salt catalyzed cyclodimerization of 1,3-cyclohexadiene.

to have a microscopic rate constant $k_{ad} = 3 \times 10^8\ M^{-1}\ s^{-1}$ in acetonitrile solvent, that is, only about a factor of 60 less than the diffusion controlled rate.[55] The highly exergonic hole transfer step is considered to have a rate constant k_{HT} approximately equal to the diffusion-controlled rate. The ionization step (k_i) is substantially endergonic but provides a low, steady-state concentration of cation radicals for the cation radical chain mechanism. Since the aminium salt catalyst progressively decomposes during the reaction to yield the corresponding neutral triarylamine, it appears likely that a catalytic mechanism also contributes, perhaps even at a rather early stage, and may well be dominant in later stages of the reaction.[56]

The use of the shelf-stable and commercially available cation radical salt **2c** as a chemical initiator/catalyst for cation radical chain reactions makes these reactions conveniently accessible for mechanistic study and synthetic applications. A number of studies have been carried out, and the results indicate that cycloaddition is an intrinsically preferred reaction mode of cation radicals. The cycloaddition of 1,3-cyclohexadiene to the three geometric isomers of 1,3-cyclohexadiene is completely stereospecific, indicating a concerted, as opposed to a stepwise, cycloaddition (Figure 5.33). *Ab initio* theoretical reaction path studies also support the proposed concerted mechanism.[55] These observations represent the first evidence for true pericyclic chemistry of cation radicals. On the other hand, it should be stressed that the feasibility of stepwise mechanisms involving distonic cation radical intermediates is undoubted, even for certain cation radical Diels–Alder reactions.[58]

The cation radical (or hole-catalyzed) Diels–Alder reaction is especially effective in two areas where the thermal Diels–Alder reaction is known to be

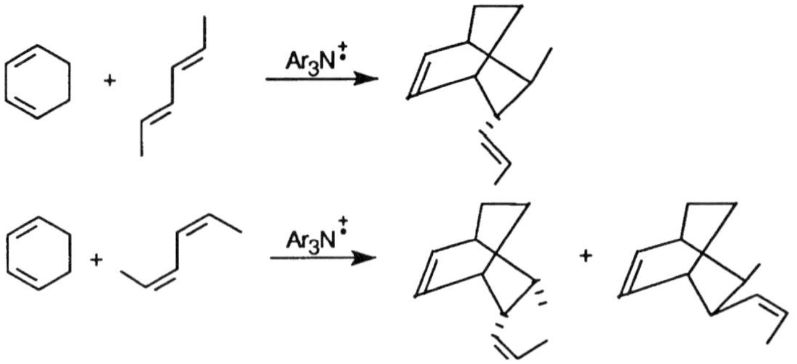

+ $Ar_3N^{+\bullet}$

+ $Ar_3N^{+\bullet}$ +

Figure 5.33 Stereospecific Diels–Alder cycloaddition of 1,3-cyclohexadiene to the 2,4-hexadienes.

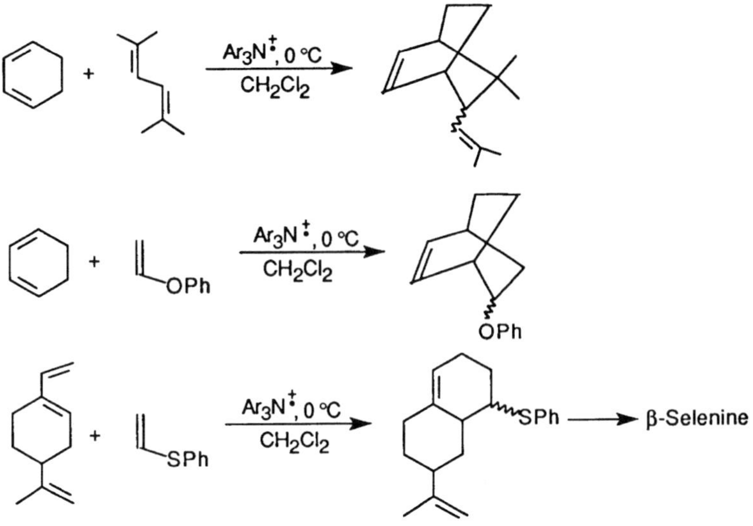

+ $Ar_3N^{+\bullet}$, 0 °C CH_2Cl_2

+ OPh $Ar_3N^{+\bullet}$, 0 °C CH_2Cl_2 OPh

+ SPh $Ar_3N^{+\bullet}$, 0 °C CH_2Cl_2 SPh → β-Selenine

Figure 5.34 Additions to sterically hindered and electron-rich dienophiles.

relatively weak, *viz*. additions involving sterically hindered or electron-rich dienophiles (Figure 5.34).[54, 59] The final reaction illustrated in Figure 5.34 has been developed as part of a synthesis of the natural product β-selenine.[60]

5.21 Role Selectivity in Diels–Alder Additions

The hypothetical Diels–Alder reaction of *s-cis*-1,3-butadiene with the ethene cation radical to give the cyclohexene cation radical is the prototype of a [4+1] Diels–Alder cycloaddition reaction, in the terminology of Woodward and Hoffmann (Figure 5.35). Orbital correlation diagrams for this reaction indicate

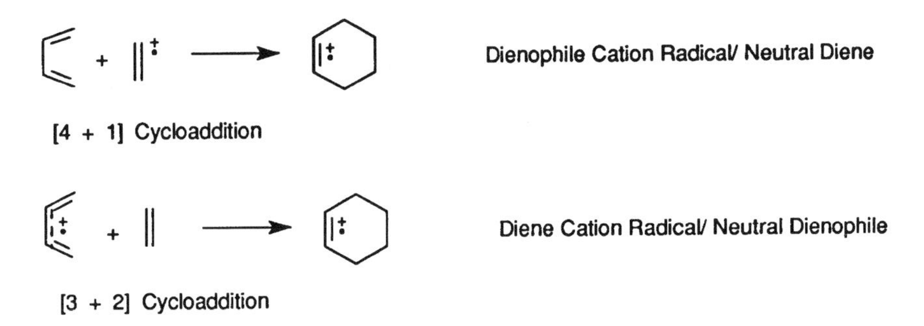

Figure 5.35 [4+1] versus [3+2] cation radical Diels–Alder reactions.

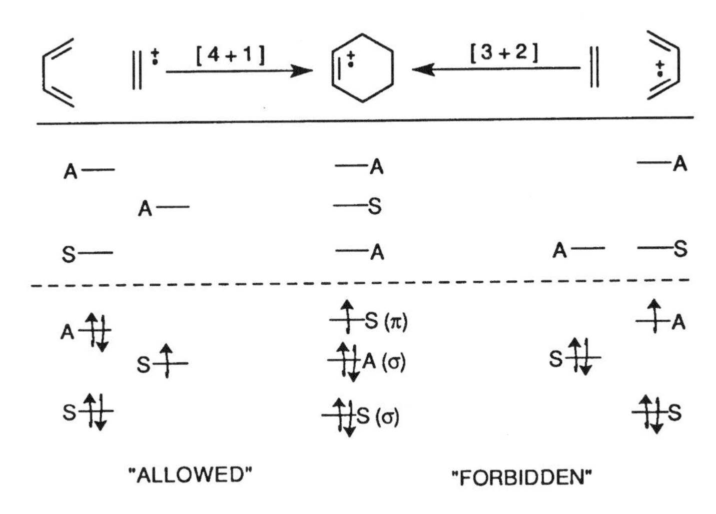

Figure 5.36 Orbital correlation diagrams for [4+1] and [3+2] cycloadditions.

that, unlike most radical or ion radical reactions, this cycloaddition is *symmetry allowed* (Figure 5.36).[60]

In contrast, the role-reversed reaction of the *s-cis*-1,3-butadiene cation radical with neutral ethene, the prototype of a [3+2] Diels–Alder cycloaddition is *symmetry forbidden*. On the other hand, orbital symmetry allowedness/forbiddenness is not necessarily expected to have a major effect on role selectivity, since activation energies on the cation radical surface are so small (a damping effect). Further, the proposed and calculated Diels–Alder transition states are highly nonsynchronous, thus also minimizing pericyclic effects. Some reactions, such as that of Figure 5.37, seem to suggest a preference for [4+1] cycloadditions. In this reaction, 1,1′-dicyclopentenyl (DCP) is far more readily oxidized to the corresponding cation radical than is 2,3-dimethyl-1,3-butadiene. In fact, the latter diene appears not to be oxidizable to the cation radical

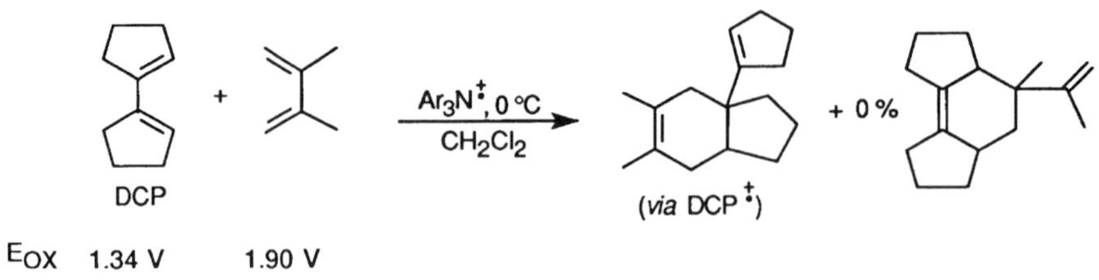

Figure 5.37 Symmetry-allowed [4+1] cycloadditions.

by **2c**, since it is stable in the presence of the catalyst. Assuming that the reactive cation radical is $DCP^{\bullet+}$, it is clear from the structure of the product that the cation radical component has cyclized exclusively as the dienophile. However, it is also possible that, owing to the conformational preference of DCP for the *s-trans* conformation, $DCP^{\bullet+}$ is formed mainly as the *s-trans* isomer, which then cannot provide the required *s-cis* dienic component. Further experiments, illustrated in Figure 5.38, indicate that there is probably little or no inherent preference for [4+1] over [3+2] cycloadditions.[61] Again, the oxidation potential of 1,2-dimethylenecyclohexane is so high that the mild aminium salt is unable to ionize it. The bis(ethylidene) derivative, on the other hand, is easily ionized and, in the absence of 1,2-dimethylenecyclohexane, rapidly undergoes cation radical Diels–Alder dimerization.

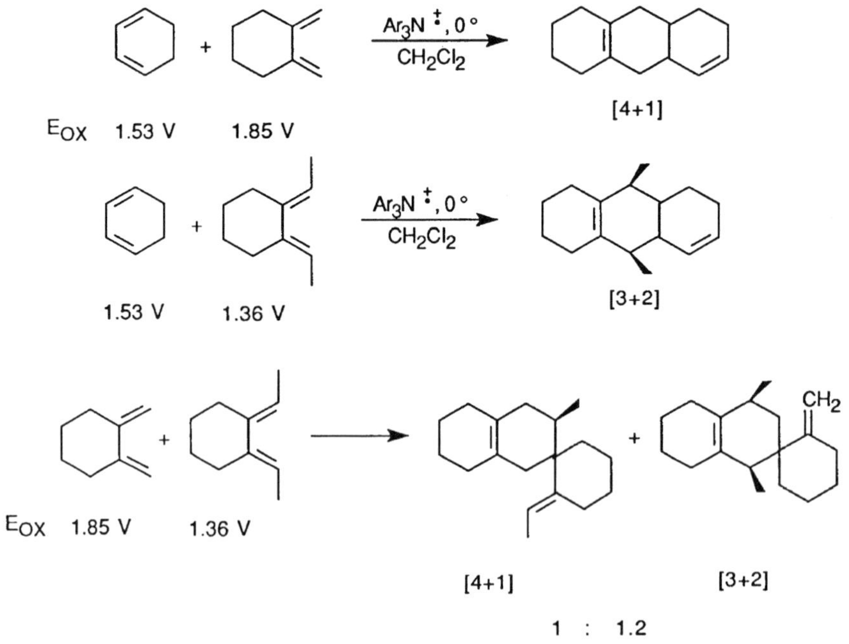

Figure 5.38 Role selectivity in the cation radical Diels–Ader reaction.

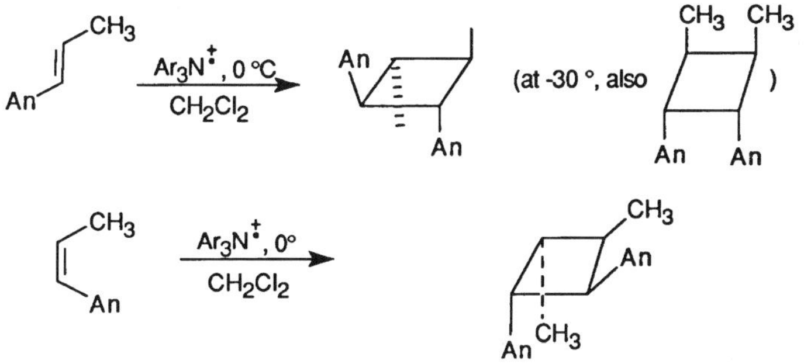

Figure 5.39 Stereospecific cyclodimerization of the anetholes.

5.22 Cyclobutanation

The original example of cation radical–neutral cycloaddition, the cyclodimerization of N-vinylcarbazole, illustrates cation radical cyclobutanation. The prototype of this reaction, viz. the cycloaddition of ethene and the ethene cation radical, if it were pericyclic (i.e., concerted rather than stepwise), would be classified as a [2+1] cycloaddition. Orbital symmetry diagrams indicate that this reaction is symmetry forbidden,[62] but the same arguments that suggest a minimal effect of symmetry forbiddenness in the [3+2] Diels–Alder are applicable here. In fact, the cyclobutanation reactions are extemely rapid and are often competitive with the Diels–Alder cycloadditions. Stereochemical studies of the cyclobutadimerization of *cis-* and *trans-*anethole find this cycloaddition to be highly stereospecific and therefore probably concerted (Figure 5.39).[63]

5.23 Periselectivity

The competition between Diels–Alder addition and cyclobutanation is herein referred to as a question of periselectivity. In many instances, the Diels–Alder reaction appears to be inherently favored, but conformational effects can favor cyclobutanation.[64] In the first reaction of Figure 5.40, Diels–Alder addition is favored. *Trans-*Anethole is the most readily ionized component, and the resulting cation radical prefers to add in the Diels–Alder fashion (specifically, [4+1] addition) rather than a possible cyclobutanation mode. On the other hand, when dicyclopentenyl (DCP) is the more ionizable component, the formation of primarily *s-trans-*$DCP^{\cdot +}$ geometrically precludes the formation of the Diels–Alder adduct in reactions with electron-rich alkenes.

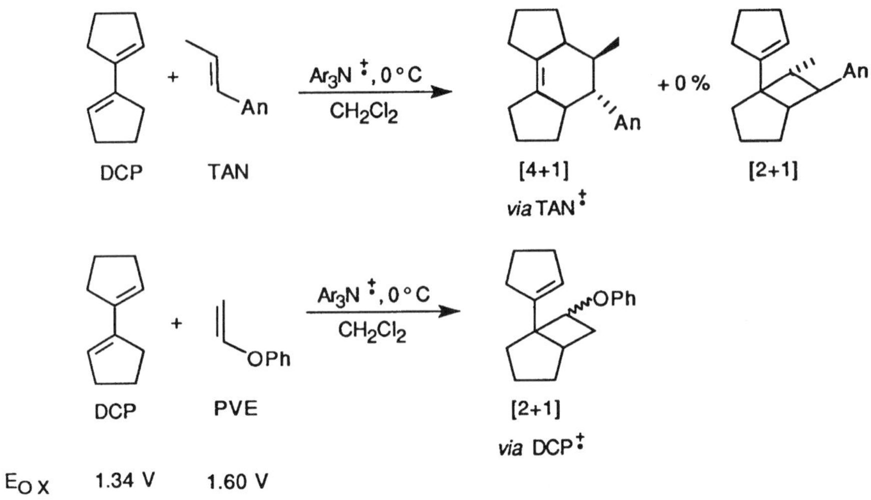

Figure 5.40 CB versus DA periselectivity.

5.24 Cation Radical versus Brønsted Acid–Catalyzed, Carbocation-Mediated Reactions

The generation of a cation radical moiety on a molecule has been seen to be structure weakening and kinetically activating. A mode of reaction that is especially strongly activated is deprotonation. Cation radicals having protons that are β to a site of charge/spin density have been seen to be superacids. Consequently, the chemistry of cation radicals and carbocations can sometimes become intermingled.[65] Further, the coupling of cation radicals produces dications, which also act as Brønsted acids through the similar loss of β protons. Consequently, attention must be given to the distinction between cation radical and carbocation mechanisms. A classic illustration is the Diels–Alder cyclodimerization of 2,4-dimethyl-1,3-pentadiene (DMP; Figure 5.41). The conventional aminium salt–catalyzed reaction of DMP yields dimer **A**.[66] A warning signal, however, is that the photosensitized electron transfer (PET) induced reaction of

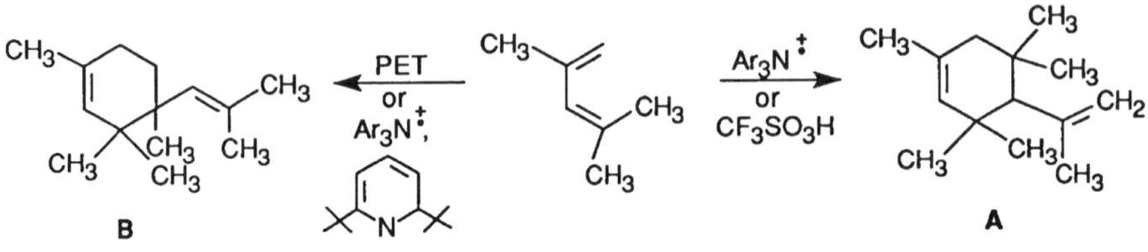

Figure 5.41 Diels–Alder cyclodimerization of 2,4-dimethyl-1,3-pentadiene.

DMP gives exclusively a different dimer (**B**). The latter method should presumably afford $DMP^{+\bullet}$ and ultimately yield the cation radical dimer. That the aminium salt–catalyzed reaction is not that of $DMP^{+\bullet}$ is further indicated by the production of dimer **A** when a strong Brønsted acid is used as the catalyst (CF_3SO_3H).[65, 66] Finally, the inclusion of the hindered amine 2,6-di-*tert*-butylpyridine with the aminium salt does produce **B**, the same dimer obtained in the PET-induced reaction.[65] These latter two reactions would appear to reflect the chemistry of $DMP^{+\bullet}$, but the reactions that yield dimer **A** are evidently acid-catalyzed, carbocation-mediated processes. A comparison of aminium salt–catalyzed reaction products with the corresponding products from the PET procedure, the hindered base-modified aminium salt procedure, and/or products of other methods that reliably furnish the desired cation radicals (including anodic oxidation) is therefore an excellent qualitative means of characterizing cation radical chemistry and of distinguishing it from carbocation chemistry induced by Brønsted acids.[65] The nature of the strong acid generated under the aminium salt conditions is still not defined, but it could be $HSbCl_6$, possibly generated by deprotonation of $DMP^{+\bullet}$.

5.25 Reactions with Dioxygen

Since ground-state dioxygen is a triplet, it should not be particularly surprising that cation radicals, like radicals, are quite reactive towards O_2. The strong retardation of the rate of cyclobutadimerization of N-vinylcarbazole by dioxygen has been mentioned previously. The reactions of cation radicals of conjugated dienes with O_2 typically lead to endoperoxides (Figure 5.42), Diels–Alder-like cycloadditions.[67] The reaction is limited to dienes that are relatively readily ionizable, but within this framework it is rather general and efficient. An especially interesting example is the endoperoxidation of ergosteryl acetate, where the persistent cation radical can be

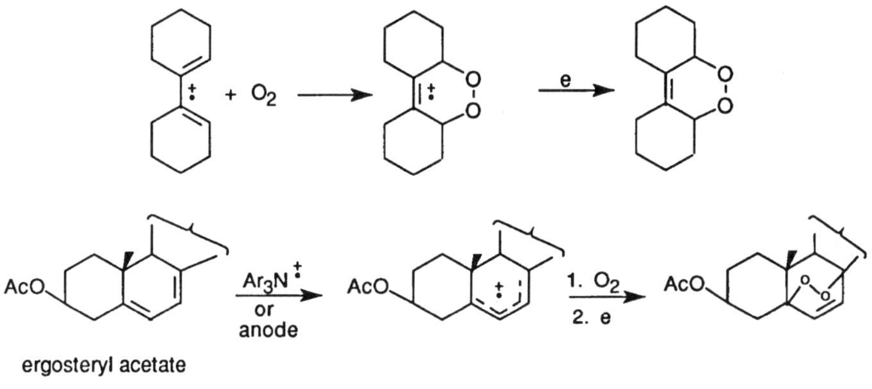

Figure 5.42 Endoperoxidation of diene cation radicals.

generated first and then subsequently treated with O_2.[68] This interesting strategy for endoperoxidation was devised by D. H. R. Barton, who reasoned that the spin forbiddenness of a concerted cycloaddition of ground-state (triplet) oxygen to a singlet diene (to give a singlet adduct) might be circumvented by employing the cation radical of the diene.[67]

Cation radicals of certain highly ionizable alkenes react with oxygen via a pathway that resembles cyclobutanation.[68] The products are 1,2-dioxetanes. The requirement for the generation of alkene cation radicals precludes the use of simple alkenes, which are not ionized by the aminium salts.

References

General References

Bard, A. J.; Ledwith, A.; Shine, H. J. *Adv. Phys. Org. Chem.* **1976**, *13*, 155.

Hammerich, O.; Parker, V. D. *Adv. Phys. Org. Chem.* **1984**, *20*, 55.

Kaiser, E. T.; Kevan, L. *Radical Ions*, John Wiley & Sons, New York, 1968.

Nelsen, S. F. in *Free Radicals*, Kochi, J. K., Ed.; John Wiley & Sons, New York, 1973, p. 565.

Roth, H. D. *Topics in Current Chemistry*, **1992**, *163*, 131.

Specific References

1. McLachlan, A. D. *Mol. Phys.* **1959**, *2*, 271; McConnell, H. M.; Robertson, R. E. *J. Chem. Phys.* **1958**, *28*, 991.
2. Bolton, J. R. in *Radical Ions*, Kaiser, E. T.; Kevan, L., Eds., John Wiley & Sons, New York, 1968, p. 12.
3. Bolton, J. R.; Fraenkel, G. K. *J. Chem. Phys.* **1964**, *40*, 3307.
4. Hoijtink, G. J.; Weijland, W. P. *Rec. Trav. Chim.* **1957**, *76*, 836.
5. Wurster, C.; Sendtner, R. *Ber.* **1879**, *12*, 1803 and 2071.
6. Gomberg, M. *J. Am. Chem. Soc.* **1900**, *22*, 757.
7. von Baeyer, A.; Villiger, V. *Ber.* **1902**, *35*, 1189.
8. Wieland, H. *Ber.* **1907**, *40*, 4260.
9. Weitz, E.; Schwechtin, H. W. *Ber.* **1926**, *59*, 2307; *idem.* **1927**, *60*, 545.
10. Bell, F. A.; Ledwith, A.; Sherrington, D. C. *J. Chem. Soc.* **1969**, 2719.
11. Walter, R. I. *J. Am. Chem. Soc.* **1955**, *77*, 5999.
12. Maslak, P.; Narvaez, J. N.; Vallombroso, T. M., Jr. *J. Am. Chem. Soc.* **1995**, *117*, 12373; Maskak, P.; Chapman, W. H., Jr.; Vallombroso, T. M., Jr.; Watson, B. A., *ibid.*, 12380.
13. Richardson, T. J.; Bartlett, N. *J. Chem. Soc., Chem. Commun.* **1974**, *427*; Richardson, T. J.; Tanzella, F. L.; Bartlett, N. *J. Am. Chem. Soc.* **1986**, *108*, 4937; Dinnocenzo, J. P.; Bannash, T. E. *J. Am. Chem. Soc.* **1986**, *108*, 4937.

14. Adams, J. Q.; Thomas, J. R. *J. Chem. Phys.* **1963**, *39*, 1904.

15. Nelsen, S. F. in *Free Radicals*, Kochi, J. K., Ed., John Wiley & Sons, New York, 1973, p. 583.

16. Hyde, J. S.; Freeman, E. S. *J. Phys. Chem.* **1961**, *65*, 1636.

17. Cole, T.; Pritchard, H.; Davidson, N.; McConnell, H. M. *Mol. Phys.* **1958**, *1*, 406; Cauggis, G.; Genies, M.; Lemaire, H.; Rassat, A.; Ravet, J. P. *J. Chem. Phys.* **1967**, *47*, 4642.

18. Ma, N. L.; Smith, B. J.; Pople, J. A.; Radom, L. *J. Am. Chem. Soc.* **1991**, *113*, 7903.

19. Mullikin, R. S.; Roothan, C. C. J. *Chem. Rev.* **1947**, *41*, 219; Meerer, A. J.; Schoonveld, L. *J. Chem. Phys.* **1968**, *48*, 522; Koppel, H.; Domcke, W.; Cederbaum, L. S.; von Niessen, W. *J. Chem. Phys.* **1978**, *69*, 4252; Bellville, D. J.; Bauld, N. L. *J. Am. Chem. Soc.* **1982**, *104*, 294.

20. Roth, H. D.; Mannion, M. L. *J. Am. Chem. Soc.* **1981**, *103*, 7210.

21. Morton, A. A.; Lanpher, E. J. *J. Org. Chem.* **1958**, *23*, 1636; Lewis, I. C.; Singer, L. S. *J. Chem. Phys.* **1965**, *43*, 2712.

22. Schmidt, W.; Steckhan, E. *Chem. Ber.* **1980**, *113*, 577.

23. Murata, Y.; Shine, H. J. *J. Org. Chem.* **1969**, *34*, 3368.

24. Bard *et al.*, p. 156.

25. *Ibid.*, p. 169.

26. *Ibid.*, p. 188; Ghosh, D.; Bauld, N. L. *J. Catalysis* **1985**, *95*, 300; Lorenz, K.; Bauld, N. L. *ibid*, 613.

27. Laszlo, P.; Lucchetti, J. *Tetrahedron Lett.* **1984**, *25*, 1567.

28. Fox, M. A.; Sackett, D. D.; Younathan, J. N. *Tetrahedron*, **1987**, *43*, 7.

29. Bauld, N. L.; Bellville, D. J. ; Gardner, S. A.; Migron, Y.; Cogswell, G. *Tetrahedron Lett.* **1982**, *23*, 825.

30. Knight, L. B., Jr.; Steadman, J.; Feller, D.; Davidson, E. R. *J. Am. Chem. Soc.* **1984**, *106*, 3700.

31. Toriyama, K.; Nunome, K.; Iwasaki, M. *J. Chem. Phys.* **1982**, *77*, 5891.

32. Hudson, C. E.; Grain, C. S.; McAdoo, D. J. Personal communication from C. E. H.

33. Bellville, D. J.; Bauld, N. L. *J. Am. Chem. Soc.* **1982**, *104*, 5700.

34. Parker, V. D.; Tilset, M. *J. Am. Chem. Soc.* **1988**, *110*, 1649.

35. Bordwell, F. G.; Cheng, J.-P. *J. Am. Chem. Soc.* **1989**, *111*, 1792.

36. Masnovi, J. M.; Sankararaman, S.; Kochi, J. K. *J. Am. Chem. Soc.* **1989**, *111*, 2263.

37. Andrulis, P. J.; Dewar, M. J. S.; Dietz, R.; Hunt, R. L., Jr. *J. Am. Chem. Soc.* **1966**, *88*, 5473; Aratani, T.; Dewar, M. J. S. *ibid.*, 5479; Andrulis, P. J.; Dewar, M. J. S. *ibid.*, 5483.

38. Kochi, J. K. in *Free Radicals*; Kochi, J.K., Ed., John Wiley & Sons, New York, 1973, p. 646.

39. Workentin, M. S.; Schepp, N. P.; Johnston, L. J.; Wayner, D. D. M. *J. Am. Chem. Soc.* **1994**, *116*, 1141.

40. Shine, H. J.; Silber, J. J. *J. Am. Chem. Soc.* **1972**, *94*, 1026.

41. Bruning, W. H.; Nelson, R. F.; Marcoux, L. S.; Adams, R. N. *J. Phys. Chem.* **1967**, *71*, 3055.

42. Kim, E. K.; Bockman, T. M.; Kochi, J. K. *J. Am. Chem. Soc.* **1993**, *115*, 3091.

43. Maslak, P.; Chapman, W. H., Jr.; Vallombroso, T. M., Jr.; Watson, B. A.; *J. Am. Chem. Soc.* **1995**, *117*, 12380.

44. Wolff, M. E. *Chem. Rev.* **1963**, *53*, 55; Neale, R. S. *Synthesis*, **1971**, 1.

45. Bauld, N.L. in *Advances in Electron Transfer Chemistry*; Mariano, P. S., Ed., JAI Press, Greenwich, 1992, p. 1; Bauld, N. L. *Tetrahedron*, **1989**, *45*, 5307.

46. Takahashi, Y.; Kochi, J.K. *Chem. Ber.* **1988**, *121*, 253.

47. Dinnocenzo, J. P.; Conlon, D. A. *J. Am. Chem. Soc.* **1988**, *110*, 2324.

48. Reynolds, D. W.; Harirchian, B.; Chiou, H.-S.; Marsh, B. K.; Bauld, N. L. *J. Phys. Org. Chem.* **1989,** *2*, 57; Bauld, N. L. *J. Computational Chem.* **1990**, *11*, 896.

49. Guo, Q.-X.; Quin, X.-Z.; Wang, J. T., Williams, F. *J. Am. Chem. Soc.* **1988**, *110*, 1974; Bauld, N. L.; Bellville, D. J.; Pabon, R.; Chelsky, R.; Green, G. *J. Am. Chem. Soc.* **1983**, *105*, 2378; Lorenz, K.; Bauld, N. L. *J. Catalysis* **1985,** *95,* 613.

50. Miyashi, T.; Konno, A.; Takahashi, Y. *J. Catalysis* **1988**, *110,* 3676.

51. Dai, S.; Pappas, R. S.; Chen, G.-F.; Guo, Q.-X.; Wang. J. T.; Williams, F. *J. Am. Chem. Soc.* **1989**, *111*, 8759.

52. Crellin, R. A.; Lambert, M. C.; Ledwith, A. *J. Chem. Soc., Chem. Commun.* **1970**, 682; Bawn, C. E. H.; Ledwith, A.; Yang, S.-L. *Chem. Ind.* **1965**, 769; Ledwith, A. *Accts. Chem. Res.* **1972**, *5*, 133.

53. Schutte, R.; Freeman, G. R. *J. Am. Chem. Soc.* **1969**, *91*, 3715; Penner, T. L.; Whitten, D. G.; Hammond, G. S. *J. Am. Chem. Soc.* **1970**, *92*, 2861.

54. Bellville, D. J.; Wirth, D. D.; Bauld, N. L. *J. Am. Chem. Soc.* **1981**, *103*, 718.

55. Calhoun, G. C.; Schuster, G. B. *J. Am. Chem. Soc.* **1984**, *106*, 6870.

56. Eberson, L.; Olofsson, B.; Svensson, J.-O. *Acta Chem. Scand.* **1992**, *46*, 1005.

57. Bauld, N. L. *J. Am. Chem. Soc.* **1992**, *114*, 5800.

58. Roth, H. D.; Schilling, M. L. *J. Am. Chem. Soc.* **1981**, *103*, 7210.

59. Pabon, R. A.; Bellville, D. J.; Bauld, N. L. *J. Am. Chem. Soc.* **1983,** *105*, 5158.

60. Bellville, D. J.; Bauld, N. L. *J. Am. Chem. Soc.* **1982**, *104*, 2665; Bauld, N. L.; Bellville, D. J.; Pabon, R.; Chelsky, R.; Green, G. *J. Am. Chem. Soc.* **1983**, *105*, 2378.

61. Chockalingam, K.; Pinto, M.; Bauld, N. L. *J. Am. Chem. Soc.* **1990**, *112*, 447.

62. Bauld, N. L.; Bellville, D. J.; Pabon, R.; Chelsky, R.; Green, G. *J. Am. Chem. Soc.* **1983**, *105*, 2378.

63. Bauld, N. L.; Pabon, R. A. *J. Am. Chem. Soc.* **1983**, *105*, 633.

64. Pabon, R. A.; Bellville, D. J.; Bauld, N. L. *J. Am. Chem. Soc.* **1984**, *106*, 2730; Kim, T.; Pye, R. J.; Bauld, N. L. J. *Am. Chem. Soc.* **1990**, *112*, 6285.

65. Gassman, P. G.; Singleton, D. A. *J. Am. Chem. Soc.* **1984**, *106*, 7993.

66. Reynolds, D. W.; Lorenz, K. T.; Chiou, H.-S.; Belleville, D. J.; Pabon, R. A.; Bauld, N. L. *J. Am. Chem. Soc.* **1987**, *109*, 4960.

67. Barton, D. H. R.; Leclerc, G.; Magnus, P. D.; Menzies, I. D. *J. Chem. Soc., Chem. Commun.* **1972**, 447.

68. Tang, R.; Yue, H. J.; Wolf, S. F.; DeMares, F. *J. Am. Chem. Soc.* **1978**, *100*, 5248.

69. Nelson, S. F. *Accts. Chem. Res.* **1987**, *20*, 276.

Exercises

5.1 Irradiation of the matrix-isolated cyclobutene cation radical with uv light at low temperature generates a new cation radical that is identical to that formed by γ radiation (radiolysis) of 1,3-butadiene. The product of this electrocyclic cleavage is therefore assumed to be the *s-trans*-1,3-butadiene cation radical. No other doublet species is formed, and the observed new cation radical is formed immediately and only upon uv radiation.

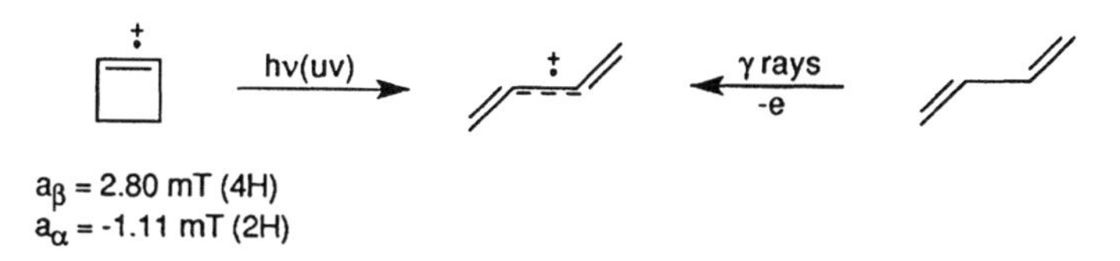

(a) Is there anything surprising about the formation of this specific cation radical upon irradiation of the cyclobutene cation radical? Propose a possible mechanism for this conversion. (b) The β hyperfine splittings in radicals such as the ethyl radical are approximately of the same magnitude as the α splittings. Explain why the beta splittings are so large in the cyclobutene cation radical. (Suggestion: Three major effects probably contribute). Is the magnitude of the α splitting unusual? (c) Of the effects mentioned above, which would you expect to be operative in the cyclobutene anion radical? Would you expect this latter species to have "anomalously high" β hyperfine splittings? Explain. (Gerson, F.; Qin, X.-Z.; Bally, T.; Aebischer, J. N. *Helv. Chim. Acta* **1988**, *71*, 1069.)

5.2 MO calculations predict that the carbene cation radical ($CH_2^{\cdot +}$) is a σ radical and that, specifically, it has the $\pi^+/\sigma^\bullet$ structure shown below, in which a $2p$ AO is vacant (π^+) and a hybrid AO is the SOMO ($\sigma^\bullet$).

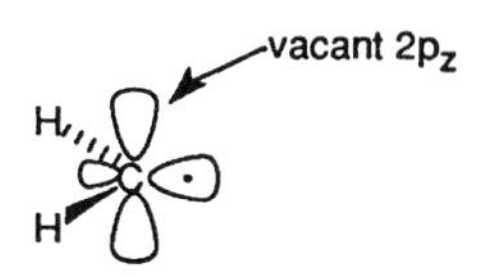

(a) Qualitatively, explain why this is a plausible structure in contrast to a $\pi^\bullet/\sigma^+$ structure. (b) The diphenylcarbene cation radical has been generated by radiolysis of matrix-isolated diphenylcarbene and also by photolysis of the diphenyldiazomethane cation radical. The ^{13}C hyperfine splitting constant from the carbene carbon in this cation radical is 9.83 mT. How can this be used to exclude the $\pi^\bullet/\sigma^+$ structure? Explain. (Bally, T.; Matzinger, S.; Truttman, L.; Platz, M. S.; Admasu, A.; Gerson, F.; Arnold, A.; Schmidlin, R. *J. Am. Chem. Soc.* **1993**, *115*, 7007.)

5.3 (a) Ionization of the phenylcyclopropane derivative shown below by the photosensitized electron-transfer method, using 1,4-dicyanobenzene as the sensitizer and acetonitrile/methanol (12:1 v/v) as the solvent, results in nucleophilic cleavage of the cyclopropane ring. Propose a mechanism for the formation of the indicated product and suggest a structure for the substrate cation radical that is consistent with the regiochemistry of the observed cleavage. What type of structure for the cation radical intermediate is definitely precluded?

DCB, hν
CH_3CN/MeOH

(b) As illustrated in the reaction above, the cleavage occurs with complete inversion of configuration at the carbon undergoing nucleophilic substitution. Reconcile this result with your proposed mechanism and again indicate what type of mechanism is clearly excluded. (Dinnocenzo, J. P.; Todd, W. P.; Simpson, T. R.; Gould, I. R. *J. Am Chem. Soc.* **1990**, *112*, 2462.)

5.4 1,4-Diazabicyclo[2.2.2]octane (dabco) has an unusually low oxidation potential for a simple tertiary amine (0.68 V versus SCE), and its cation radical has a half-life on the order of several seconds (unusually long for a simple amine cation radical).

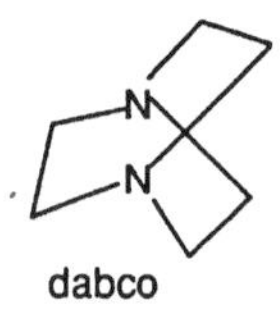

dabco

(a) Use FO theory to explain the ease of formation and relatively long lifetime of this cation radical. What does the bicyclic nature of the cation radical have to do with its stability? (b) The ESR spectrum of the dabco cation radical shows a 1.70-mT splitting from two equivalent ^{14}N atoms and a 0.73-mT splitting from 12 equivalent protons. Indicate the significance of the magnitude and the multiplicity (2 N's) of the ^{14}N hyperfine splitting in regard to the structure of the cation radical. (McKinney, T. M.; Geske, D. H. *J. Am. Chem. Soc.* **1965**, *87*, 3013.)

5.5 (a) The ionization of benzene at 4.2 K by X-radiolysis yields a cation radical having the following hyperfine splittings: $a_H = -0.82$ (2H), $a_H = -0.24$ (4H). When allowed to warm to 100 K, the ESR spectrum becomes an evenly spaced septet with a binomial intensity distribution ($a_H = -0.43$ mT). Depict the SOMO of the benzene cation radical observed at 4.2 K and explain the transition to a simple septet spectrum.

X-rays
$CFCl_3$
4.2K

(b) In $CFCl_2CF_2Cl$ solvent, the septet spectrum mentioned above is also observed at 77–100 K, but at and above 100 K, the spectrum changes to a 13-line one with $a = -0.21$. Explain. (Iwasaki, M.; Toriyamo, K.; Nunone, K. *J. Chem. Soc. Chem. Commun.* **1983**, 320.)

5.6 A direct method for studying cation radical structures is potentially available via detailed analysis of the vibrational structure of the photoelectron spectrum (PES) of the corresponding neutral molecule. When studied in this way, the structure of the cyclobutadiene cation radical is found to be rectangular (D_{2h}), and qualitatively similar to the ground state of cyclobutadiene itself.

(a) Explain in terms of an HMO energy-level diagram why the cyclobutadiene cation radical would not be expected to have a square (D_{4h}) structure.

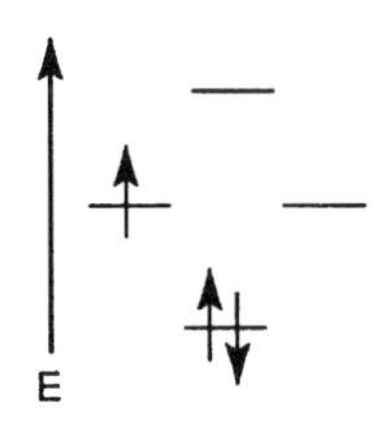

(b) The HMOs of the open-shell singlet of neutral cyclobutadiene are depicted below. Based upon the arguments for a distorted cation radical, would this singlet necessarily be expected to undergo distortion?

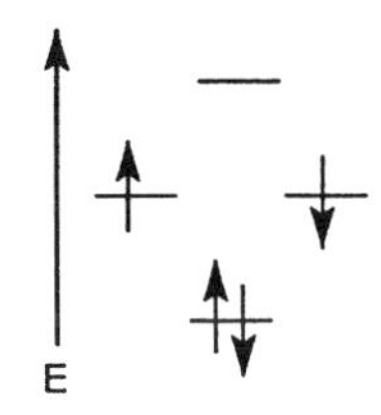

(Note: The distortion observed in the cation radical is referred to as a first-order Jahn–Teller distortion; that in the neutral is a second-order or pseudo-Jahn–Teller distortion.) (Kohn, D. W.; Chen, P. *J. Am. Chem. Soc.* **1993**, *115*, 2844.)

5.7 Upon γ irradiation of semibullvalene, a cation radical is produced that has the following ESR spectrum: $a_{1,5}$ = 3.62 mT (2H); $a_{2,4,6,8}$ = −0.77 mT (4H).

The structure is evidently a bis(allylic) cation radical, but could conceivably be either a rapidly equilibrating distonic cation radical or a bis(homobenzene) type of cation radical:

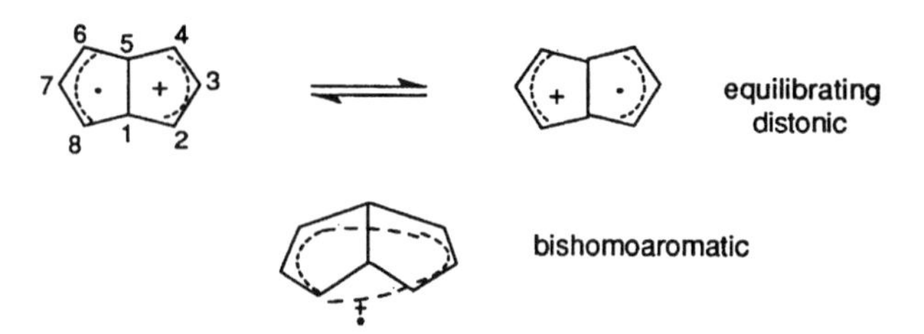

(a) Consider whether the (α) splittings of the hydrogens at C_2, C_4, C_6, and C_8 are qualitatively consistent with splittings from hydrogens at a terminal allylic position. (b) Is the absence of resolvable splittings from protons at C_3 and C_7 consistent with protons at a central allylic position? (c) Does the magnitude of the β hyperfine splitting suggest a choice between equilibrating distonic and bishomoaromatic structures? Explain.

5.8 The radiolytic oxidation of either 1,5-hexadiene or bicyclo[2.2.0]hexane yields the same binomial seven-line ESR pattern [a = 1.20 mT (6H)] that is assigned to the cyclohexane-1,4-diyl cation radical.

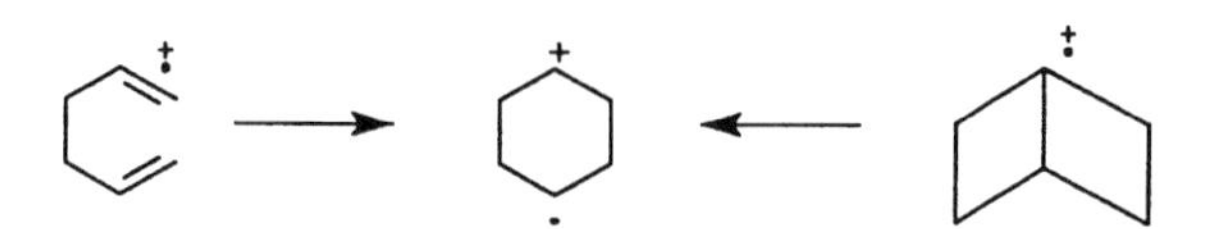

(a) Reconcile this simple septet spectrum with the proposed structure of the cation radical. (b) The cyclohexane-1,4-diyl cation radical has been found to have a chair as opposed to a possible boat structure by ionizing the deuterated bicyclohexane substrate shown below. The resulting cation radical is found to have a six-line spectrum with the same splitting constant as before. Explain how this experiment rules out a boat structure but is consistent with a chair structure.

D D —(- e)→ D D → cyclohexane-1,4-diyl cation radical (6 lines)

5.9. Monoamine oxidase (MAO), a flavin-dependent enzyme, catalyzes the oxidation of amine neurotransmitters to aldehydes in a series of reactions that is considered to involve initial electron transfer from the amine to the enzyme, giving an aminium cation radical. The latter is usually considered to lose an α proton to yield an α-aminoalkyl radical, which is then oxidized to an immonium ion. The latter is ultimately hydrolyzed to give an aldehyde.

$$RCH_2\ddot{N}H_2 \underset{}{\overset{Fl \quad Fl^{\bar{\cdot}}}{\rightleftharpoons}} RCH_2\overset{+\cdot}{N}H_2 \xrightarrow{-H^+} R\dot{C}HNH_2 \xrightarrow{Fl \quad Fl^{\bar{\cdot}}} RCH{=}\overset{\oplus}{N}H_2 \rightarrow RCHO$$

When *trans*-2-phenylcyclopropylmethylamine is used as a substrate, the enzyme functions normally and oxidizes the amine to *trans*-2-phenylcyclopropanecarbaldehyde.

Ph–C₃H₄–$CH_2\ddot{N}H_2$ $\xrightarrow{MAO}$ Ph–C₃H₄–CHO

(a) How could this result be rationalized in terms of the radical mechanism above, considering the rapid cleavage expected of the cyclopropylcarbinyl radical intermediate? (b) How could the result be rationalized within a simple mechanistic framework that involves the aminium ion and the immonium ion but circumvents the radical intermediate? (c) When cubylcarbinylamine is used as a radical trap ($k_{probe} = 3 \times 10^{10}\,s^{-1}$), the enzyme is irreversibly inactivated, and the oxidation products include two in which the cubane ring has undergone cleavage. What conclusions does this result suggest concerning the mechanism of MAO oxidation of amines?

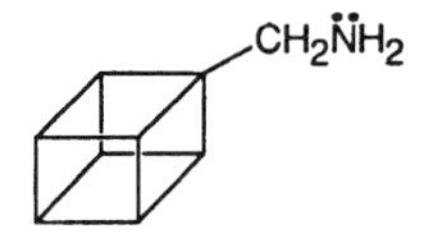

(d) Is there a property, other than the rapid rate of the cleavage reaction, that could make the cubylcarbinyl radical cleavage an especially good probe reaction for enzyme-catalyzed reactions? (Silverman, R. B.; Zhon, J. P.; Eaton, P. C. *J. Am. Chem. Soc.* **1993**, *115*, 8841.)

5.10 The cytochrome P-450 catalyzed oxidation of amines is analogous to the MAO oxidation in the previous question in that the first step is well established as involving electron transfer to yield an aminium cation radical. The reaction proceeds efficiently for a wide structural variety of amines, but when cyclopropylamines are used, the enzyme is irreversibly inactivated–a phenomenon referred to as mechanism-based inactivation. Propose an explanation for this inactivation specifically by cyclopropylamines.

R-cyclopropyl–$\ddot{N}HCH_2Ph$ $\xrightarrow{P\text{-}450}$ R-cyclopropyl–$\overset{+\cdot}{N}HCH_2Ph$ $\xrightarrow{inactivation}$

(Macdonald, T. L.; Zirvi, K.; Burka, L. T.; Peyman, P.; Guengerich, F. P. *J. Am. Chem. Soc.* **1982**, *104*, 2050.)

5.11 The Diels–Alder cycloadditions of a series of mono- and disubstituted stilbenes to 2,3-dimethyl-1,3-butadiene, catalyzed by triarylaminium salts, have been studied via competition kinetics. The relative rates (log k_{rel}) correlate very well with σ^+ ($\rho = -4.16$, $r^2 = 0.988$). The peak oxidation potentials of these same stilbene substrates also correlate well with σ^+ (slope 0.272, $r^2 = 0.989$), and the slope corresponds to $\rho = -5.02$. The kinetic studies and the oxidation potential studies were both carried out in acetonitrile at 0°C, and both mono- and symmetrically disubstituted stilbenes fit well on both plots.

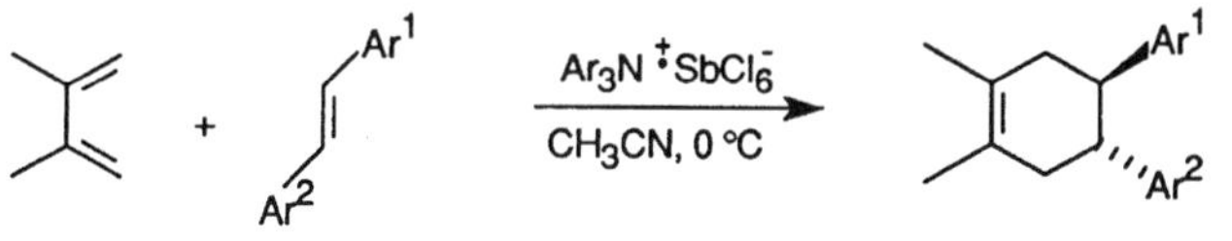

$Ar_3N^{\cdot +}SbCl_6^-$

CH_3CN, 0 °C

Ar[1] Ar[2] Ar[1] Ar[2]

(a) Propose a detailed mechanism for these reactions that is consistent with these data. (b) Are these data consistent with an electrophilic mechanism involving rate-determining addition of a proton or other electrophile to stilbene? Explain. (c) Are these data consistent with the involvement of the 2,3-dimethyl-1,3-butadiene cation radical? Explain. (d) Propose a plausible reason why the observed ρ value is less than the ρ value for reversible ionization to the cation radical (from the oxidation potentials). Under what conditions could the observed ρ be equal to that for reversible ionization? (e) The absolute rates of these reactions are found to be linearly dependent upon the concentration of the diene. Interpret this result in terms of your proposed mechanism.

CHAPTER

6

Ion Radical Pairs and Electron Transfer

Electron transfer (ET) between two neutral molecules yields an *ion radical pair*, a unique chemical entity that contains both a cation radical and an anion radical moiety (Figure 6.1). Depending upon the distance between the donor and acceptor molecules at the instant of electron transfer, either a *contact* (also called intimate) or a *solvent-separated* (also called "solvent-penetrated") ion radical pair could be formed. Given a sufficiently long lifetime, and a relatively polar solvent, these ion radical pairs may dissociate into free cation and anion radicals. Alternatively, or competitively, they may undergo back electron transfer (BET) to regenerate the original neutral molecules in their ground electronic state. In the case of contact ion radical pairs, the ion radicals may also react with each other by radical coupling, proton transfer, cycloaddition, or other covalent bond-forming reactions. Both types of ion radical pairs and, most especially, the free ion radicals may react with external molecules, especially via the very reactive cation radical moiety. The reactions of cation radicals and of anion radicals have already been considered in the preceding

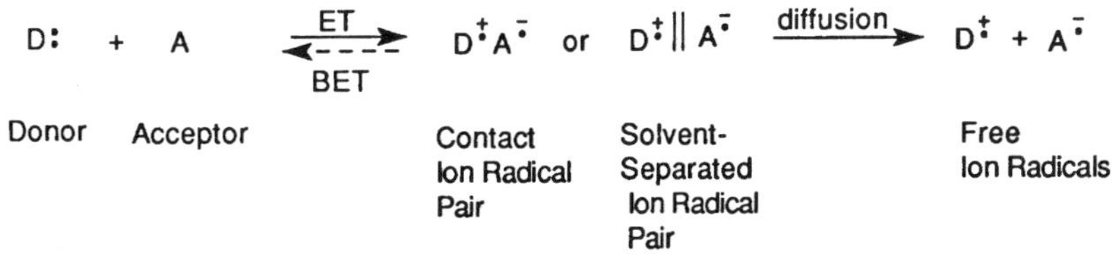

$$\underset{\text{Donor}}{D\!:} + \underset{\text{Acceptor}}{A} \underset{\text{BET}}{\overset{\text{ET}}{\rightleftharpoons}} \underset{\substack{\text{Contact}\\ \text{Ion Radical}\\ \text{Pair}}}{D^{\dot{+}}A^{\dot{-}}} \text{ or } \underset{\substack{\text{Solvent-}\\ \text{Separated}\\ \text{Ion Radical}\\ \text{Pair}}}{D^{\dot{+}} \| A^{\dot{-}}} \xrightarrow{\text{diffusion}} \underset{\substack{\text{Free}\\ \text{Ion Radicals}}}{D^{\dot{+}} + A^{\dot{-}}}$$

Figure 6.1 Ion radical pairs and ion radicals.

$$\Delta G^0 = [E_{ox}(D) - E_{red}(A)] - E_{excit}$$

where $E_{ox}(D)$ = oxidation potential of D (Donor)

$E_{red}(A)$ = reduction potential of A (Acceptor)

Figure 6.2 The Weller equation.

two chapters. In this chapter we shall focus primarily on the unique aspects of ion radical *pairs* and upon the electron-tansfer (ET) processes that generate them and, in part, consume them.

6.1 The Energetics of Electron Transfer

The energetic feasibility of electron transfer can be readily predicted using the Weller equation (Figure 6.2).[1] The quantity inside the brackets represents the free energy required to generate the ion radicals $D^{\dot{+}}$ and $A^{\dot{+}}$ from the corresponding neutrals. Since most organic molecules have substantially positive oxidation potentials and substantially negative reduction potentials (both relative to SCE, for example), the electron-transfer process that generates ion radical pairs from neutral molecules in their ground electronic states is normally quite strongly endergonic. However, if either A or D is in an electronically excited state, the excitation energy (E_{excit}) represents a powerful thermodynamic driving force for electron transfer. Consequently, photosensitized (or photoinduced) electron transfer (PET) is a much more general method than thermal electron transfer (TET) for generating ion radical pairs.

6.2 Rates of Electron Transfer

The relative rates of a series of closely related electron-transfer reactions can be predicted from the Marcus equation, which is given in its simplest form in Figure 6.3.[2] In this equation, the activation free energy for an electron-transfer reaction under consideration is $\Delta G^{\ddagger}$, while $\Delta G_0^{\ddagger}$ (the intrinsic activation energy) is the corresponding activation energy for a reference reaction that is identical in all respects to the reaction under consideration but that has no thermodynamic driving force (i.e., $\Delta G_0^0 = 0$). The free energy change (thermodynamic driving force) corresponding to the reaction under consideration is given by ΔG^0. Essentially, the

$$\Delta G^{\ddagger} = \Delta G_0^{\ddagger}\,[1 + \Delta G^0 / 4\,\Delta G_0^{\ddagger}]^2$$

Figure 6.3 The Marcus equation: Simplified version.

$$\Delta G^{\ddagger} = \Delta G_0^{\ddagger} + 1/2\,\Delta G^0 + (\Delta G^0)^2/16\Delta G_0^{\ddagger}$$
$$\Delta\Delta G^{\ddagger} = \Delta G^{\ddagger} - \Delta G_0^{\ddagger} = 1/2\,\Delta G^0 + (\Delta G^0)^2/16\Delta G_0^{\ddagger}$$

Figure 6.4 An expanded Form of the Marcus equation.

Marcus equation predicts the form of the dependence of the activation free energy ($\Delta G^{\ddagger}$) on the thermodynamic free energy change (ΔG^0) of the reaction. An equivalent expression for the Marcus equation (Figure 6.4) makes clear, for example, that for relatively small driving forces (ΔG^0 small relative to $\Delta G_0^{\ddagger}$), only one-half of the driving force (ΔG^0) is useful for decreasing the activation energy. However, to diminish the activation energy to zero requires $\Delta G^0 = -4\Delta G_0^{\ddagger}$, as is seen most readily from the first form of the Marcus equation. Perhaps most interesting of all, it is a consequence of the quadratic form of the Marcus equation that when $-\Delta G^0 > 4\Delta G_0^{\ddagger}$, the activation energy ($\Delta G^{\ddagger}$) once again becomes positive and continues to increase as ΔG^0 becomes progressively more negative. This so-called Marcus "inverted region" has been observed for the highly exergonic back-electron-transfer reactions of ion radical pairs.

6.3 The Marcus Equation: Derivation

The derivation is based simply upon intersecting reactant (R) and product (P) free energy parabolas on a reaction coordinate diagram. The reactant system is assumed to be described by the free energy function $G(R) = \lambda x^2$, where x is the value of the reaction coordinate, taken to be 0 for the energy minimum of the reactant system and 1 for the product system energy minimum (Figure 6.5). First a reference reaction is considered that has no thermodynamic driving force ($\Delta G^0 = 0$). The free energy function for the product system is then given by $G(P) = \lambda(x - 1)^2$. The transition state for the reaction is considered to occur at the intersection of the two parabolas, at which point $G(R) = G(P) = G(\ddagger)$. By setting $\lambda x^2 = \lambda(x - 1)^2$, the value of the reaction coordinate at the transition state is obtained ($x = 0.5$). Substituting this value of x into either $G(R)$ or $G(P)$ gives $G(\ddagger) = \lambda/4$ and thus $\Delta G_0^{\ddagger} = G(\ddagger) - G(R) = \lambda/4$. The parameter λ is easily seen to be the value of $G(R)$ at $x = 1$ or of $G(P)$ at $x = 0$. Considering the former case, λ can be construed as the energy of the reactants when distorted to the equilibrium geometry of the products. For a reaction in solution this would also include the solvation shell.

In the case of electron transfer between two neutral molecules A and D, affording an ion radical pair $A^{\overline{\bullet}}/D^{\overset{+}{\bullet}}$, λ would refer to the energy of A and D distorted to the equilibrium geometry of the anion radical and the cation radical ($A^{\overline{\bullet}}$ and $D^{\overset{+}{\bullet}}$, respectively) and in the equilibrium solvation shell(s) of $A^{\overline{\bullet}}/D^{\overset{+}{\bullet}}$. The parameter λ is thus referred to as the reorganization energy of the reaction and is composed of internal (λ_i) and solvent (λ_s) contributions.

To complete the derivation, an analogous reaction is considered that has the same value of λ, but that has a thermodynamic driving force ($\Delta G^0 \neq 0$, Figure 6.6). The

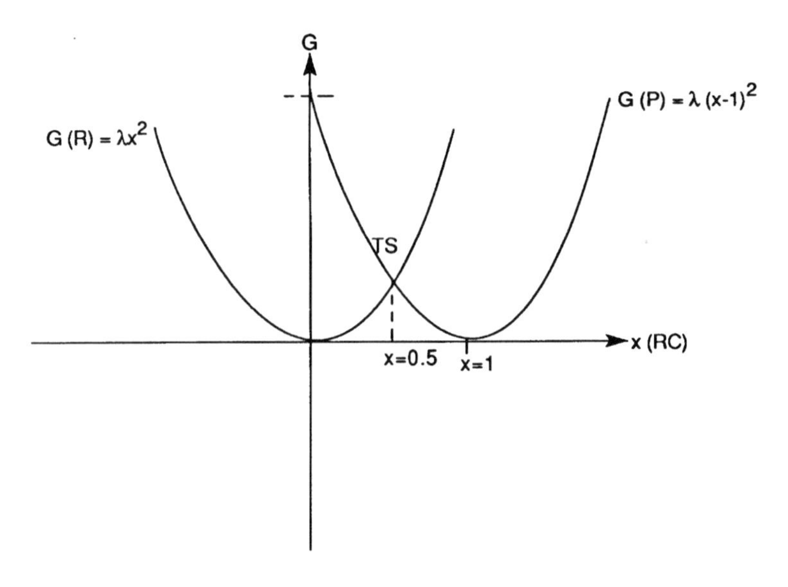

Figure 6.5 Marcus parabolas for the reference reaction with $\Delta G^\circ = 0$.

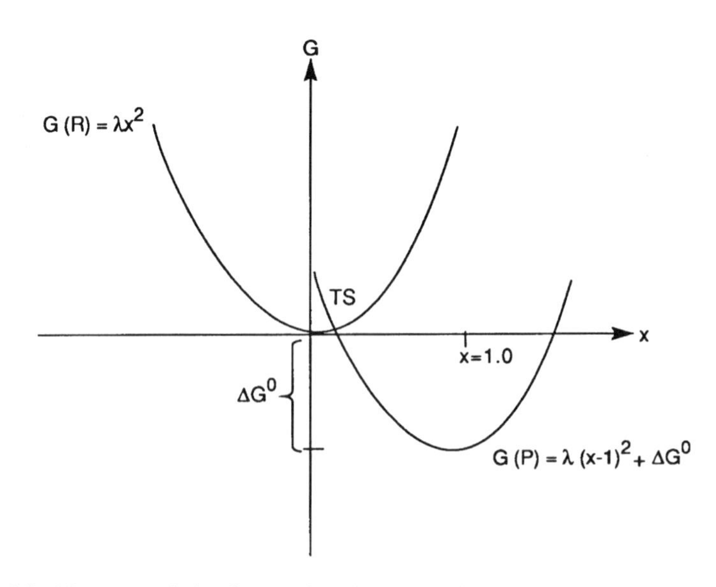

Figure 6.6 Marcus parabolas for reactions involving thermodynamic driving force.

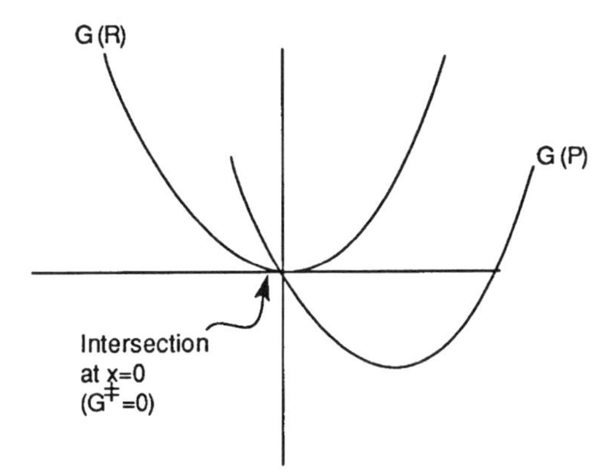

Figure 6.7 Intersecting Marcus parabolas for the case where $\Delta G^0 = -\lambda$.

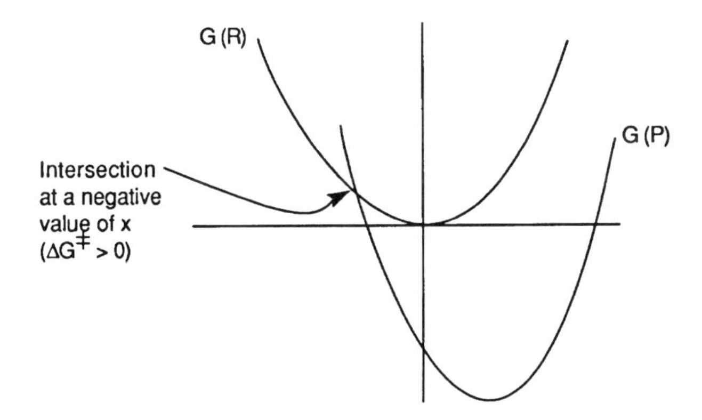

Figure 6.8 Intersecting Marcus parabolas for the case where $-\Delta G^0 > \lambda$.

product free energy function now becomes $G(P) = \lambda(x - 1)^2 + \Delta G^0$. Once again setting $G(P) = G(R)$, the reaction coordinate value corresponding to the transition state is obtained: $x = \frac{1}{2}(1 + \Delta G^0/\lambda)$. Substituting this value of x into $G(R) = \lambda x^2$ affords the Marcus equation (recognizing that $\lambda = 4\ \Delta G_0^{\ddagger}$). It is easily seen that the intersection (TS) occurs at $x = 0$ for a reactant parabola that has $\Delta G^0 = -\lambda$. In this case, $\Delta G\ddagger = 0$ (Figure 6.7). When $-\Delta G^0 > \lambda$, the intersection occurs at a negative value of x, corresponding to $\Delta G\ddagger > 0$ (Figure 6.8).

6.4 Electrostatic Effects: The Full Marcus Equation

When both reactants or both products are ionic, as in thermal electron-transfer reactions between neutral molecules or in back electron transfer between ion radical pairs, electrostatic effects arise that, in principle, affect the Marcus equation in two

$$\Delta G^{\ddagger} = \Delta G_0^{\ddagger} [1 + \Delta G^{0'}/\lambda]^2 + Z_1 Z_2 e^2 f/Dr_{12}$$

where:

$$\Delta G^{0'} = \Delta G^0 + (Z_1 - Z_2 - 1)(e^2 f/Dr_{12})$$

Z_1, Z_2 = charges on the two reactants

f = ionic strength

D = dielectric constant of the solvent

r_{12} = distance (between reactants) at which electron transfer occurs

Figure 6.9 The Marcus equation with electrostatic effects.

ways (Figure 6.9). First, the relevant free energy change ($\Delta G^{0'}$) is no longer ΔG^0 when the reactant or product state is stabilized or destabilized by electrostatic effects. For example, when neutral molecules undergo electron transfer to give an ion radical pair, the product state is stabilized by electrostatic attractions between the oppositely charged ions. This corresponds to the case where $Z_1 = Z_2 = 0$ (uncharged reactant states). The free energy change is decreased by e^2f/Dr_{12}. Conversely, for back electron transfer within an ion radical pair, the standard free energy change is *increased* by e^2f/Dr_{12}. These effects appear to be rather small when the solvent involved is polar (large D), as in the case of acetonitrile. It is noted that the electrostatic correction to ΔG^0 vanishes when one reactant is uncharged *and* one product is uncharged.

The electrostatic effect on the activation free energy (the second term in Figure 6.9) is nil when either reactant is uncharged. The term therefore vanishes for ET between neutral molecules, but contributes to the lowering of the activation barrier for back electron transfer ($Z_1 = +1$, $Z_2 = -1$). However, it should be noted that the activation free energy now refers to the dissociated ions as the initial reactant state, not to the activation energy of the back-electron-transfer step itself, which is given by the first term in Figure 6.9.

6.5 Limitations of the Marcus Equation

In a very strict sense, the $\Delta G^{\ddagger}$ and $\Delta G_0^{\ddagger}$ of the Marcus equation could be understood to refer to reactions that are identical except for the standard free energy changes, which are, respectively, ΔG^0 and zero. Obviously, it would be impossible to vary ΔG^0 without affecting the reaction in some way. More practically, the Marcus equation should hold so long as the reorganization energy, λ, remains reasonably constant for the series. This milder restriction nevertheless places serious limitations on the range of application of the Marcus equation. An excellent example is provided by the effect of methyl substituents attached to an aromatic ring.[3] In the back-electron-transfer reaction between the 1,2,4,5-tetracyanobenzene anion radical ($TCB^{\bar{\bullet}}$) and various alkylated benzene cation radicals ($D^{\dot{+}}$) in acetonitrile, the value of λ_s varies from 0.70 eV for *p*-xylene to 0.59 for durene (1,2,4,5-tetramethylbenzene) and to 0.45 for hexamethylbenzene. The decrease in λ_s as a consequence of progressive alkyl substitution apparently reflects the increased ability of the substituents to stabilize the cation radical moiety and the concomitant diminished solvation of the cation radical. Results such as this make the strong point that even when the basic substrate structures remain constant, too great a variation of substituent effects could result in deviation from the Marcus equation. These deviations can even include obscuring the Marcus inverted region. Finally, it may be worthwhile to point out that the Marcus equation can be, and has been, applied to a wide range of reactions other than electron transfer, including S_N2 reactions.[4]

6.6 Back Electron Transfer in Contact Ion Radical Pairs

Contact ion pairs can be generated by irradiation of charge-transfer (CT) complexes at the wavelength corresponding to the charge-transfer absorption (Figure 6.10). Since back electron transfer lowers the quantum yield of free ions, higher quantum yields of the latter are obtained when back electron transfer is slow. By measuring the quantum yield of free ions (Φ_{ions}) and having rather accurate estimates of k_{sep} and k_{solv}, the rate constants for back electron transfer (k_{-et}) have been obtained for a variety of donors.[5a] The free energy changes (ΔG^0) for the back-electron-transfer reactions were obtained from electrochemical measurements [$\Delta G^0 = E_{red}(A) - E_{ox}(D)$]. The log k_{-et} versus $-\Delta G^0$ data were originally fitted to an equation of the Marcus type in order to obtain the apparent value of the reorganization energy (λ = 0.85 eV). Very recently, λ has been measured independently (0.73 eV), permitting the rates of back electron transfer to be *predicted* by the Marcus equation.[5b] The plot in Figure 6.11 labeled $A^{\bullet -}D^{\bullet +}$ shows the fit of the calculated curve (solid curve) with the experimental data for back electron transfer within the contact ion radical pair (squares). It is noted that the fit is excellent and that increasing the thermodynamic driving force (greater $-\Delta G^0$) corresponds to a decreased rate of back electron transfer. The requirement for $\Delta G^{\ddagger} = 0$ (maximum rate) is $-\Delta G^0 = \lambda$

$$D + A \rightleftharpoons \underset{\text{CT complex}}{D \longrightarrow A} \underset{k_{-et}}{\overset{h\nu_{CT}}{\rightleftharpoons}} D^{\dot{+}}A^{\bar{\bullet}} \underset{k_{-solv}}{\overset{k_{solv}}{\rightleftharpoons}} D^{\dot{+}}//A^{\bar{\bullet}} \xrightarrow{k_{sep}} \underset{\text{free ion radicals}}{D^{\dot{+}} + A^{\bar{\bullet}}}$$

A = 2,6,9,10-tetracyanoanthracene
D = alkylated benzenes (3-6 alkyl groups)
λ_{CT} = 460 nm
solvent = acetonitrile

Figure 6.10 Generation of contact ion radical pairs by irradiation of charge transfer complexes.

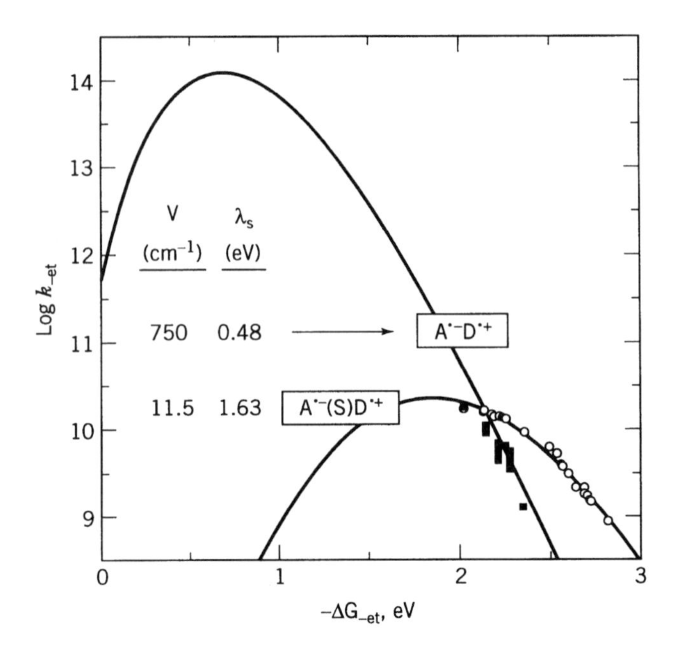

Figure 6.11 A plot of log k_{-et} for back electron transfer versus the thermodynamic driving force, $-\Delta G_{-et}$, for back electron transfer. The plot labeled $\bar{A^{\bullet}}$ $\overset{+}{D^{\bullet}}$ corresponds to back-electron-transfer in contact ion radical pairs, whereas the plot labeled $\bar{A^{\bullet}}$ (S)$\overset{+}{D^{\bullet}}$ corresponds to back electron transfer from solvent separated ion radical pairs. Figure kindly supplied by Dr. Samir Farid of the Eastman Kodak Company, Rochester, New York.

= 0.73 eV. Consequently, the Marcus inverted region includes those reactions that are at least 0.73 eV (16.8 kcal/mol) exergonic.

6.7 Back Electron Transfer in Solvent-Separated Ion Radical Pairs

As noted previously, a more general method for generating ion radical pairs is provided by photosensitized electron transfer (PET). In this approach, acceptor molecules, rather than donor–acceptor complexes, are selectively excited by radiation of the appropriate wavelength. Donor molecules then transfer an electron to the acceptor excited state in a highly exergonic process (Figure 6.12). In contrast to the excitation of a charge–transfer complex, this electron transfer could, *a priori*, lead to either contact or solvent-separated ion radical pairs. A PET study parallel to the one previously described for back electron transfer in contact pairs was carried out, and the results conclusively rule out the formation of contact ion radical pairs in the PET process in acetonitrile solvent.[6] Once again the quantum yield of free ion radicals is observed to increase as the reaction exergonicity ($-\Delta G^0$) increases,

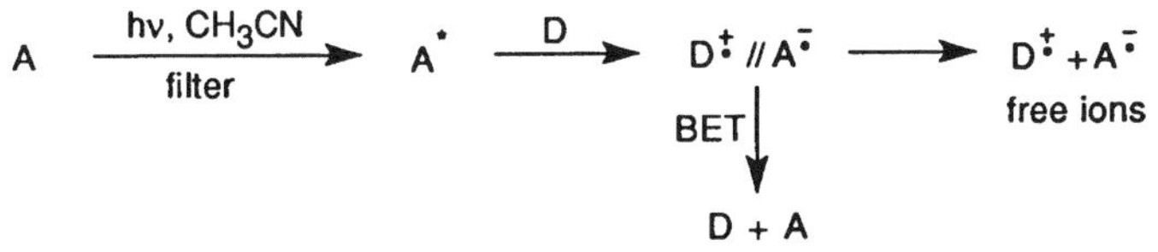

$$A \xrightarrow[\text{filter}]{h\nu,\ CH_3CN} A^* \xrightarrow{D} D^{\bullet +}//A^{\bullet -} \longrightarrow \underset{\text{free ions}}{D^{\bullet +} + A^{\bullet -}}$$
$$D^{\bullet +}//A^{\bullet -} \xrightarrow{\text{BET}} D + A$$

Figure 6.12 Generation of solvent-separated ion radical pairs by photosensitized electron transfer.

indicating that the back-electron-transfer rate constants (k_{-et}) decrease as the thermodynamic driving force increases (the Marcus "inverted region"). However, the dependence of the quantum yields of free ion radicals and also the back-electron-transfer rate constants upon $-\Delta G^0$ is very different from that observed for the contact ion pairs. Specifically, a fit of the data to the Marcus equation yields a much larger reorganization energy, $\lambda = 1.9$ eV. The large solvent contribution to this ($\lambda_s = 1.6$ eV) suggests that the pairs from which back electron transfer occurs are much more highly solvated than the contact ion pairs considered previously. Presumably solvent-separated ion pairs are formed directly by electron transfer between A^* and D, through a solvent molecule (acetonitrile).

The recent independent determination of the reorganization energy for back electron transfer within solvent-separated ion pairs ($\lambda = 1.88$) is in excellent agreement with the value previously obtained by curve fitting ($\lambda = 1.9$). The plot of the calculated Marcus curve and its excellent fit to the experimental data (circles) are illustrated in Figure 6.11 [the curve labeled $A^{\bullet -}(S)D^{\bullet +}$]. The Marcus plots in Figure 6.11 provide a clear basis for the preference for electron transfer over the longer distance (ca. 7 Å) required to provide a solvent-separated ion pair. Although the intrinsic activation energy ($\lambda/4$) and the reorganization energy (λ) for BET are smaller in contact than in solvent-separated ion radical pairs, the extremely large exergonicities of these BET reactions (2–3 eV) exceed λ by much more for contact ion radical pairs ($\lambda = 0.73$ eV) than for solvent-separated ion radical pairs ($\lambda = 1.88$ eV). The former are therefore much further into the inverted region than are the latter, and are more adversely affected by the extremely large $-\Delta G^0$.

In the reaction system under consideration, electron transfer to give contact ion radical pairs would be preferred for reactions having exergonicities of less than ca. 2 eV (46 kcal/mol). This is also valuable information, since it makes it clear that mildly exergonic or endergonic electron transfers (e.g., thermal ET) should occur via contact ion radicals, since these have much smaller reorganization energies (λ) and therefore smaller barriers ($\Delta G_0^{\ddagger} = \lambda/4$) to electron transfer.

6.8 Production of Triplets via Back Electron Transfer

Since back electron transfer from an anion radical to a cation radical is often exergonic to the extent of 2–3 eV (46–69 kcal/mol), it should not be too surprising if, in certain cases, one of the neutral reactants is generated in an excited triplet

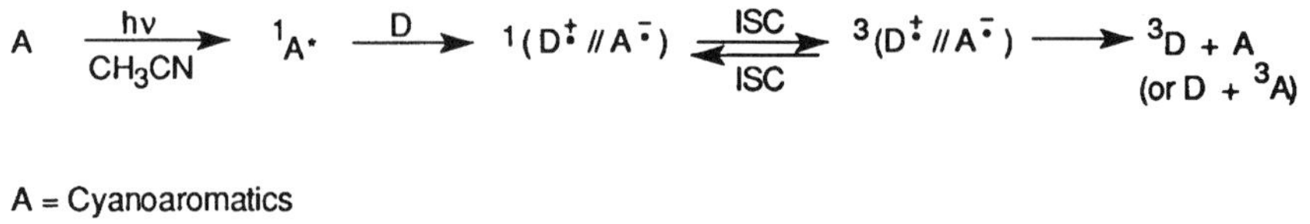

$$A \xrightarrow[CH_3CN]{h\nu} {}^1A^{\bullet} \xrightarrow{D} {}^1(D^{+\bullet}//A^{-\bullet}) \underset{ISC}{\overset{ISC}{\rightleftarrows}} {}^3(D^{+\bullet}//A^{-\bullet}) \longrightarrow {}^3D + A \ (\text{or } D + {}^3A)$$

A = Cyanoaromatics

Figure 6.13 Formation of triplets via back electron transfer in ion radical pairs.

electronic state. If the ion radical pair is initially generated in a singlet state (i.e., if A^* is an excited singlet state), this requires an intersystem crossing (ISC) to the triplet ion radical pair, followed by back electron transfer to afford either ${}^3D$ or ${}^3A$, or both (Figure 6.13).

On the other hand, if the excited-state acceptor in the PET procedure is a triplet state, the ion radical pair is formed directly in its triplet state. In the latter state, back electron transfer to give the neutral molecules D and A in their ground states is spin forbidden. If sufficient energy is not available from back electron transfer to give either ${}^3D$ or ${}^3A$, the triplet ion radical either dissociates to free ion radicals or crosses over to the singlet state (Figure 6.14).

As an example of singlet excited-state photosensitized electron transfer leading to triplet chemistry, consider the PET reaction of methyl 2,3-diphenylcyclopropene-carboxylate (CP), sensitized by 9,10-dicyanoanthracene (DCA; Figure 6.15). The singlet excitation energy of DCA is 68 kcal/mol (2.95 eV), easily sufficient to provide efficient electron transfer from CP to *DCA, giving $CP^{+\bullet}//DCA^{-\bullet}$ [$\Delta G^0 = E_{ox}(CP) - E_{red}\ (DCA) - E_{excit}\ (DCA) = 1.69 - (-0.88) - 2.95 = 2.57 - 2.95 = -0.38$ eV]. The back-electron-transfer reaction to provide ground-state CP and DCA is 2.57 eV exergonic and is well into the Marcus inverted region ($\lambda \approx 1.9$ eV), thus allowing diffusive separation and intersystem crossing to compete extensively. Back electron transfer from the triplet ion radical pair is also exergonic and leads to ^{3}CP [$\Delta G^0 = E_{red}\ (DCA) - E_{ox}\ (CP) + E_{excit} = -2.57 + 2.3 = -0.27$ eV, where $E_T(CP) =$ 53 kcal/mol or 2.3 eV].

The intermediacy of ^{3}CP* in dimer formation is confirmed by triplet quenching experiments in which a low-energy triplet quencher is shown to diminish the yield

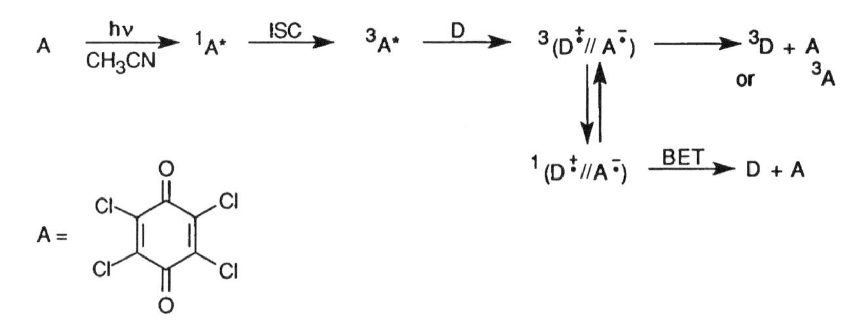

Figure 6.14 Direct formation of triplet ion radical pairs via excited triplet acceptors.

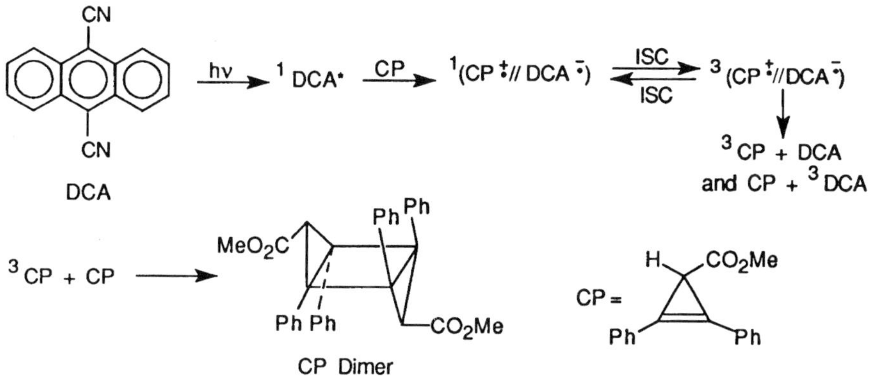

Figure 6.15 A Case of triplet formation via BET.

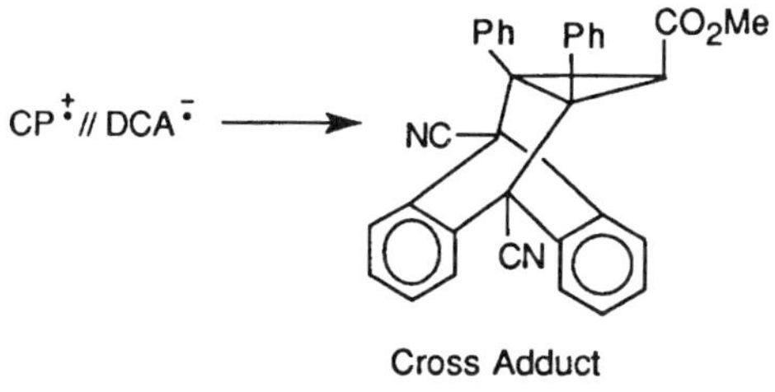

Figure 6.16 Cycloaddition of an ion radical pair.

of dimer but not to affect the yield of the cross adduct, which arises from an interesting cycloaddition of $CP^{+\bullet}/DCA^{-\bullet}$ (Figure 6.16).

6.9 Ion Radical Chemistry via Photosensitized Electron Transfer

The cyclodimerization of N-vinylcarbazole, as described in Chapter 5, is now considered the classic example of a cation radical/neutral cycloaddition reaction. The reaction was originally carried out by Ellinger using chloranil/*hν*, in what is presumably the prototype of the photosensitized electron-transfer method.[8] The cation radical chain mechanism was later identified and rigorously characterized by Ledwith's group. The method has subsequently been developed and used extensively for generating ion radical pairs and especially for effecting cation radical chemistry analogous to that discovered by Ellinger and Ledwith. Most of the cation radical reactions that have been described in Chapter 5 can be carried out under PET conditions. In a number of cases, the PET method has been found to be preferable to the aminium salt catalytic method as a consequence of the strong acid generated in the course of the latter reactions. As an example, the acid-sensitive

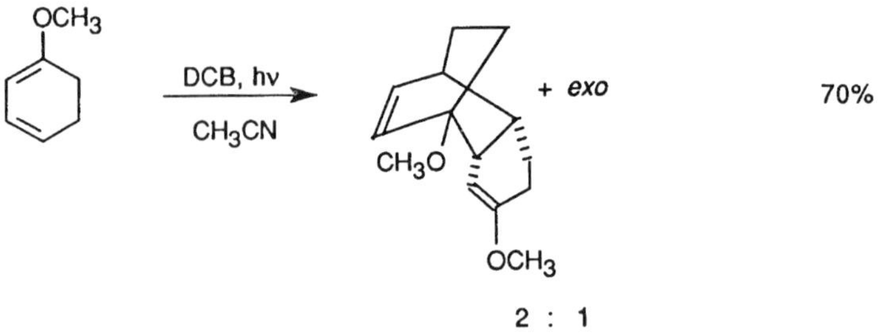

OCH_3 DCB, hν CH_3CN CH_3O + *exo* 70% OCH_3 2 : 1

Figure 6.17 Cation radical Diels–Alder reactions under PET conditions.

substrate 1-methoxycyclohexa-1,3-diene undergoes the cation radical Diels–Alder reaction smoothly under PET conditions (1,4-dicyanobenzene sensitizer, 70% yield), but the reaction fails under aminium salt conditions (Figure 6.17).[9] Both the reactant and product are decomposed by the strong acids present under aminium salt conditions. On the other hand, rapid back electron transfer can limit the range of applicability of the PET method. For example, the Diels–Alder cross addition of 2,5-dimethyl-2,4-hexadiene and 1,3-cyclohexadiene proceeds efficiently under aminium salt conditions; (Figure 6.18), but fails completely under PET conditions.[10, 11] Apparently the cycloaddition of the cation radical of this hindered diene to cyclohexadiene is relatively slow, allowing back electron transfer from the

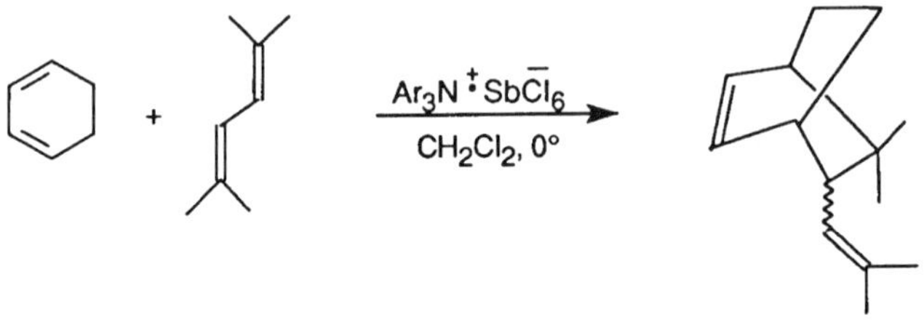

$Ar_3N^{+\bullet}SbCl_6^-$ CH_2Cl_2, 0°

Figure 6.18 Aminium salt catalyzed Diels–Alder addition.

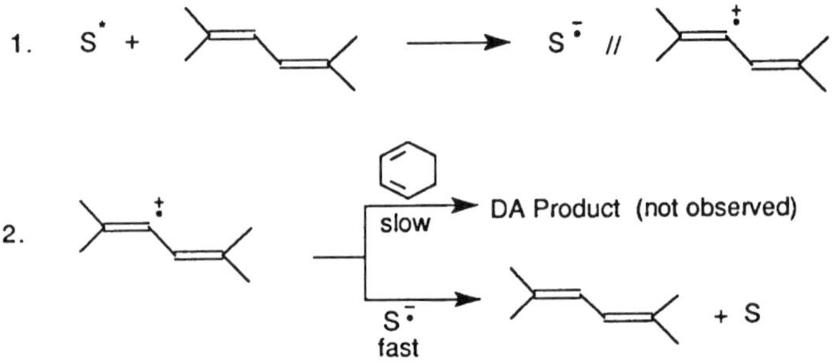

1. S^* + → $S^{\bullet-}$ // $^{+\bullet}$

2. slow → DA Product (not observed)

$S^{\bullet-}$ fast → + S

Figure 6.19 Back electron transfer versus cycloaddition.

sensitizer anion radical to dominate. The latter reaction returns only the ground state of the reactants (Figure 6.19). The usually facile cyclodimerization of 1,3-cyclohexadiene is also completely suppressed.

6.10 Thermal Electron Transfer

Electron transfer between neutral organic molecules in their electronic ground states is relatively much less favorable energetically than photosensitized electron transfer but is nevertheless well exemplified by reactions such as that of chloranil with 1,4-bis(dimethylamino)benzene (Figure 6.20).[12] Even in this highly favorable instance, a polar solvent is required. In relatively nonpolar media (such as ether or chloroform) only charge-transfer complex formation is observed. The possibility of electron-transfer mechanisms for certain pericyclic reactions seems to have been introduced by Kosower, who proposed the mechanism of Figure 6.21 for the reaction betwen tetracyanoethylene (TCNE) and methyl vinyl ether (MVE).[13] The essential aspects of this mechanism are, first, ET to yield the ion radical pair $MVE^{\dot{+}}$ $TCNE^{\dot{-}}$, followed by coupling of the two ion radicals to yield an intermediate zwitterion, and finally closure of the zwitterion to yield a cyclobutane adduct. The generally accepted mechanism for the reaction involves nucleophilic addition of MVE to TCNE (electrophilic addition to MVE), giving the zwitterionic intermediate directly. The possibility of ET mechanisms as an alternative to traditional polar

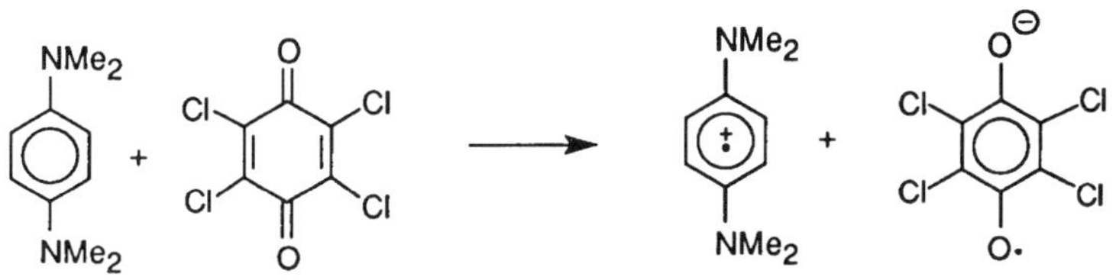

Figure 6.20 A facile thermal electron transfer yielding ion radical pairs.

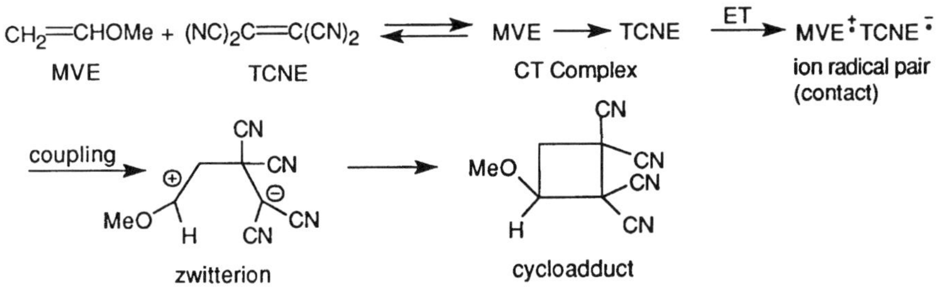

Figure 6.21 ET versus polar mechanisms for cycloaddition.

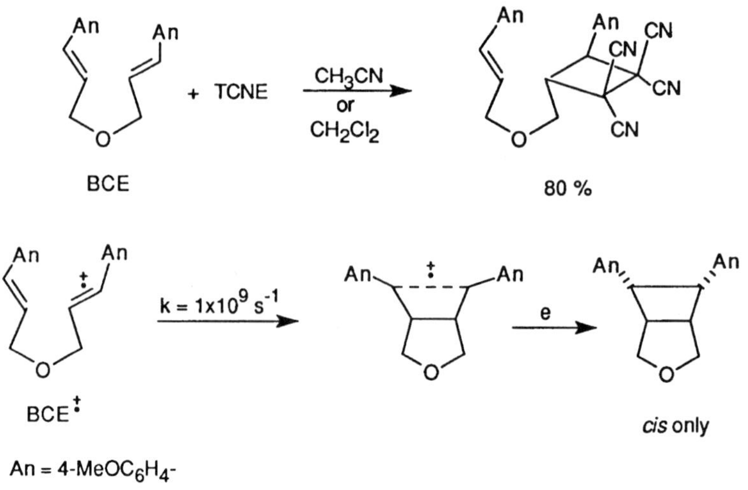

Figure 6.22 A cation radical probe for ET.

mechanisms has attracted much attention, but the number of well-established TET mechanisms remains relatively small. Recent evidence appears to rule out the ET possibility in at least one reaction of TCNE with an electron-rich alkene through the use of a cation radical trap (Figure 6.22).[14] The substrate (BCE) reacts smoothly with TCNE in acetonitrile or dichloromethane solvent at room temperature to yield the expected cyclobutane adduct. The cation radical of the substrate molecule is known to cyclize at a rate of $10^9\,s^{-1}$, but none of the cation radical cyclization product is found in the reaction with TCNE. Since the latter product would have been detectable at the 0.1% level, it is clear that if an ET mechanism were operative, the reaction of the substrate cation radical with the TCNE anion radical must be at least 1 000 times faster than the intramolecular cyclization. This would require a rate of at least $10^{12}\,s^{-1}$ for the ion radical coupling, a rate that appears improbable for a covalent bond forming reaction. The lifetime of the hypothetical cation radical intermediate would then be $< 10^{-12}$ s, a lifetime that is at least very close to that expected of a transition state ($\sim 10^{-13}$ s). To rule out the ET mechanism more conclusively, the ion radical pair $BCE^{\bullet+}TCNE^{\bullet-}$ was generated by irradiation of the $BCE^{\bullet}TCNE$ CT complex at the CT band. Under these conditions the cation radical cyclization product is readily observed.[15] Finally, it was shown that ^{3}BCE yields a mixture of the *cis* and *trans* isomers of the cyclization products, while the cation radical cyclization stereoselectively affords only the *cis* isomer. Consequently, ^{3}BCE cannot be the precursor of the cyclization product in the CT irradiation experiments. These results support the operation of a traditional polar mechanism in this cyclobutanation reaction.

Another potentially useful means of probing for cation radicals, and therefore for ET mechanisms that generate ion radical pairs is the use of substituent effects in reactions of *m*- and *p*-substituted aryl vinyl sulfides and aryl vinyl ethers. The

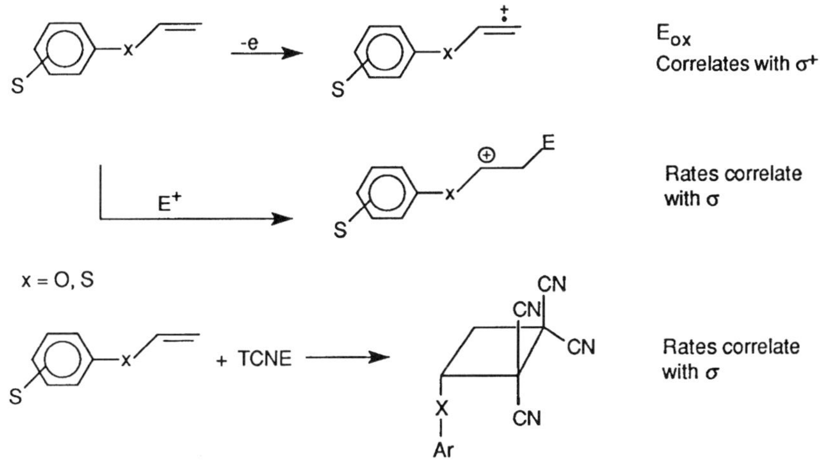

Figure 6.23 A simple substituent probe for dinstiguishing ET versus polar mechanisms.

ionization of these substrates to cation radicals nicely correlates with the Brown σ^+ substituent parameter,[16] while electrophilic additions to the vinyl double bond correlate equally well with the Hammett σ parameter (Figure 6.23).[17] The addition of TCNE to both sets of substrates correlates uniquely with σ, again confirming a polar mechanism involving electrophilic addition to the vinyl ether and vinyl sulfide moiety.[19]

6.11 Stable Cation Radical–Anion Radical Pairs

Stable examples of all the major types of radical species are now known, and the cation radical–anion radical pair is no exception. The initial example was provided by the exergonic thermal electron transfer between tetracyanoquinodimethane and tetrathiafulvalene (Figure 6.24).[19] The product is an organic conductor. Another impressive example is the ion radical pair derived from decamethylferrocene and tetracyanoethylene.[20] The latter is the first reported molecule-based magnet, being ferromagnetic below a critical temperature (T_c) of 4.8 K. As such it represents an important advance towards the ultimate goal of achieving the synthesis of organic ferromagnets that have T_cs well above room temperature. The synthesis of stable ion radical pairs is an important aspect of this research, but since ferromagnetism is a bulk property, it is currently not easy to predict which stable ion radical pairs will be ferromagnetic, that is, have all spins unpaired in the bulk solid. For example, the decamethylferrocene-TCNQ ion radical pair is metamagnetic, that is, it is ferromagnetic, but only in the presence of strong magnetic fields.[21]

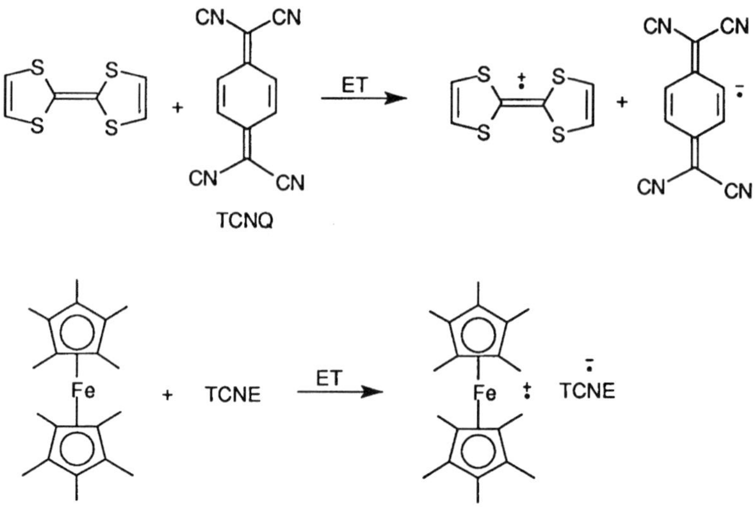

Figure 6.24 Stable ion radical pairs.

6.12 Intramolecular Electron Transfer

Exergonic intermolecular electron transfer between an anion radical and a neutral molecule is pervasive in anion radical chemistry and plays an important mechanistic role in such reactions as the $S_{RN}1$ reaction (Chapter 4). Intramolecular electron transfer (ET) is of especially great interest theoretically. In molecules in which the donor (anion radical) and acceptor (neutral) moieties are widely separated by a relatively rigid molecular framework (spacer), ET must occur without the benefit of a contact interaction or even close approach of the relevant functionalities. Under these circumstances many fundamental questions arise, including the following: (1) Is long range intermolecular ET efficient enough to compete with intermolecular ET? (2) What is the mechanism of intramolecular ET, for example, does it occur preferentially through bonds or through space? (3) Is a specific stereochemical relationship preferred (e.g., an *anti* relationship of bonds)? and (4) Is Marcus-type behavior observed?

A number of basic studies have begun to address these issues. A classic example is the study of long-range ET between a 4-biphenylyl anion radical moiety ($B^{\bar{\bullet}}$) and various neutral acceptor moieties (A) via a 5α-androstane spacer (Sp; Figure 6.25).[22] The rates of intramolecular ET (k_{et}) are relatively fast and range over more than three orders of magnitude ($5.6 \times 10^5\ s^{-1}$ for A = 4-biphenylyl to $> 10^9\ s^{-1}$ for 1-pyrenyl and hexahydronapthoquinon-2-yl). The observation of a Marcus inverted region in this study represented the first experimental confirmation of the theoretical prediction of this unique effect. The inverted region spans the four most readily reduced quinone moieties, for which $\Delta G^0 = -1.93$ to -2.40 eV. The value $\lambda = 1.20$ is found from the Marcus plot.

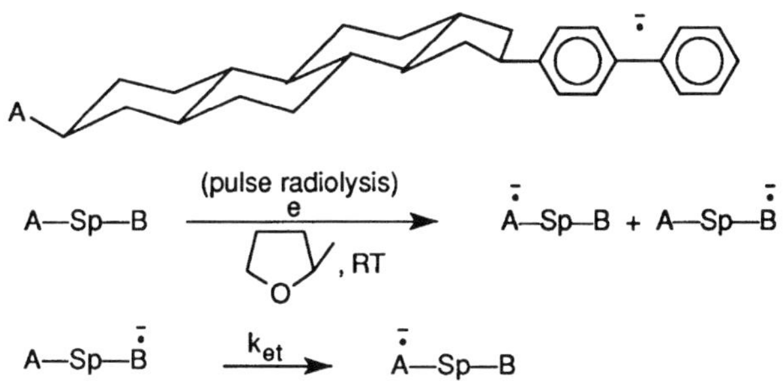

A

A—Sp—B —(pulse radiolysis) e, THF, RT→ $\dot{A}^{-}$—Sp—B + A—Sp—$\dot{B}^{-}$

A—Sp—$\dot{B}^{-}$ —k_{et}→ $\dot{A}^{-}$—Sp—B

A = 4-biphenyl (ΔG^0 =0); 2-naphthyl (ΔG^0 = -0.05 eV); 9-phenanthryl (ΔG^0 = -0.16); 1-pyrenyl (ΔG^0 = -0.52); hexahydronaphthoquinon-2-yl (ΔG^0 = -1.23), 2-napthoquinon-2-yl (ΔG^0 = -1.93), 2-benzoquinonyl (ΔG^0 = -2.10); 5-chloro-2-benzoquinonyl (ΔG^0 = -2.29); 5,6-dichlorobenzoquinon-2-yl (ΔG^0 = -2.40)

Figure 6.25 Long-range intramolecular electron transfer.

The former studies were also extended to include a range of different spacers, including 1,3- and 1,4-cyclohexanediyl and 2,7- and 2,8-*trans*-decalindiyl, maintaining constant donor (4-biphenylyl anion radical) and acceptor (2-naphthyl) functions.[22] The rates of ET vary from *ca.* $10^6\,s^{-1}$ (androstane spacer) to nearly $10^{10}\,s^{-1}$ (1,3-cyclohexanediyl). A plot of log k_{rel} versus the number of bonds separating the two functionalities (4–10) is nicely linear, whereas the correlations between log k_{et} and various distance parameters are of lower quality. These data are consistent with a preference (of unknown magnitude) for through-bond ET. Further, the ET rates are lower for axial–axial and equatorial–axial than for equatorial–equatorial donor and acceptor groups, suggesting that an *anti* relationship between bonds is the prefered stereochemistry for ET. Analogous studies of hole transfer from a cation radical moiety to a neutral moiety and of triplet energy transfer from a 4-benzophenonyl moiety have yielded very similar results.[23, 24]

6.13 Electron Transfer via Tunneling

Careful temperature dependence-studies of the reactions of Figure 6.24 reveal the startling fact that the rates of ET are esentially *temperature independent.*[25] Consequently, the activation barriers must be entropic rather than enthalpic. This indicates that efficient tunneling is responsible for high rates of ET and relatively inefficient tunneling for lower rates, including the Marcus inverted region.

References

1. Leonhardt, H.; Weller, A. *Ber. Bunsen ges. Phys. Chem.* **1963,** *67*, 791.
2. Marcus, R. A. *Faraday Disc.* **1964**, *29*, 21; Marcus, R.A. *J. Phys. Chem.* **1968**, *72*, 891.
3. Gould, I. R.; Noukakis, D.; Gomez-Jahn, L.; Goodman, J. L.; Farid, S. *J. Am. Chem. Soc.* **1993**, *115*, 4405.
4. Albery, W. J.; Kreevoy, M. M. *Adv. Phys. Org. Chem.* **1978**, *16*, 87.
5. (a) Gould, I. R.; Moody, R.; Farid, S. *J. Am. Chem. Soc.* **1988**, *110*, 7242. (b) Gould, I. R.; Farid, S. *Acct. Chem. Res.* **1996**, *29*, 522–528.
6. Gould, I. R., Moser, J. E.; Ege, D.; Farid, S. *J. Am. Chem. Soc.* **1988**, *110*, 1991.
7. Brown-Wensley, K. A. S.; Mattes, S. L.; Farid, S. *J. Am. Chem. Soc.* **1978**, *100*, 4162.
8. Ellinger, L. P. *Polymer* **1964,** *5*, 559.
9. Pabon, R. A.; Bellville, D. J.; Bauld, N. L. *J. Am. Chem. Soc.* **1983**, *105*, 5158.
10. Bellville, D. J.; Wirth, D. D.; Bauld, N. L. *J. Am. Chem. Soc.* **1981**, *103*, 718.
11. Jones, C. R.; Allman, B. J.; Mooring, A.; Spahic, B. *J. Am. Chem. Soc.* **1983**, *105*, 652.
12. Bijl, D.; Kainer, H.; Rose-Innes, A. C. *Naturwiss.* **1954**, *41*, 303; Kosower, E. M. in *Progress in Physical Organic Chemistry*, Vol. 3, Cohen, S. G.; Streitwieser, A., Jr.; Taft, R. W.; Eds., John Wiley & Sons, New York, 1965, p. 115.
13. Kosower, E. M. in *Progress in Physical Organic Chemistry*, Vol. 3, Cohen, S. G.; Streitwieser, A., Jr.; Taft, R. W.; Eds., John Wiley & Sons, New York, 1965, p. 124.
14. Kim, T.; Mirafzal, G. A.; Bauld, N. L. *Tetrahedron Lett.* **1993**, *34*, 7201.
15. Kim, T.; Sarker, H.; Bauld, N. L. *J. Chem. Soc. Perkin Trans. 2* **1995**, 577.
16. Aplin, T.; Bauld, N. L. Unpublished results.
17. McClelland, R. A. *Can. J. Chem.* **1977**, *55*, 548; Fueno, T.; Matsumura, I.; Okuijama, T.; Furukawa, J. *Bull. Chem. Soc. Jpn.* **1968**, *41*, 818.
18. Solomonov, B. N.; Arkhirieva, I. A.; Konovalov, A. I. *Zh. Org. Khim.* **1980**, *16*, 1666 and 1670.
19. Krief, A. *Tetrahedron* **1986**, *42*, 1209.
20. Miller, J. S.; Epstein, A. J.; Reiff, W. M. *Isr. J. Chem.* **1987**, *27*, 363; Miller, J. S.; Epstein, A. J., Reiff, W. M. *Chem. Rev.* **1988**, *88*, 201; Miller J. S.; Calabrese, J. C.; Harlow, R. L.; Dixon, D. A.; Zhang, J. H.; Reiff, W. M.; Chittipeddi, S.; Selover,M. A.; Epstein, A. J. *J. Am. Chem. Soc.* **1990**, *112*, 5496.
21. Candela, G. A.; Swartzendruber, L.; Miller, J. S.; Rice, M. J. *J. Am. Chem. Soc.* **1979**, *101*, 2755.
22. Miller, J. R.; Calcaterra, L. T.; Closs, G. L. *J. Am. Chem. Soc.* **1984**, *106*, 3047; Closs, G. L.; Calcaterra, L. T.; Green, N. J.; Penfield, K. W.; Miller, J. R. *J. Phys. Chem.* **1986**, *90*, 3673.
23. Johnson, M. D.; Miller, J. R.; Green, N. D.; Closs, G. L. *J. Phys. Chem.* **1989**, *93*, 1173.
24. Closs, G. L.; Johnson, M. D.; Miller, J. R.; Piotrowiak, P. *J. Am. Chem. Soc.* **1989**, *111*, 3751.
25. Liang, N.; Miller, J. R.; Closs, G. L. *J. Am. Chem. Soc.* **1990**, *112*, 5353.

Exercises

6.1 (a) In a system consisting of hexamethylbenzene (HMB) as the donor, 2,6,9,10-tetracyanoanthracene (TCA) as the acceptor, and acetonitrile as the solvent, $HMB^{+\bullet}/TCA^{-\bullet}$ contact ion radical pairs were generated by irradiation of the HMB/TCA charge-transfer complex at the CT band using a pulsed dye laser (410 nm). The quantity $E_{ox}(HMB) - E_{red}(TCA) = 2.03$ eV in this system. Using the Marcus equation and the appropriate λ value given in this text, calculate the free energy of activation for back electron transfer ($\Delta G^{\ddagger}_{-et}$). Assuming that the maximum rate of BET (corresponding to $\Delta G^{\ddagger} = 0$) is $10^{13}\ s^{-1}$, calculate the rate constant (k_{-et}) for back electron transfer in this system. Is this in the Marcus inverted region? Explain. (b) For an analogous system in which the ion radical pairs are formed by the photosensitized electron-transfer (PET) method, calculate $\Delta G^{\ddagger}_{-et}$ and compare this value to that obtained in the previous question. Which is smaller and why? Assuming that the maximum rate of back electron transfer in this system (for $\Delta G^{\ddagger} = 0$) is $1.8 \times 10^{10}\ s^{-1}$, calculate k_{-et}. Why is the maximum rate constant less than in the previous question, and why does it have this particular value?

6.2 The following data refer to the 1,2,4,5-tetracyanobenzene anion radical/hexamethylbenzene cation radical contact ion pair generated by irradiation of the corresponding charge-transfer (CT) complex at the CT wavelength:

TCB · HMB (Me6) $\xrightarrow[\text{solvent}]{h\nu_{CT}}$ $TCB^{-\bullet}HMB^{+\bullet}$ $\xrightarrow{\Delta G_{-et}}$ TCB + HMB (ground states)

Solvent	$-\Delta G^0_{-et}$ (eV)	λ_s	
cyclohexane	2.57	0.14	
CCl_4	2.50	0.16	$\lambda_i = 0.31$ eV
diethyl carbonate	2.46	0.36	
$CHCl_3$	2.43	0.44	
$ClCH_2CH_2Cl$	2.45	0.53	

(a) Rationalize the qualitative sense of the dependence of $-\Delta G^0$ upon solvent polarity. (b) Rationalize the sense of the dependence of λ_s upon solvent polarity. (c) In which solvent is back electron transer (–et) likely to be the slowest? Explain. (Gould, I. R.; Noukakis, D.; Goodman, J. L.; Young, R. N.; Farid, S. *J. Am. Chem. Soc.* **1993**, *115*, 3830.)

6.3 The efficiency of reactions of ion radicals generated via photosensitized elecron transfer with external reactants is often diminished by the occurrence of competing back electron transfer (BET). The conditions indicated below have been suggested as providing optimum efficiency for the separation of ion pairs and minimizing BET:

$$\text{N-methylacridinium } PF_6^{\ominus} + D \xrightarrow{h\nu,\ CH_2Cl_2} D^{\cdot +}$$

For example, the quantum yield of dissociated biphenyl cation radicals is Φ = 0.89 under these conditions. (a) The quantum yield of biphenyl cation radicals in acetonitrile is only $\Phi = 0.33$ when the same sensitizer is used. Clarify the basis for the enhanced quantum yield of free cation radicals in dichloromethane, which is much less polar (ε = 8.93) than acetonitrile (ε = 35.4). (b) The N-methylacridinium ion sensitizer also represents an important improvement over neutral sensitizers such as 1,4-dicyanobenzene. Explain why a cation sensitizer would be expected to promote more efficient generation of free ion radicals. (Todd, W. P.; Dinnocenzo, J. P.; Farid, S.; Goodman, J. L.; Gould, I. R. *J. Am. Chem Soc.* **1991**, *113*, 3601.)

6.4 When 1,3-cyclohexadiene is irradiated in the presence of 1,4-dicyanobenzene (DCB) in degassed acetonitrile solution, using an appropriate filter so that only DCB absorbs, both the Diels–Alder dimers (DA) described in the text and the cyclobutane dimers (CB) illustrated below are formed. When 1,3,5-trimethoxybenzene (E_{ox} = 1.46 V) or 4,4′-dimethoxybiphenyl (E_{ox} = 1.21 V) is added as a quencher, the formation of the DA dimers is suppressed, but CB dimer formation is essentially unaffected.

$$\text{1,3-cyclohexadiene} \xrightarrow[CH_3CN]{DCB,\ h\nu} \text{DA } (\textit{endo}>\textit{exo}) + \text{CB } (\textit{anti} + \textit{syn})$$

E_{OX} 1.63V *vs.* Ag/Ag^+

(a) Propose a mechanism for the formation of the CB dimers that is consistent with the above data. (b) Propose a mechanism for the formation of the DA dimers that is consistent with supression of the reaction by added quenchers. (c) Would indene (E_{ox} = 1.65 V) be expected to quench DA dimer formation? Explain. (Calhoun, G. C.; Schuster, G. C. *J. Am. Chem. Soc.* **1984,** *106*, 6870.)

6.5 Dialkoxytrimethylenemethanes, generated reversibly by thermal cleavage of methylenecyclopropanes, undergo stereospecific cycloaddition to a variety of mildly electron-deficient alkenes to give ketene acetal (KA) dericatives, as illustrated below (the E-alkene derivative was also studied).

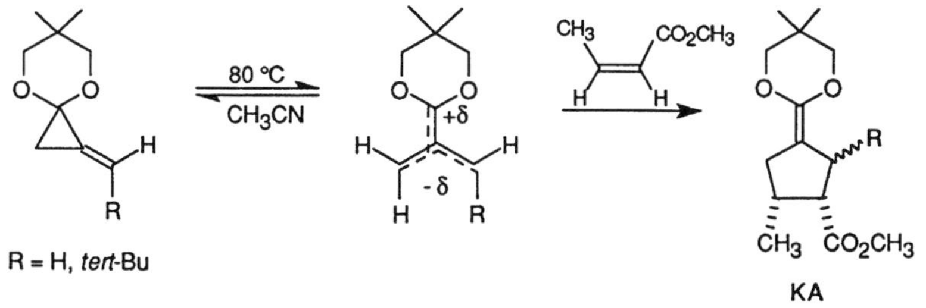

O O H R

R = H, *tert*-Bu

80 °C

CH_3CN

O O +δ H H -δ H R

CH_3 CO_2CH_3 H H

O O R CH_3 CO_2CH_3

KA

However, when the alkene has a reduction potential less negative than −1.8 V, the regiochemistry of addition is altered and *exo*-methylenecyclopentanone ketals are formed as the major products.

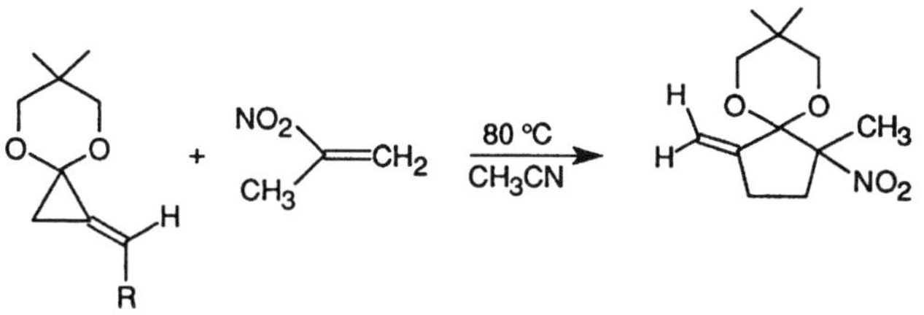

O O H R + NO_2 CH_3 $=CH_2$

80 °C

CH_3CN

H H O O CH_3 NO_2

When E and Z alkene isomers are used, the reaction is found to be nonstereospecific, and the unreacted alkene is partially isomerized. The smaller amounts of ketene acetals (KA) produced are formed largely stereospecifically.

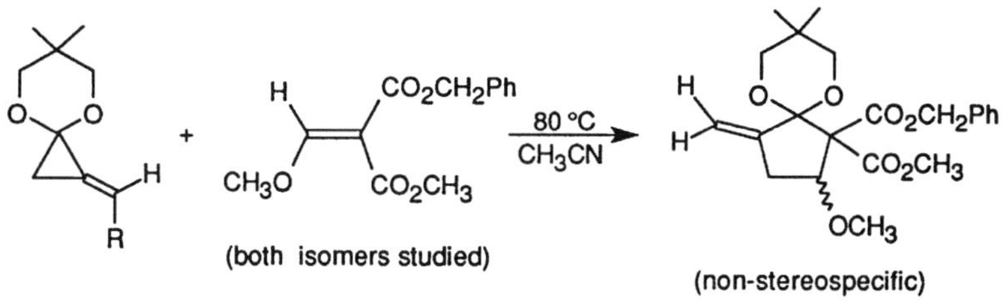

O O H R + H CO_2CH_2Ph CH_3O CO_2CH_3

(both isomers studied)

80 °C

CH_3CN

H H O O CO_2CH_2Ph CO_2CH_3 OCH_3

(non-stereospecific)

Propose a mechanism for the formation of the *exo*-methylene adducts that is consistent with these data and rationalize the change in regiochemistry in terms of this mechanism. Explain, also, the isomerization of the alkene in reactions in which the *exo*-methylene adducts are formed, but not in those reactions that give only the ketene acetal adducts. (Yamago, S.; Ejiri, S.; Nakamura, M.; Nakamura, E. *J. Am. Chem. Soc.* **1993**, *115*, 5344.)

CHAPTER 7

Triplets and Higher Multiplets

The triplet spin states (especially T_0) of radical pairs have already been seen (Chapter 3) to play an indispensable role (via spin sorting) in chemically induced dynamic nuclear polarization (CIDNP) and the mechanistic applications of this phenomenon. Precisely the same can be said of CIDNP phenomena arising from ion radical pairs. It has just been seen (Chapter 6) that ion radical pair triplets may undergo back electron transfer to yield excited triplet electronic states of neutral molecules. In the realm of chemical reactions, excited triplet-state intermediates have an important role in direct photochemistry (via intersystem crossing) and in triplet sensitized photoreactions. Ground-state triplets are involved in the chemistry of dioxygen and some carbenes and carbene analogues. Certain cyclic conjugated π electron systems (antiaromatics) have ground-state or near-ground-state triplets. Examples of stable triplets are even available. Finally, spin multiplets higher than triplet are of both theoretical and practical interest, the latter in connection with the development of organic ferromagnets.

7.1 Spin Functions of the Triplet State

Triplet molecules can be defined as molecules that have two unpaired electrons. Since the electron spins are not paired, they must, in molecular-orbital terminology, occupy two different MOs. The Pauli Exclusion Principle requires that spin functions be either symmetric or antisymmetric with respect to electron exchange, so that the permissible spin functions for a system having two electrons in different MOs are as shown in Figure 7.1.

Triplet States
$$\left\{\begin{array}{l}\alpha(1)\,\alpha(2) \equiv T_1 \\ \beta(1)\,\beta(2) \equiv T_{-1} \\ \alpha(1)\,\beta(2) + \alpha(2)\,\beta(1) \equiv T_0\end{array}\right.$$

Singlet State
$$\alpha(1)\,\beta(2) - \alpha(2)\,\beta(1) \equiv S$$

Figure 7.1 Triplet spin functions.

In the T_1 state, both electrons have α spin, and in the T_{-1} state both electrons have β spin. The plus and minus designations are derived from the convention that the α electron has spin $m_s = +\frac{1}{2}$ and the β electron spin $m_s = -\frac{1}{2}$. Consequently the total spin is +1 for the state that has two α electrons and −1 for two β electrons. Note that electron exchange (interchange of the labels 1 and 2) in $\alpha(1)\,\alpha(2)$ and $\beta(1)\,\beta(2)$ leaves the spin function unchanged. Thus these simple triplet spin functions are both symmetric with respect to electron exchange. The simple function $\alpha(1)\beta(2)$ cannot represent a valid singlet or triplet spin function, since electron exchange converts it into a different function, viz. $\alpha(2)\,\beta(1)$. However, the additive combination of these two functions does provide the third symmetric (triplet) spin function. The subtractive combination corresponds to an antisymmetric (singlet) spin function. The overall wave function of the triplet state is then the product of the appropriate spin function (T_1, T_{-1}, or T_0) and the spatial part of the wave function (ψ_{space}). Since the overall wave function (the product of the space and spin functions) is necessarily antisymmetric with respect to electron exchange, the spatial part of the triplet wave function must be antisymmetric, while the spatial part of the singlet wave function is symmetric (Figure 7.2). The value $m_s = 0$, which characterizes the T_0 state specifically, corresponds to a relative orientation of the spin magnetic moment vectors (μ_1, μ_2) of the electrons that affords a zero component of the net magnetic moment along the axis of quantization (assumed here to be the Z axis). The T_0 state does, however, have a nonzero component of the magnetic moment in the XY plane, while the singlet does not (Figure 7.3). In a simple model, the singlet is considered to have magnetic moment vectors that are equal and opposite and that precess around the Z axis. The magnetic moment vectors of T_0, however, are additive in the XY plane. As the magnetic moment vectors μ_1 and μ_2 precess, they

Triplet: $\quad {}^3\Phi = \psi^{A}_{space} \cdot \psi^{S}_{spin} \qquad \psi^{S}_{spin} = T_{-1}, T_0 \text{ or } T_{+1}$

Singlet: $\quad {}^1\Phi = \psi^{S}_{space} \cdot \psi^{A}_{spin} \qquad \psi^{A}_{spin} = S$

Figure 7.2 Overall (space and spin) wave functions for triplet states.

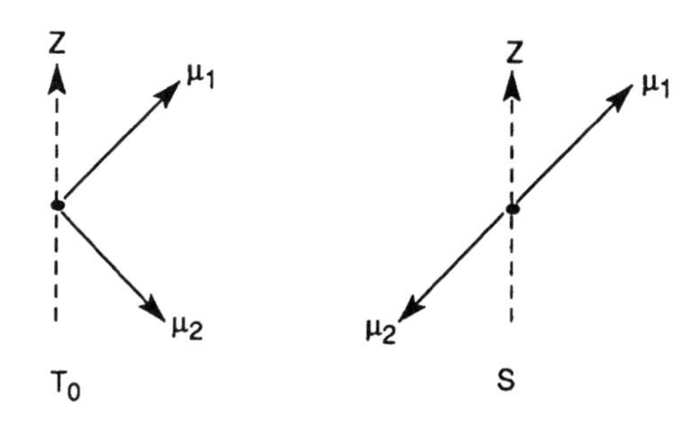

Figure 7.3 Classical picture of T_0 and S magnetic moment vectors.

tend to lose phase coherence. Eventually, in this way, T_0 and S are interconverted. Singlet–triplet interconversions therefore commonly involve the T_0 state specifically.

7.2 Relative Stability of Triplets and Singlets

In the simple case where two molecular orbitals of equal energy (i.e., doubly degenerate MOs) are available for occupancy by two electrons, three distinct kinds of states are possible (Figure 7.4). Besides the triplet state, there is the possibility of an *open-shell singlet* state, in which each MO is singly occupied as in the triplet, and also the possibility of closed-shell singlet states, in which either one of the MOs is doubly occupied and the other vacant. The latter states are, however, subject to geometric distortion (called Jahn–Teller distortion) that removes the orbital

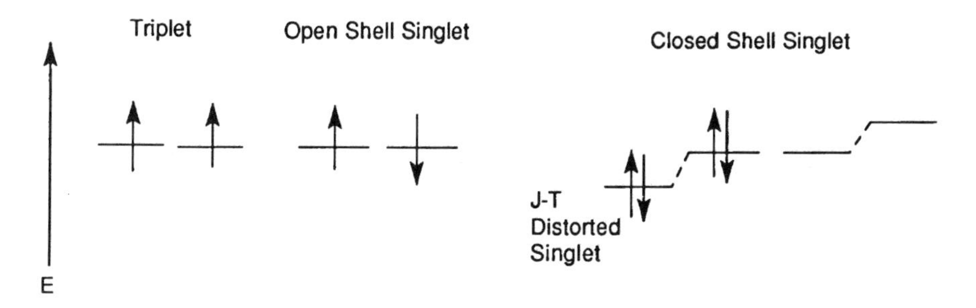

Figure 7.4 Triplets, open-shell singlets, and closed-shell singlets.

degeneracy, stabilizing the occupied MO and destabilizing the vacant one. Since only one of the MOs is occupied, the result is net stabilization of the Jahn–Teller distorted singlets relative to the undistorted closed-shell singlets. On the other hand, an extension of Hund's Rule for the ground states of atoms suggests that the triplet state should be more stable than either the open-shell singlet or the undistorted closed-shell singlet. Hund's Rule does not apply to a comparison of the relative stabilities of the triplet and the distorted singlets, since these states have different geometries. Consequently, it is not possible to make a general statement concerning the ground state of such a system except to say that a low-lying triplet state is usually available and may be the ground state. The energetic advantage of the triplet state over the undistorted singlet state derives essentially from the Pauli Exclusion Principle. Electrons having the same spin (as in triplets) are less repulsive toward each other than are electrons of opposite (paired) spin, because the motions of the former are automatically correlated by Pauli Exclusion to minimize repulsions. In molecular orbital calculations, the relative stabilization of the triplet is based upon the fact that the exchange integral (K_{ij}) between two electrons (in the molecular orbitals labeled i and j) exists only when the electrons have the same spin (Figure 7.5). The spin part of K_{ij} (the second double integral) is unity only when both electrons have the same spin. Since the α and β spin functions are orthogonal, the spin part of K_{ij} is zero for electrons with paired spins (Figure 7.6). The exchange integrals themselves are positive, but enter into the calculation subtractively; that is, they reduce the repulsions between the electrons.

$$K_{ij} = \iint \psi_i(1)\psi_j(2)e^2/r_{12}\psi_j(1)\psi_i(2)d\tau_1 d\tau_2 \iint \psi_{spin}(1)\psi_{spin}(2)d\tau_1 d\tau_2$$

K_{ij} = Exchange Integral

e^2/r_{12} = electron repulsion Hamiltonian operator

r_{12} = distance between electrons 1 and 2

$\psi_{spin} = \alpha$ or β

Figure 7.5 Exchange stabilization of the triplet state.

$$\iint \alpha(1)\,\beta(2)\,d\tau_1 d\tau_2 = 0$$

$$\iint \alpha(1)\,\alpha(2)\,d\tau_1 d\tau_2 = 1$$

$$\iint \beta(1)\,\beta(2)\,d\tau_1 d\tau_2 = 1$$

Figure 7.6 Orthogonality of the α and β spin functions.

7.3 Noninteracting or Weakly Interacting Triplets

When the AO compositions of the MOs (ψ_i, ψ_j) occupied by the two electrons are similar, i.e., when the electron clouds ψ_i^2 and ψ_j^2 overlap substantially, the repulsions between the electrons are relatively large. The repulsion is represented quantitatively by the coulomb replusion integral J_{ij}:

$$J_{ij} = \int\int \psi_i^2(1)e^2/r_{12}\ \psi_j^2(2)d\tau_1 d\tau_2$$

When the repulsion integral is large, the corresponding exchange integral (K_{ij}) is usually also relatively large. A large exchange integral engenders a large stabilization of the triplet state relative to the undistorted singlet. It is in instances of this type that triplet ground states are especially probable.

On the other hand, if the MOs ψ_i and ψ_j have essentially no AOs in common (i.e., they are situated on different atomic centers), the exchange integral is small, and triplet and undistorted singlet states are of nearly equal energy. Such triplets and their corresponding singlets are often referred to as *diradicals* or *double doublets*.

7.4 Excited-State Triplets

Triplet exchange stabilization is perhaps most evident in the stabilization of excited-state triplets relative to the corresponding excited singlets (Figure 7.7). The first excited singlet state of benzophenone has E_S = 75 kcal/mol, while the triplet-state energy is E_T = 69 kcal/mol.[1] The singlet–triplet separation in aromatics is typically much larger, for example, E_S = 76 and E_T = 47 for anthracene.[1] In the latter case,

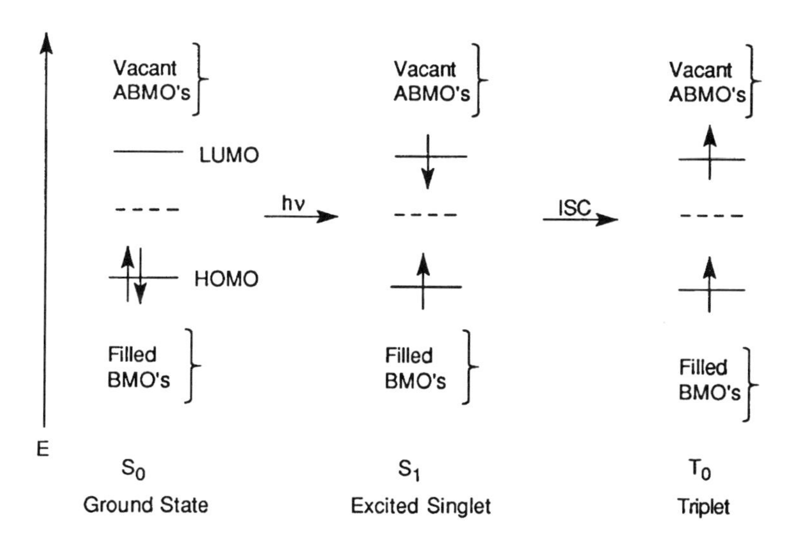

Figure 7.7 MO diagrams of ground and excited singlet and triplet states.

ψ_i and ψ_j are the HOMO and LUMO of anthracene, which are "paired" orbitals according to the Pairing Theorem. The electron distributions in these MOs are very similar and give rise to a large exchange integral and the attendant large stabilization of the triplet state relative to the excited singlet. In sharp contrast, the excited state of benzophenone is of the $n \rightarrow \pi^*$ type. As a result, one of the 2 MOs (say ψ_i) is essentially localized on oxygen in a nonbonding n-type AO, while the other (ψ_j) is a π^* MO that is delocalized over the carbonyl group and both aryl rings. Further, to the extent that the electron density in ψ_j is on the carbonyl group in the π^* MO, it is more highly concentrated on the carbonyl carbon than on oxygen. Consequently, the exchange integral K_{ij} is quite small, and the triplet state is only modestly stabilized relative to the singlet. However, there is an important benefit associated with small S–T energy separations, viz., the intersystem crossing (ISC) rate is much faster. Accordingly, the rate of ISC from S_1 to T is $10^{11}\ s^{-1}$ for benzophenone and $5 \times 10^9\ s^{-1}$ for anthracene.[2] The primary benefit of the extremely fast ISC rate for benzophenone excited singlets is the ready accessibility of benzophenone triplets. Since the ISC rate is faster than the diffusion rates ($\sim 10^{10}\ s^{-1}$), triplet benzophenone can be generated cleanly without the complications that might arise from having a mixture of excited singlet and triplet states.

7.5 Dioxygen: A Stable Triplet Ground State

Undoubtedly the most common example of a stable triplet ground-state molecule is dioxygen. This important molecule can be thought of as having two three-electron π bonds (Figure 7.8). Of the twelve valence electrons required for two oxygen atoms, two electrons occupy an MO that corresponds (approximately) to the O–O σ bond. Two electron pairs occupy (linear combinations of) oxygen nonbonding MOs involving the O_{2S} and O_{2px} AOs (essentially *sp* AOs). The remaining six electrons must be accommodated by the π MOs. This requires that the π_y and π_z bonding MOs be filled and that single electrons occupy the π_y^* and π_z^* ABMOs. In a valid sense, dioxygen can be said to have a partial *triple* bond; that is, it has a σ bond and two very partial π bonds. Obviously the strength of these three-electron

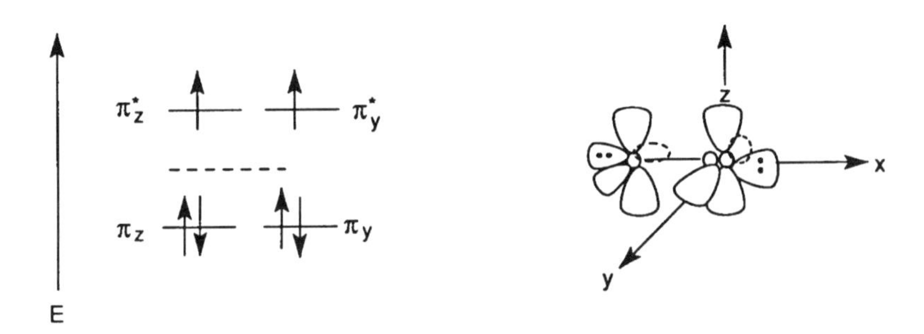

Figure 7.8 The π MOs of dioxygen.

bonds is much less than that of a comparable two-electron bond. As a triplet, dioxygen is an excellent and highly reactive radical trap and also a trap for other more reactive triplets. The autoxidation chemistry of dioxygen has already been considered in Chapter 2 and its reactions with cation radicals in Chapter 5.

The open-shell singlet state of dioxygen (the $^1\Sigma$ state) is 37.6 kcal/mol higher in energy than the ground-state triplet ($^3\Sigma$). However, the closed-shell singlet ($^1\Delta$) state is actually the first excited state of dioxygen, lying 22.5 kcal/mol above the $^3\Sigma$ ground state. The chemistry of singlet oxygen in the $^1\Delta$ state has been developed extensively, since this state can be produced chemically from singlet precursors and photochemically from triplet dioxygen by triplet sensitization or photosensitized electron transfer, but is not considered further here.

7.6 Stable Organic Triplets

A number of surprisingly stable triplet molecules have been prepared by the simple strategy of duplicating or simulating a stable radical moiety. The triplet of Yang and Castro (**1**), which is based upon the structure of galvinoxyl, is especially conveniently accessible from a synthetic standpoint, and is impressively stable (Figure 7.9).[3] Another impressively stable triplet is the perchlorinated Schlenk hydrocarbon (**2**).[4] This and the Rajca triplet (**3**) are based upon the trityl structure.[5] Finally, various elaborations of the TEMPO radical structure have been prepared (e.g., **4**).[6]

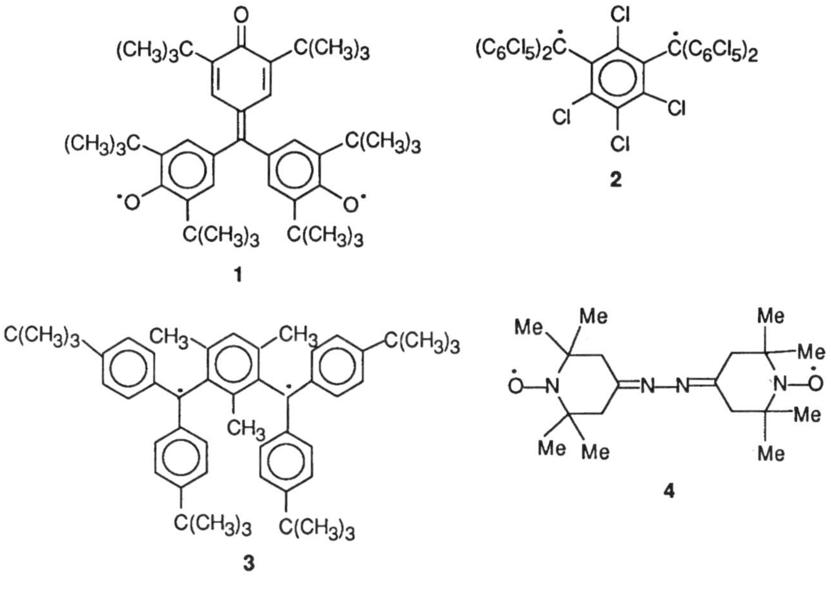

Figure 7.9 Stable organic triplets.

7.7 Persistent Triplets in Antiaromatic Systems

The classic examples of ground-state triplets in antiaromatic systems are provided by several cations of the cyclopentadienyl family. The parent cation has the HMO energy levels displayed in Figure 7.10. The ground state of this 4π-electron system is expected to be either a triplet or a Jahn–Teller distorted, closed-shell singlet. The initial attempts to study a cyclopentadienyl cation system involved the pentaphenyl derivative. Treatment of the corresponding alcohol with boron trifluoride in methylene chloride at –60°C apparently does yield the pentaphenylcyclopentadienyl cation, since the ESR spectrum confirms the formation of a triplet (vide infra for triplet ESR spectra).[7] These observations are gratifying, but they do not address the question of whether the observed triplet is actually the ground state of the cation, as opposed to a thermally populated, low-lying excited state. A critical test in this regard is the relationship of the ESR signal intensity to the temperature. For a thermally populated excited state, an increase in the temperature should increase the concentration of triplets and, therefore, tend to increase the signal intensity. The Curie–Weiss law, however, indicates that the signal intensity of a ground-state triplet should increase as the temperature is decreased (i.e., a plot of intensity vs. $1/T$

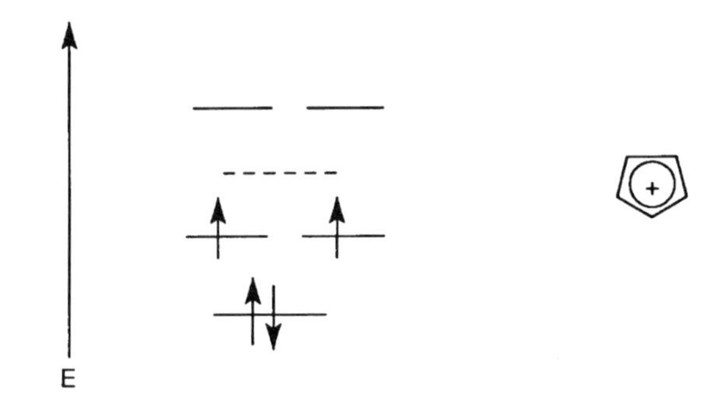

Figure 7.10 MOs of the cyclopentadienyl cation.

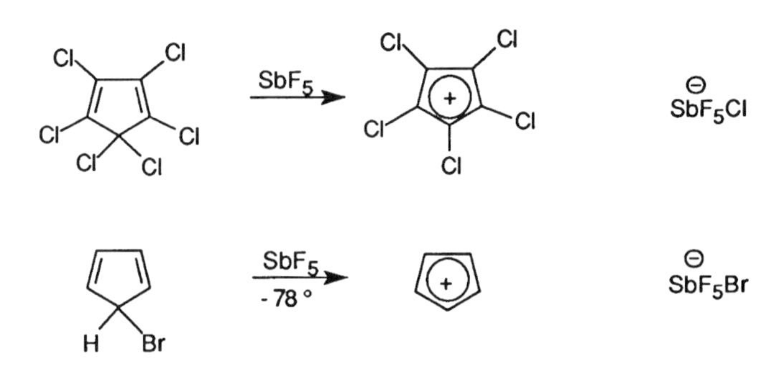

Figure 7.11 Ground-state cyclopentadienyl cation triplets.

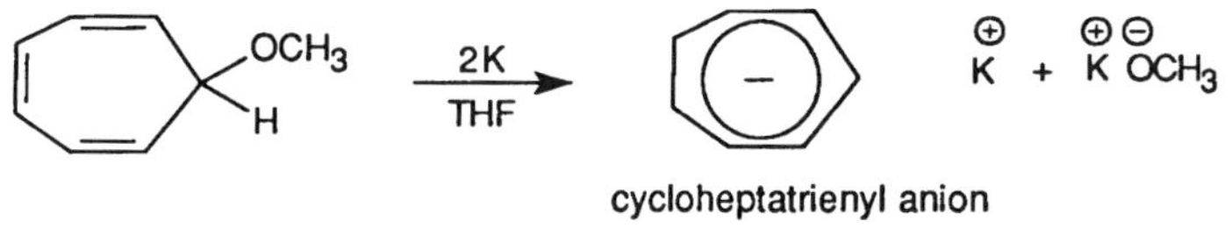

OCH_3 / H —2K, THF→ $K^{\oplus}$ + $K^{\oplus}\,{}^{\ominus}OCH_3$

cycloheptatrienyl anion

Figure 7.12 Antiaromatic systems with closed-shell singlet ground states.

is linear with positive slope). Consequently, when thermal population is occurring, deviations from the Curie–Weiss law are observed. In the present case, the Curie–Weiss law is not obeyed, and the triplet is found to be just slightly higher in energy than the ground-state singlet. In contrast, the pentachlorocyclopentadienyl cation (Figure 7.11) does obey the Curie–Weiss law and has a ground triplet state.[7] Finally, the parent cation has also been found to have a triplet ground state.[7]

In rather sharp contrast to these observations, the parent cycloheptatrienyl anion (Figure 7.12), an 8π-electron system, is relatively stable at room temperature in ethereal solvents and reveals no triplet ESR absorptions at all.[8] Similarly, the ground state of 1,3-cyclobutadiene is the rectangularly distorted closed-shell singlet.[9]

7.8 Carbenes: Reactive Ground-State Triplets

The chemistry of divalent carbon intermediates (carbenes) is an especially fruitful area for studying and exploiting the chemistry of ground-state triplets. The parent carbene (methylene) has a triplet ground state that is approximately 9 kcal/mol below the singlet state (Figure 7.13).[10] Both states are bent, but the triplet has a

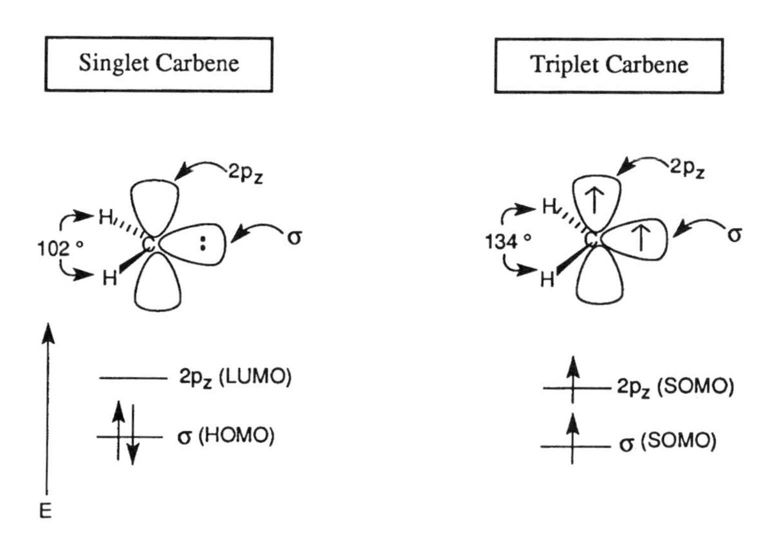

Figure 7.13 Singlet versus triplet carbenes.

$$Ph_2C{=}O \xrightarrow{h\nu} {}^1Ph_2C{=}O^* \xrightarrow{ISC} {}^3Ph_2C{=}O^* \xrightarrow{CH_2N_2} {}^3CH_2N_2^* + Ph_2C{=}O$$

$$^3CH_2N_2^* \longrightarrow N_2 + {}^3CH_2$$

Figure 7.14 Generation of triplet carbene.

wider valence angle (134°) than the singlet (102°). The bending of the singlet state is clearly expected, since the unshared pair should prefer to occupy an MO with the highest feasible "*s* content" (lowest energy), *viz*. a hybrid (sp^2) orbital, leaving the higher-energy $2p_z$ AO vacant. However, in triplet carbene, an *sp* hybridization state with a valence angle of 180° might have been expected, since only this hybridization state would provide degenerate SOMOs ($2p_z$,$2p_y$). Since triplet carbene is also bent, it is clear that the hybridization state is not *sp* and that the two SOMOs are nondegenerate. Evidently the exchange stabilization of the triplet is greater than the energy difference between the SOMOs; otherwise both electrons would prefer to be in the σ orbital, providing a singlet ground state.

Although the triplet state of carbene is clearly the ground state, its production in solution is not necessarily straightforward. Given that spin is conserved in organic reactions, the more usual singlet carbene precursors afford singlet carbene directly. The subsequent reactions of these extremely reactive intermediates are normally faster than ISC to the triplet. An effective approach to triplet carbene is the triplet photosensitized decomposition of diazomethane (Figure 7.14).[11] As noted previously, intersystem crossing in singlet excited-state benzophenone is extremely fast, affording triplet benzophenone. The triplet energy of benzophenone is quite high (E_T = 69 kcal/mol), so that triplet energy transfer to diazomethane is energetically feasible. The triplet diazomethane then loses dinitrogen to afford triplet carbene, in a process that conserves spin. The subsequent reactions of triplet carbene provide an instructive contrast to those of singlet carbene.

7.9 Reactions of Triplet Carbene

Both singlet and triplet carbene add readily to carbon–carbon π bonds such as alkene double bonds. While the addition of singlet carbene is concerted and stereospecific, triplet carbene adds in a nonstereospecific, stepwise manner (Figure 7.15).[11,12] Since spin is expected to be conserved, the hypothetical concerted addition of triplet carbene to an alkene would have to generate the triplet excited state of the cyclopropane, and this would be energetically prohibitive. Instead, addition occurs to give a triplet diradical intermediate, which must be converted to a singlet diradical state before closure to the ground (singlet) state of the cyclopropane adduct. The diradical has sufficient time to undergo rotation around the carbon–carbon bond that was formerly the double bond, but is now single. The stereochemical integrity of the original alkene is therefore lost, and both *cis*- and

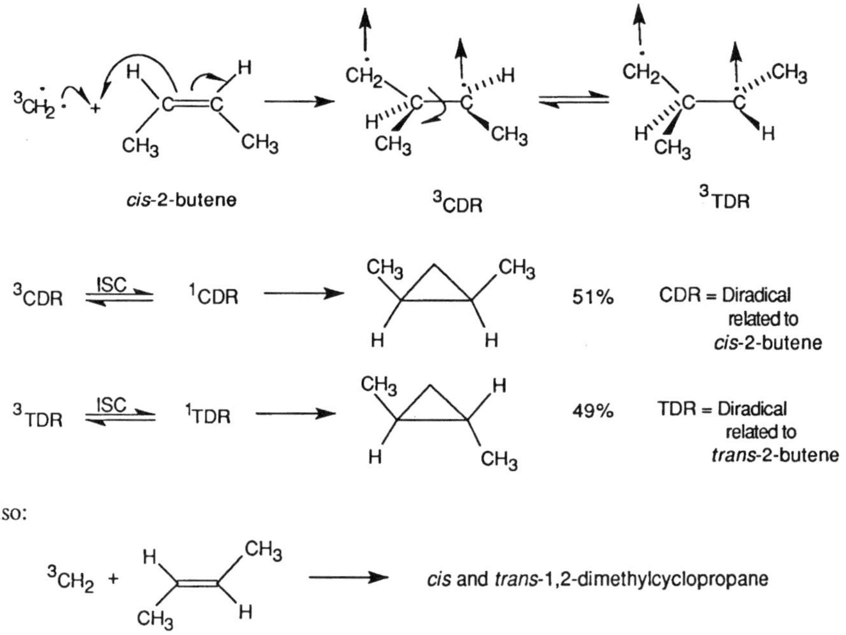

Figure 7.15 Nonstereospecific addition of triplet carbene to the 2-butenes.

trans-1,2-dimethylcyclopropane are formed from either *cis*-2-butene or *trans*-2-butene, in sharp contrast to the results with singlet carbene.

7.10 Diphenylcarbene

The unpaired electrons in this carbene are of course far more delocalized than in the parent carbene. Consequently the exchange stabilization of the diphenylcarbene triplet is much less than that of the parent carbene, and the triplet–singlet energy separation is decreased. The diphenylcarbene triplet ground state is just 5.1 kcal/mol below the singlet state.[13] At least partially as a consequence of the smaller energy separation of the triplet and singlet states, the conversion of singlet diphenylcarbene, generated thermally from singlet precursors, to the triplet ground state is fast enough to compete with the addition reactions of diphenylcarbene. The latter are, of course, somewhat slower than for the parent carbene. The result is that, typically, even when singlet diphenylcarbene is generated initially, the addition chemistry observed reflects a mixture of competing singlet (stereospecific) and triplet (nonstereospecific) additions (Figure 7.16). Not surprisingly, triplet diphenylcarbene is more highly bent ($\theta = 150°$) than the parent triplet methylene.[14] Forced constriction of the valence angle, as in the case of fluorenylidene, tends to disfavor the triplet state. In

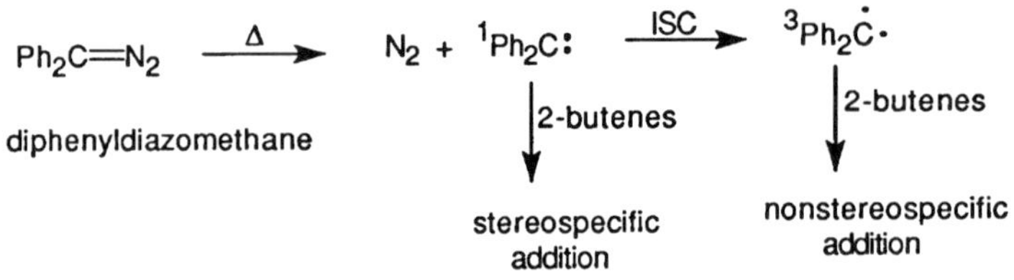

$Ph_2C{=}N_2$ —Δ→ N_2 + $^1Ph_2C:$ —ISC→ $^3Ph_2\dot{C}\cdot$

diphenyldiazomethane

$^1Ph_2C:$ —2-butenes→ stereospecific addition

$^3Ph_2\dot{C}\cdot$ —2-butenes→ nonstereospecific addition

Figure 7.16 Interconversion of singlet and triplet diphenylcarbene.

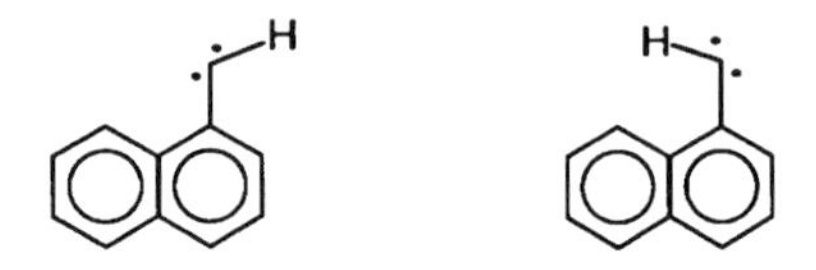

Figure 7.17 Isomeric carbenes resulting from bent structures.

the latter carbene, however, the triplet state is still the ground state but is only 1.1 kcal/mol more stable than the singlet state.[13] The bent valence angle in triplet carbenes is reflected in a novel way in α-naphthylmethyl carbene, where two geometric isomers are observed (Figure 7.17).[15]

7.11 Triplet Carbenes That Are Not Ground States

The singlet state of carbenes can be preferentially stabilized in several ways. As has been mentioned, increased delocalization of the unpaired electrons and constriction of the valence angle tend to increase the relative stability of the singlet state. In the case of carbenes related to diphenylcarbene, planarization of the aryl rings also tends to enhance the relative stability of the singlet state.[16] Probably the most potent effect is to increase the HOMO–LUMO energy gap in the singlet (the SOMO–SOMO gap in the triplet) to the point at which it surpasses the exchange stabilization of the triplet.

Dichlorocarbene, for example, has a singlet ground state, which is ca. 12 kcal/mol more stable than the triplet state as a result of the increase in the energy of the LUMO resulting from its interaction with the unshared pairs of the chlorine substituents (Figure 7.18).[17] Other resonance-electron-donating substituents such as alkoxy exert similar effects.

7.12 Nitrenes

Divalent nitrogen diradicals, i.e. nitrenes, are the nitrogen equivalent of carbenes (Figure 7.19). Triplet states of several nitrenes, especially phenylnitrene and

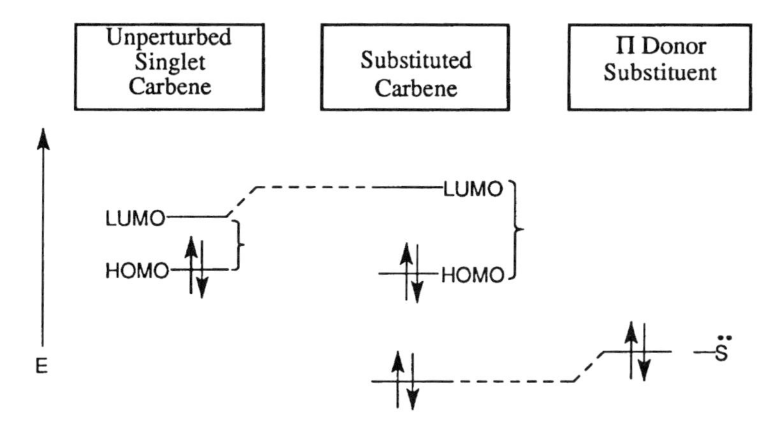

Figure 7.18 Increasing the HOMO–LUMO gap in carbenes by electron pair donor substituents.

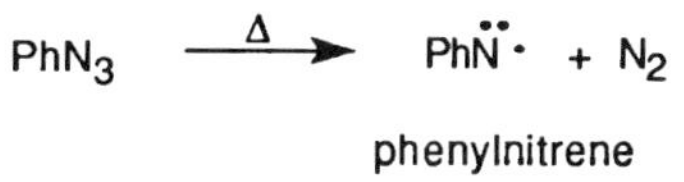

$$PhN_3 \xrightarrow{\Delta} Ph\ddot{N}\cdot + N_2$$

phenylnitrene

Figure 7.19 Nitrenes.

carbethoxynitrene, are well known and are believed to be the ground states of these species.[18]

7.13 Triplet Ground States of Non-Kekulé Hydrocarbon Diradicals

The trimethylenemethane (TMM) and tetramethyleneethane (TME) systems have attracted much attention, since no reasonable canonical structures can be written for these molecules other than diradical structures (Figure 7.20). The HMO energy levels also reveal degenerate frontier orbitals, and triplet states are naturally expected to be at least close to the ground states. These triplets have in fact been observed by ESR spectroscopy (see the section on triplet ESR at the end of this chapter).[19] Both triplets obey the Curie–Weiss law and are therefore ground-state triplets. The ESR spectrum of TMM generated by irradiating a single crystal of the precursor ketone (Figure 7.20) even reveals hyperfine structure. Specifically, the six equivalent protons of TMM give rise to a septet hyperfine splitting pattern (a = 8.9 G). Similarly, when TME is generated by irritating a diazo precursor at 10 K in a 2-methyltetrahydrofuran glass (Figure 7.20), the characteristic ESR spectrum of a triplet is generated. Nonet (9-line) hyperfine splittings are observed, indicating the presence of eight equivalent protons (a = 10.5 G). A Curie–Weiss plot confirms that this is the ground state of TME. Recent calculations suggest that *planar* TME is

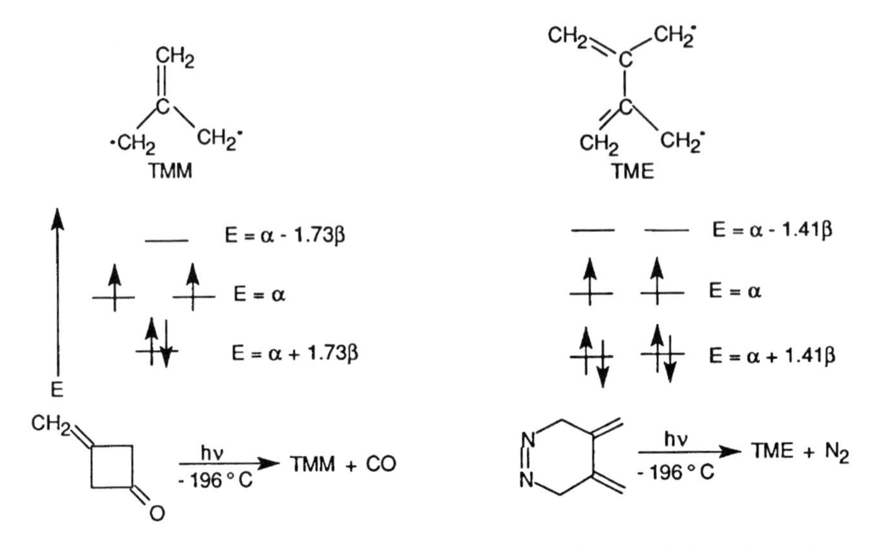

Figure 7.20 MO diagrams of the trimethylenemethane and tetramethyleneethane triplets.

more stable in the *singlet* state and that the observed ground-state *triplet* has the two allylic moieties twisted out of planarity by about 50°.[20]

7.14 Triplet Sensitization: Diene Triplets

The production of excited triplets by photoexcitation to an excited singlet state, followed by intersystem crossing to the more stable triplet, has already been illustrated for benzophenone. This method, however, lacks generality since excited singlets often have reaction modes that occur more rapidly than ISC. The selective generation of excited triplets is better accomplished by triplet sensitization, as has been illustrated in the generation of triplet carbenes. The primary requirements for an appropriate sensitizer are a chromophore that absorbs in the ultraviolet region in an area where the substrate is essentially transparent, an excited singlet state that undergoes extremely rapid and efficient intersystem crossing, and a triplet energy that is higher than that of the desired triplet. Triplet energy transfer from the sensitizer triplet to the substrate, under the latter conditions, is exergonic and efficient. An excellent illustration of triplet sensitization involves the generation of diene triplets and their subsequent cycloadditions to ground-state diene molecules.[21] An ether solution of 1,3-butadiene and benzophenone is irradiated in a Pyrex vessel (or using a Pyrex filter). Since butadiene has a uv λ_{max} = 218 nm, and since Pyrex effectively absorbs all the shorter-wavelength uv light quanta ($\lambda < 290$ nm), light is absorbed exclusively by the sensitizer. The benzophenone triplet energy (E_T = 69 kcal/mol) is substantially greater than that of 1,3-butadiene (E_T = 60 kcal/mol),

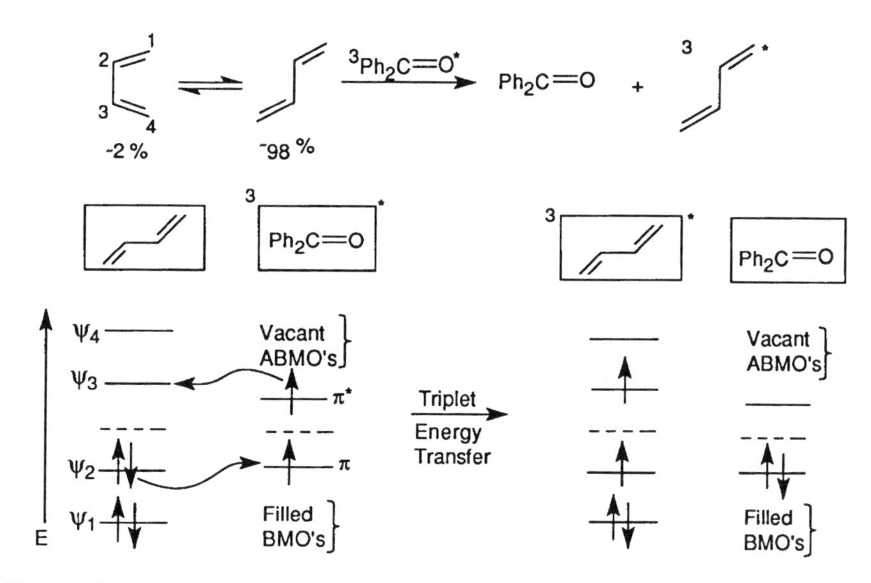

Figure 7.21 Triplet energy transfer from benzophenone to 1,3-butadiene.

and butadiene triplets are formed rather efficiently. The triplet energy transfer process is depicted in Figure 7.21. Because butadiene exists predominantly in the *s-trans* conformation, the diene triplets are preponderantly of the *s-trans* geometry. Because the HOMO of butadiene (ψ_2) is antibonding between C_2–C_3 and the LUMO is bonding between C_2–C_3, the removal of an electron from ψ_2 and its replacement into ψ_3 sharply increases the C_2–C_3 bond order (Figure 7.22). Consequently, the *s-trans* triplets are unable to equilibrate with the corresponding *s-cis* triplets. It is noted in passing that the C_1–C_2 and C_3–C_4 bond orders are sharply decreased for parallel reasons. The diene triplets then effect homolytic addition to ground-state 1,3-butadiene, also in the preferred *s-trans* form, affording the E,E-bis(allylic) triplet diradical (Figure 7.23). After conversion to the singlet diradical (ISC), the latter can cyclize only to a divinylcyclobutane. Cyclization to

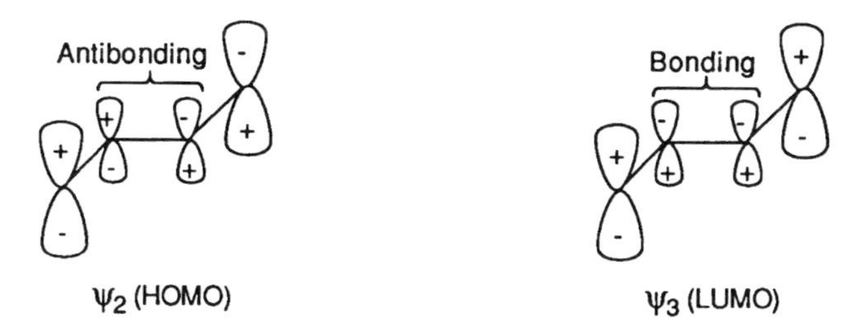

Figure 7.22 Orbital picture of the butadiene HOMO and LUMO.

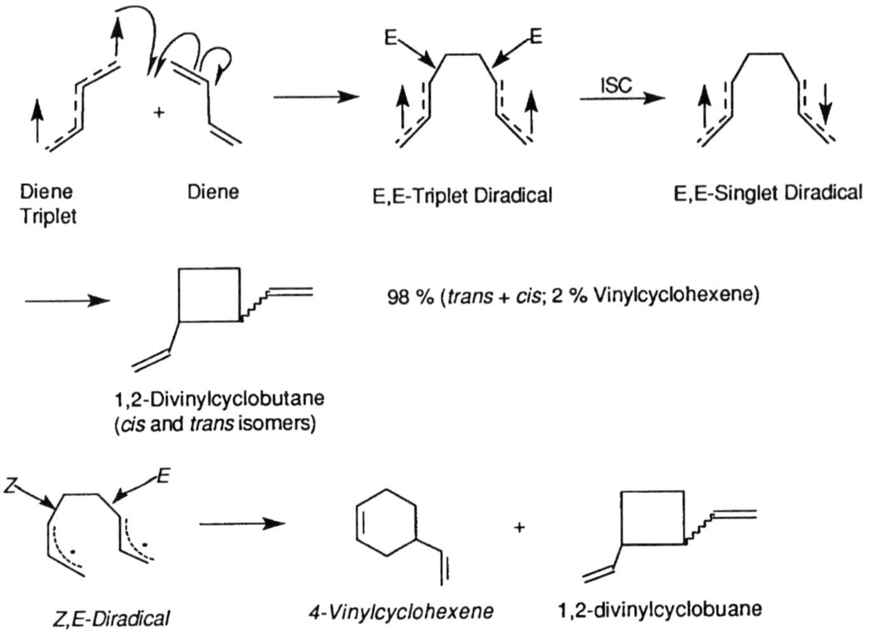

Diene Triplet
Diene
E,E-Triplet Diradical
ISC
E,E-Singlet Diradical
98 % (*trans* + *cis*; 2 % Vinylcyclohexene)
1,2-Divinylcyclobutane (*cis* and *trans* isomers)
Z,E-Diradical
4-Vinylcyclohexene
1,2-divinylcyclobuane

Figure 7.23 Triplet sensitized cyclodimerization of 1,3-butadiene.

vinylcyclohexene requires at least one *s-cis* diene component (either derived from the diene triplet or the ground-state diene) to produce the Z,E-diradical, which can cyclize to either 1,2-divinylcyclobutane or 4-vinylcyclohexene.

7.15 Geometric Isomerization of Triplet States

As noted in the previous discussion, the C_1–C_2 (and C_3–C_4) bond order of dienes is sharply diminished in the triplet state produced by Franck–Condon (i.e., vertical) excitation of the ground-state diene. Relaxation of this triplet state to its global energy minimum generates a perpendicular triplet diradical in which one of the methylene groups is rotated by 90° with respect to an allylic radical moiety.[22] The

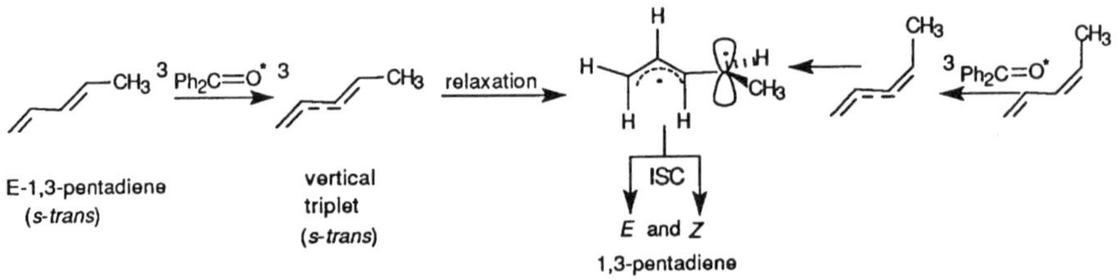

$^3Ph_2C{=}O^{\bullet}$
relaxation
ISC
E-1,3-pentadiene (*s-trans*)
vertical triplet (*s-trans*)
E and *Z* 1,3-pentadiene

Figure 7.24 Geometric isomerization of diene triplets.

same relaxed triplet state is thus produced from either *cis*- or *trans*-1,3-pentadiene, and this triplet eventually undergoes ISC to yield a mixture of these isomers (Figure 7.24). Styrene and alkene triplets isomerize in the same fashion and for the same reasons. Alkene triplets have been referred to as "1,2-diradicals."

7.16 Thermally Generated Excited Triplets

When singlet molecules undergo highly exergonic reactions, are they precluded by the spin conservation principle from forming energetically accessible triplet excited-state product molecules? A number of interesting instances are now available that demonstrate that the spin conservation principle as it applies to chemical reactions is not at all inviolate. The classic example is the cycloreversion of 1,2-dioxetanes, which generates acetone triplets with high efficiency (Figure 7.25).[23] The retroelectrocyclic reaction of Dewar benzene is a case where triplets (benzene) are formed, but only relatively inefficiently (*ca.* 0.1–0.01% yield).[24]

7.17 Triplet ESR Spectra

The ESR spectra of triplets in which there are two localized, remote, and noninteracting radical sites are essentially the same as those of a corresponding monoradical. Molecules of this kind are most appropriately termed *diradicals* or *double doublets*. As an example, the ESR spectrum of the bis(nitroxyl) radical **5** is very similar to that of TEMPO, its close monoradical analogue (Figure 7.26).[25] The

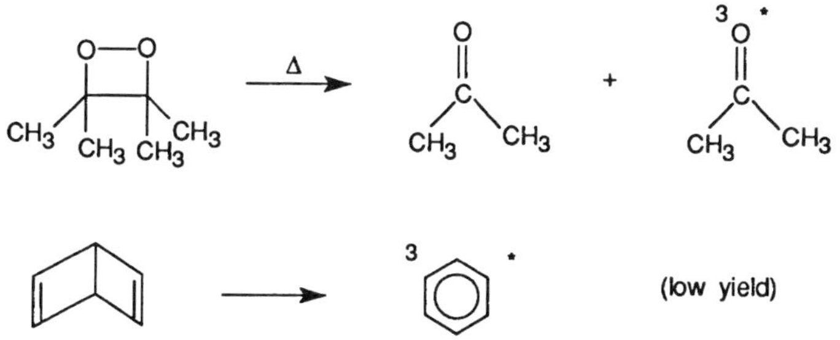

(low yield)

Figure 7.25 Thermal generation of excited triplets from ground-state singlets.

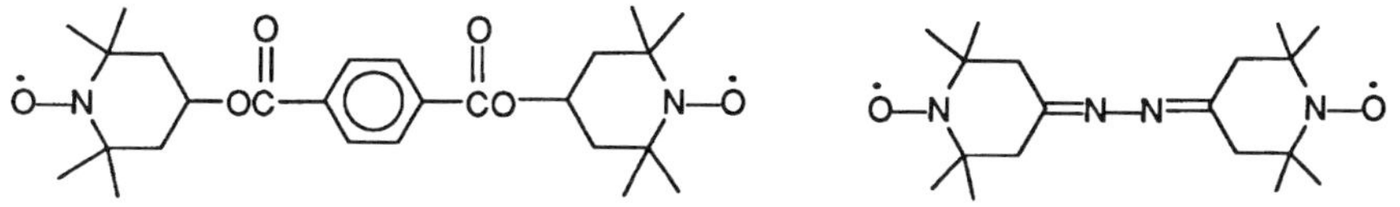

5 **6**

Figure 7.26 ESR spectra of noninteracting and weakly interacting triplets.

spectrum consists of three major lines of equal intensity (not 1:2:1) derived from splitting by a single nucleus of spin $S = 1$ (i.e., from a single N nucleus). If the distance between the nitroxyl centers is decreased, as in **6**, spin exchange occurs even though the interaction between the two spins is very weak. The result is that each odd electron is delocalized over both radical sites, and the spectrum consists of a quintet (splittings from two equivalent nitrogen nuclei). The hyperfine splitting (0.74 mT) is almost exactly one-half of that observed in **5** (1.56 mT). Except for delocalization by spin exchange, the triplet ESR spectrum is virtually unchanged from that of a monoradical analogue. The distances between the nitroxyl radical sites in **5** and **6** are, respectively, 17 and 10 Å.

When the odd electrons occupy π MOs in the same conjugated system or, even more, when they occupy AOs on the same atom, the magnetic dipolar interactions between the unpaired electrons are much larger and give rise to characteristic triplet ESR spectra that are easily distinguished from monoradical spectra.[26] In the most typical form, triplet ESR spectra observed in frozen glassy matrices consist of seven lines (but only five lines for molecules of C_{3v} or higher symmetry). Additional and much smaller electron–nuclear *hyperfine* splittings may be observed, but frequently these are not resolved. The total width of the spectrum, corresponding to the magnitude of the electron–electron magnetic dipolar interactions is typically 1–2 orders of magnitude greater than the width of organic monoradical spectra (e.g., 1000–4000 versus 25–50 G). As will be seen in the more detailed discussion that follows, the width of the spectrum is a quantitative measure of the proximity of the two odd electrons and varies as $1/r^3$, where r is the average distance between the unpaired electrons. More specifically, the total width of the spectrum is equal to $2D$, where D is one of the so-called zero field parameters that characterize a specific triplet molecule (vide infra).

The seven lines of a typical ESR spectrum of randomly oriented triplets in frozen glassy media arise essentially as follows. In the limiting (hypothetical) situation where the electron spins are accurately quantized with respect to the external applied magnetic field (H_0), that is, when the magnetic dipolar interaction between the electrons is much smaller than their interactions with H_0, the spin states are $m_s = -1$, 0, +1, as noted previously. The excitations from $m_s = -1$ to 0 and from $m_s = 0$ to +1 represent single quantum jumps (i.e., flipping of a single electron spin $\Delta m_s = 1$) and are allowed, but they have the same energy and would give rise to a single, unsplit absorption. The transition $m_s = -1$ to +1 represents a double quantum jump ($\Delta m_s = 2$) and would be forbidden. In this hypothetical situation, then, there would be no fine splitting (as distinguished from electron–nuclear hyperfine splitting) of the ESR signal by the electron–electron magnetic dipolar interaction and nothing to distinguish the triplet from a monoradical. However, the magnetic dipolar interaction between the electrons is actually of the same order of magnitude as the interaction with the external magnetic field, so that the electron spins are not accurately quantized with respect to the latter. Further, the three spin states are nondegenerate even in the absence of an external field (shown at the left of Figure 7.28). The result is that the spacing between the adjacent energy levels is different, giving rise to two "$\Delta m_s = 1$" absorption lines, as shown in Figure 7.27. Since the

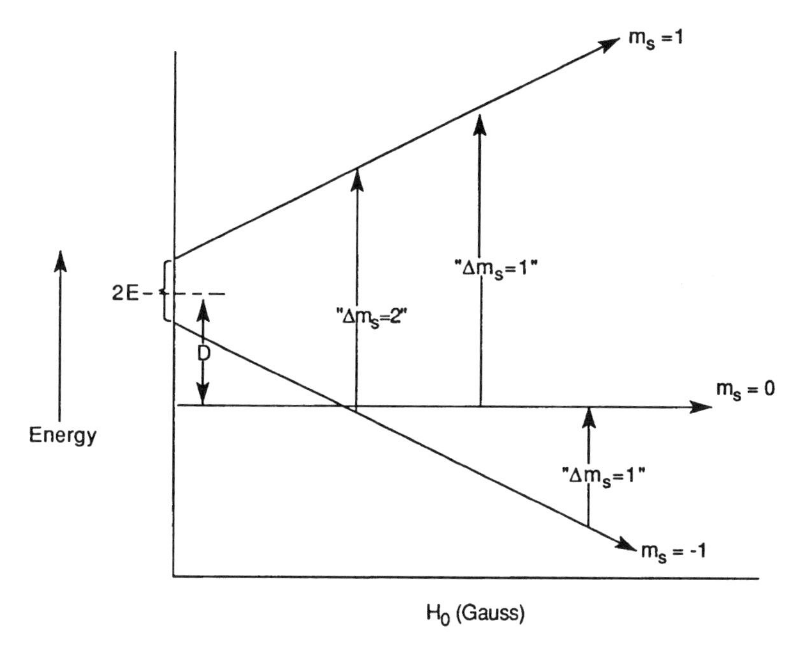

Figure 7.27 Triplet ESR absorptions.

electron spins are not rigorously quantized with respect to the external magnetic field, the designation $\Delta m_s = 1$ is not rigorous and is placed in quotes. For the same reason, the "$\Delta m_s = 2$" transition is no longer forbidden. Although the $m_s = 0$ state remains a quantum state, the other two states are now linear combinations of the $m_s = -1$ and $+1$ states. At zero field, the $m_s = 0$ state is the state of lowest energy. The upper two states are separated by the energy quantity $2E$ (vide infra), and the average energy of these two states is greater than that of the $m_s = 0$ state by the energy quantity D. The quantities D and E are called the *zero-field parameters* and are the key spectral parameters that characterize a particular triplet state.

In the previous discussion, it was assumed that the triplet molecule was oriented with a specific molecular axis (say the Z axis) parallel to the external field. This is critical in triplet ESR spectroscopy because the dipolar interactions between electrons are anisotropic; that is, they are highly directional. Although the direct (through-space) electron–nuclear dipolar interactions of monoradical spectra are also anistropic, these relatively weak interactions are averaged out by tumbling in solution spectra, leaving only isotropic, through-bond interactions. Electron–electron dipolar magnetic interactions, being much larger, are not averaged out even in solution spectra. That is, irrespective of the orientation of the molecule with respect to H_0, the electron spins tend to remain fixed relative to each other. Furthermore, most triplet spectra are measured in frozen media, where tumbling is absent anyway. We have so far established that two "$\Delta m_s = 1$" transitions and one

"$\Delta m_s = 2$" transition occur for molecules with (let us assume) their *Z* axes aligned with H_0. Two additional transitions are observed when the molecule has each of the perpendicular (*X*,*Y*) axes oriented along the magnetic field. Since the "$\Delta m_s = 2$" transition is essentially isotropic, there is only one such absorption for any molecular orientation. In short, there are two "$\Delta m_s = 1$" absorptions each for molecules having their *X*, *Y*, and *Z* axes, respectively, aligned with the external field, plus a single isotropic $\Delta m_s = 2$ transition, making a total of seven absorption lines. Molecules with intermediate geometries apparently contribute relatively little to the absorption spectra. The separation between the outermost lines of the spectra is twice the zero-field parameter *D*. The other two pairs of corresponding lines are separated by $D + E$ and $D - E$. The zero-field parameter *E* is a measure of the distinction between the *X* and *Y* molecular axes. For molecules in which these two directions are equivalent, $E = 0$. Therefore, molecules that have threefold or higher symmetry have only five-line triplet ESR spectra.

7.18 Triphenylene Triplet

The photochemically generated triplet of triphenylene in 2-methyl-tetrahydrofuran glass at 77 K (Figure 7.28) shows a five-line ESR spectrum.[27] The "$\Delta m_s = 1$" absorptions are found at 1835, 2449, 3914, and 4698 G. The separation between the outermost lines (2863 G) corresponds to 2*D*, and that between the inner lines (1431 G) to *D*. Converted to the usual units, $D = 0.1338\ \text{cm}^{-1}$. The "$\Delta m_s = 2$" absorption (the so-called half-field absorption) occurs at 1411 G.

7.19 Naphthalene Triplet

In the same manner, naphthalene triplets show a seven-line spectrum with $D = 0.10046$ and $E = 0.01536\ \text{cm}^{-1}$.[27] The "$\Delta m_s = 2$" absorption occurs at 1508.1 G, and the six "$\Delta m_s = 1$" absorptions at 2198.2, 2451.7, 2922.0, 3515.2, 4036.5, and 4350.0 G, respectively ($H_0 = 3278.9$ G). The large spectral width compared to monoradical spectra and the low-field "$\Delta m_s = 2$" absorptions clearly distinguish

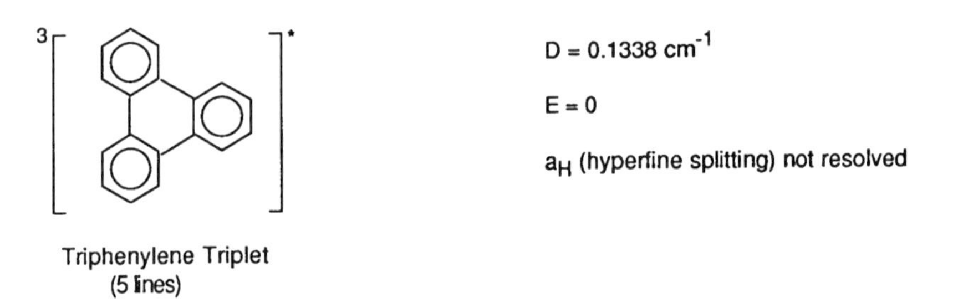

Figure 7.28 Zero-field parameters of the triphenylene triplet.

these as triplet-state spectra, while $E \neq 0$ is appropriate for naphthalene. Although hyperfine splittings were not resolved in the randomly oriented triplet spectra, these appropriate hyperfines have been resolved for oriented naphthalene triplets.[2]

The D values and hence the spectral widths are even greater for more localized triplets such as diphenylcarbene ($D = 0.4055\ \text{cm}^{-1}$, $E = 0.0194$) and cyanonitrene ($D = 1.544$, $E = 0$).[27] In the spectra of such triplets, where dipolar interactions are especially strong, the symmetry of the spectra about H_0 and also some transitions are lost.

7.20 Trimethylenemethane (TMM) Triplet

The triplet state of this important non-Kekulé hydrocarbon diradical has been generated by two routes, as illustrated in Figure 7.29.[19] Irradiation of an appropriate diazo precursor at –196°C yields a five-line spectrum without hyperfines. However, a much better resolved spectrum is obtained from a single crystal of a ketonic precursor. The five principal lines are found at 1590 G ("$\Delta m_s = 2$"), 3005, 3129, 3407, and 3531 G ("$\Delta m_s = 1$"). A seven-line hyperfine splitting arising from six equivalent protons is also observed ($a = 8.9$ G). The Curie–Weiss law is obeyed, and this triplet appears to be the ground state of TMM.

7.21 Higher Multiplets

Molecules having three unpaired spins are designated as quartets ($2S + 1$, $S = \frac{3}{2}$). Similarly, quintets have four unpaired spins, etc. Relatively stable tri- and tetraradicals based upon meta linked trityl radicals have been prepared, and their ESR spectra reveal quartet and quintet spin states, respectively. The radicals are generated from the corresponding trianions and tetranions by iodine oxidation.[28] As a specific example the triradical of Figure 7.30 shows five equally spaced lines in the "$\Delta m_s = 1$" region as expected for a triradical. The line separation, $D = 0.0066\ \text{cm}^{-1}$. Also, the "$\Delta m_s = 2$" region consists of a broad quartet, and a $\Delta m_s = 3$" absorption is also observed (unsplit).

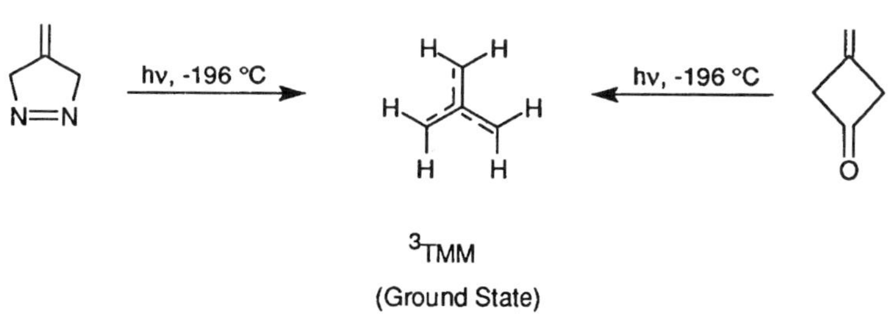

hv, -196 °C

hv, -196 °C

^{3}TMM

(Ground State)

Figure 7.29 Generation of trimethylenemethane.

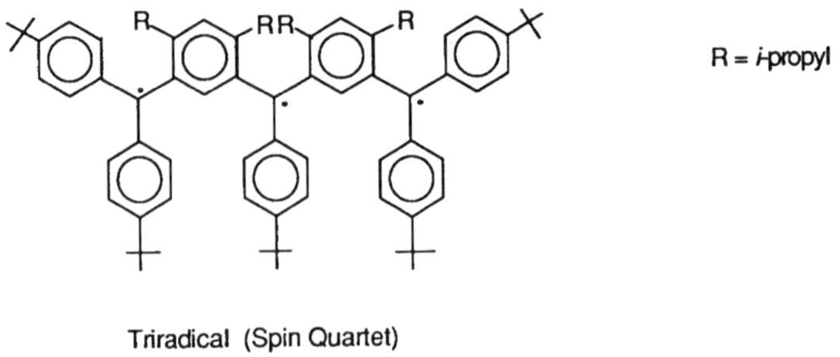

R = *i*-propyl

Triradical (Spin Quartet)

Figure 7.30 A persistent organic triradical.

References

General Reference

Turro, N. J. *Modern Molecular Photochemistry*, Benjamin/Cummings, Menlo Park, 1978.

Specific References

1. Turro, p. 352.
2. Turro, p. 186.
3. Yang, N. C.; Castro, A. J. *J. Am. Chem. Soc.* **1960**, *82*, 6208.
4. Veciana, J.; Rovira, C.; Crespo, M. I.; Armet, O.; Domingo, V. M.; Palacio, F. *J. Am. Chem. Soc.* **1991**, *113*, 2552.
5. Rajca, A.; Utamapanya, S.; Xu, J. *ibid.*, 9235.
6. Briere, R.; Dupeyre, R.-M.; Lemaire, H.; Morat, C.; Rassat, A.; Rey, P. *Bull. Soc. Chim. France* **1965**, 3290; Rozantsev, E. G.; Golubev, V. A. *Izuest. Akad. Nauk. SSSR, Ser. Khim.* **1965**, *4*, 718.
7. Breslow, R. in *Topics in Nonbenzenoid Chemistry*, Vol. 1, Nozoe, T.; Breslow, R.; Hafner, K.; Itô, S.; Murata, I., Eds., Hirokawa, Tokyo, 1973, p. 81; Garratt, P. J. *Aromaticity*, John Wiley & Sons, New York, 1986, p. 166.
8. Bauld, N. L.; Brown, M. S. Unpublished results.
9. Masamune, S.; Nakamura, N.; Suda, M.; Ona, H. *J. Am. Chem. Soc.* **1973**, *95*, 8481.
10. Schaefer, H. F., III *Science*, *231*, 1100.
11. Kopecky, K. R.; Hammond, G. S.; Leermakers, P. *J. Am. Chem. Soc.* **1962**, *84*, 1015.
12. Skell, P. S.; Woodworth, R. C. *J. Am. Chem. Soc.* **1956**, *78*, 4496; Skell, P. S.; Garner, A. Y. *ibid.*, 5430.
13. Sitzman, E. V.; Langan, J.; Eisenthal, K. B. *J. Am. Chem. Soc.* **1984**, *106*, 1868.
14. Wasserman, E.; Trozzolo, W. A.; Yager, W. A.; Murray, R. W. *J. Chem. Phys.* **1964**, *40*, 2408; Brandon, R. W.; Closs, G. L.; Davoust, C. E.; Hutchinson, C. A., Jr.; Kohler, B. E.; Silbey, R. *J. Chem. Phys.* **1965**, *43*, 2006.
15. Trozzolo, A. M.; Wasserman, E.; Yager, W. A. *J. Am. Chem. Soc.* **1965**, *87*, 129.
16. Metcalf, J.; Halevi, E. A. *J. Chem. Soc. Perkin II* **1977**, *634*.
17. Davidson, E. R. in *Diradicals*, Borden, W. T., Ed., John Wiley & Sons, New York, 1982, p. 73.

18. Waserman, E.; Smolinsky, G.; Yager, W. A. *J. Am. Chem. Soc.* **1964**, *86*, 3166; Moriarty, R. M.; Rahman, M.; King, G. J. *J. Am. Chem. Soc.* **1966**, *88*, 842.

19. Dowd, P. *Accts. Chem. Res.* **1972**, 5, 242; Dowd, P. *J. Am. Chem. Soc.* **1970**, *92*, 1066; Dowd, P.; Chang, W.; Paik, Y. N. *J. Am. Chem. Soc.* **1986**, *108*, 7416.

20. Nash, J. J.; Dowd, P.; Jordan, K. D. *J. Am. Chem. Soc.* **1992**, *114*, 10071.

21. Dilling, W. L. *Chem. Rev.* **1969**, *69*, 845.

22. Unett, D. J.; Caldwell, R. A. *Res. Chem. Intermed.* **1995**, *21*, 665.

23. Turro, N. J. *Modern Molecular Photochemistry*, Benjamin Cummings, Menlo Park, 1978, p. 597.

24. Lechtkin, P.; Breslow, R.; Schmidt, A. H.; Turro, N. J. *J. Am. Chem. Soc.* **1973**, *95*, 3025.

25. Briere, R.; Dupeyre, R.-M.; Lemaire, H.; Mora, C.; Ramsat, A.; Rey, P. *Bull. Soc. Chim. Fr.* **1965**, 3290.

26. Carrington, A.; McLachlan, A. D. *Introduction to Magnetic Resonance*, 1967, Chapter 8.

27. Wasserman, E.; Snyder, L.-C.; Yager, W. A. *J. Chem. Phys.* **1964**, *41*, 1763.

28. Rajca, A.; Utamapanya, S.; Xu, J. *J. Am. Chem. Soc.* **1991**, *113*, 9235; Rajca, S.; Rajca, A. *J. Am. Chem. Soc.* **1995**, *117*, 9172.

Exercises

7.1 The dication of hexachlorobenzene was generated as indicated below and is a ground-state triplet.

$$C_6Cl_6 \xrightarrow[Cl_2]{SbF_5} C_6Cl_6^{+\cdot} \xrightarrow[h\nu]{77K} C_6Cl_6^{2+}$$

(a) Using the HMO energy levels of benzene, depict by means of energy-level diagrams the conversion to a cation radical and then to a dication. Explain why the dication might reasonably be expected to have a triplet ground state.

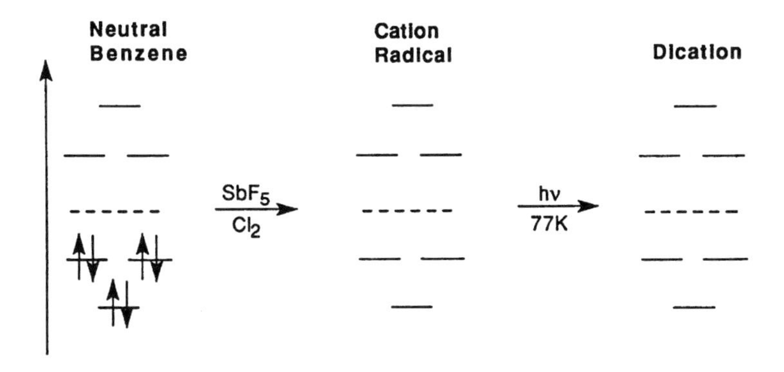

(b) How would it be evident, qualitatively, that the spectrum observed is that of a triplet as opposed to a doublet? (c) How could it be established experimentally that the triplet is the ground state of the dication, as opposed to a thermally populated excited state? (d) The zero-field parameter E $< 0.0003\ cm^{-1}$. What information does this provide about the structure of the dication? Is it consistent with the symmetry expected from the dication MO energy-level diagram? (e) The zero-field parameter *D*, interestingly, is found to be significantly less than for the triplet state of *neutral* hexachlorobenzene and also that of benzene itself. Explain this in terms of differential substituent effects in the dication and neutral.

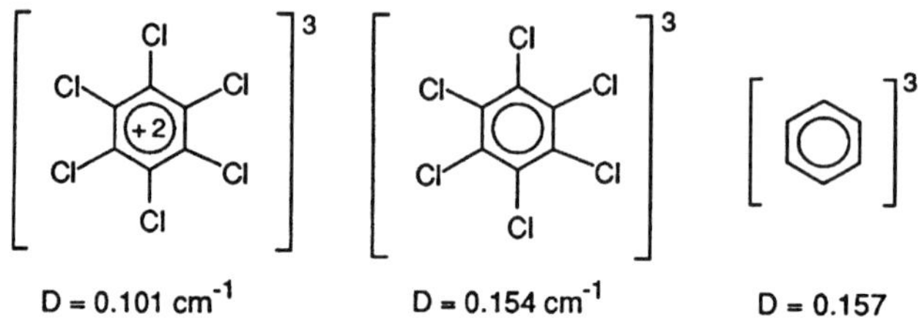

D = 0.101 cm^{-1} D = 0.154 cm^{-1} D = 0.157

(f) The hexachlorobenzene cation radical signal still remains strong in the samples for which the triplet spectra are obtained. This signal occurs at 3283 G. One of the lines of the triplet spectrum occurs at 1534 G. To which transition does this signal correspond? Explain. (g) The remaining four lines of the triplet spectrum occur at 2232, 2687, 3776, and 4399 G. From these data, calculate the zero-field parameter, *D*, in gauss. (h) How is it evident from this spectrum that the zero-field parameter *E* is negligible? (Wasserman, E.; Hutton, R. S.; Kuck, V. J.; Chandross, E. A. *J. Am. Chem. Soc.* **1974**, *96*, 1965.)

7.2 Buckminsterfullerene (C_{60}) has triply degenerate LUMOs and is easily reduced electrochemically to the anion radical ($C_{60}^{\bar{\bullet}}$), dianion (C_{60}^{-2}), trianion (C_{60}^{-3}), and tetranion (C_{60}^{-4}). The dianion is found to have a triplet ground state, while the trianion is only a doublet. (a) Illustrate the electronic configurations of the dianion and trianion on the abbreviated energy-level diagrams below.

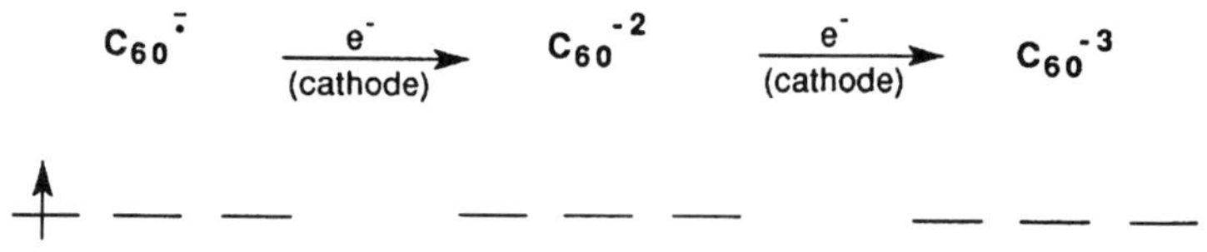

$C_{60}^{\bar{\bullet}}$ $\xrightarrow[\text{(cathode)}]{e^-}$ C_{60}^{-2} $\xrightarrow[\text{(cathode)}]{e^-}$ C_{60}^{-3}

(b) The *D* values of the dianion triplet and of the neutral C_{60} triplet correspond roughly to 6.1 Å and 12.6 Å average separations of the electron spins. Which species, the dianion or the neutral, would be expected to have the larger separation of spins (smaller *D* value)? Explain. (Dubois, D.; Jones, M. T.; Kadish, K. M. *J. Am. Chem. Soc.* **1992**, *114*, 6446).

7.3 Propargylene (ethynyl carbene) can be generated by photolysis of 3-diazopropyne. When this carbene is generated in the presence of the 2-butenes, the expected cyclopropane products are obtained in the indicated ratios.

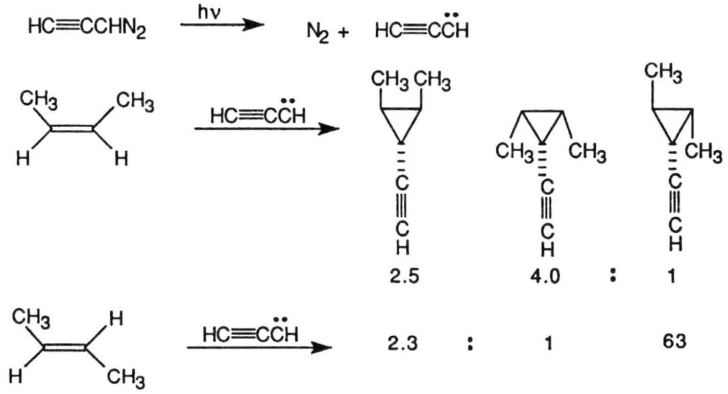

$$HC{\equiv}CCHN_2 \xrightarrow{h\nu} N_2 + HC{\equiv}C\ddot{C}H$$

cis-2-butene $\xrightarrow{HC\equiv C\ddot{C}H}$ 2.5 : 4.0 : 1

trans-2-butene $\xrightarrow{HC\equiv C\ddot{C}H}$ 2.3 : 1 : 63

(a) Assuming that propargylene is generated initially as the singlet, provide a mechanistic interpretation of these data. (b) What stereochemical result would be expected if these reactions involved triplet propargylene exclusively? Explain. (c) The observed value of the zero-field parameter E is approximately zero for triplet propargylene in frozen solutions. Suggest a likely structure for this species and depict its π-electron systems. (d) Suggest a likely structure for singlet propargylene. (e) Assuming that C_1 and C_3 of the diazopropyne precursor can be distinguished by isotopic labeling, propose a means of distinguishing the cycloaddition reactions of singlet and triplet propargylene that utilizes such a labeling approach. (Skell, P. S.; Klebe, J. *J. Am. Chem. Soc.* **1960**, *82*, 247; Bernheim, R. A.; Kempf, R. J.; Gramas, J. V.; Skell, P. S. *J. Chem. Phys.* **1965**, *43*, 196.)

7.4 Atomic carbon has a triplet ground state (^3P) that is 1.3 eV below the lowest-energy singlet state (^1D). The ^3P state can be considered essentially as a dicarbene in which one carbene moiety is a triplet and the other a singlet. When atomic carbon is produced via a carbon arc and deposited first on a neopentane surface at –196°C in order to permit the decay of the singlet to the ground-state triplet and then *cis*- or *trans*-2-butene is added as a substrate after a brief delay, the following results are obtained:

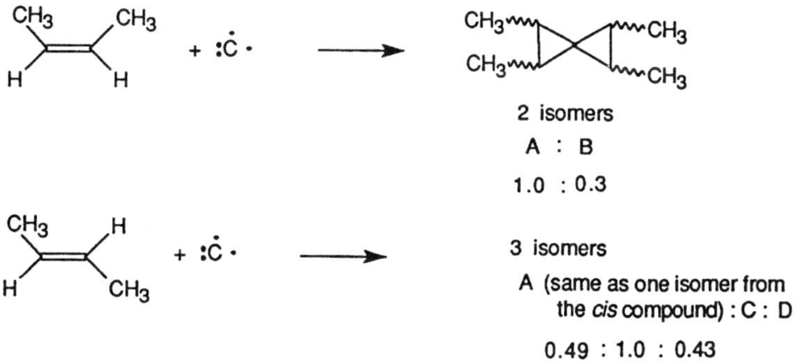

cis-2-butene + :Ċ· ⟶ (spiropentane product)
2 isomers
A : B
1.0 : 0.3

trans-2-butene + :Ċ· ⟶
3 isomers
A (same as one isomer from the *cis* compound) : C : D
0.49 : 1.0 : 0.43

Suggest specific structures for the tetramethylspiropentane isomers *A*, *B*, *C*, and *D* and rationalize their formation mechanistically. Be sure to explain, for example, why *B* is not formed from *trans*-2-butene and why *C* and *D* are not formed from *cis*-2-butene, but *A* is formed from both alkenes.

7.5 2,3-Dimethylenebicyclo[2.2.0]hexane decomposes thermally in the temperature range 40–60°C. The decompositions are strictly first order ($\Delta H^{\ddagger} = 17.5$ kcal/mol, $\Delta S^{\ddagger} = -22.7$ eu). The sole products of the reaction are the dimers shown below. Even when an excess of diethyl fumarate is included in the solvent (tetrachloroethylene), the dimers are still formed in excellent yield, and no cross adducts involving the fumarate are formed. However, when the decomposition is carried out in neat, dilute *trans*, *trans*-2,4-hexadiene, 50% of the cross adducts indicated below are formed, in addition to 50% of the usual dimers.

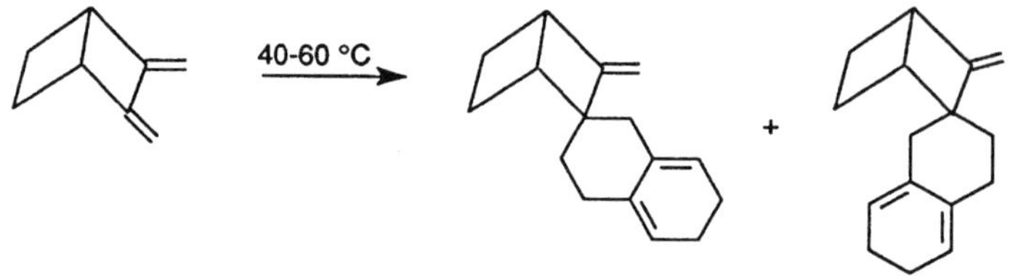

40-60 °C

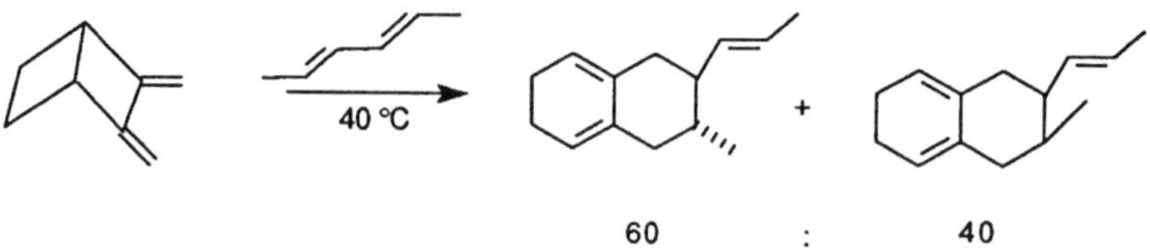

+

40 °C

+

60 : 40

(a) Propose a detailed mechanism for the formation of the dimers that is consistent with the kinetic data and that explains the absence of any reaction with diethylfumarate. (b) Rationalize the observation that, even in neat and dilute 2,4-hexadiene solution, 50% of the dimer is still formed. Write a mechanism for the formation of the two cross adducts. (Chang, C.-S.; Bauld, N. L. *J. Am. Chem. Soc.* **1972**, *94*, 7593; Bauld, N. L.; Chang, C.-X. *ibid.*, 7594.

Index